Cellular and Molecular Methods in Neuroscience Research

Springer
*New York
Berlin
Heidelberg
Barcelona
Hong Kong
London
Milan
Paris
Singapore
Tokyo*

Adalberto Merighi
Giorgio Carmignoto
Editors

Cellular and Molecular Methods in Neuroscience Research

Foreword by A. Claudio Cuello

With 66 Illustrations

 Springer

Adalberto Merighi DVM
Department of Veterinary Morphophysiology
Rita Levi Montalcini Center Brain Repair
University of Torino
Gruglasco I-10095, Italy

Giorgio Carmignoto
Department of Experimental and Biomedical Sciences
CNR Center for the Study of Biomembranes
University of Padova
Padova I-35121, Italy

Cover illustration: The cover image is an art-work composition from original tables of the book that emphasizes the range of novel analysis methods now available in cellular and molecular neuroscience research, in contrast to the classical neuroanatomical approach of Golgi staining, which represented a milestone in neurohistology during the first half of the last century. The center image is a photograph from an original Golgi preparation by Giovanni Godina, Emeritus Professor of Veterinary Anatomy at the University of Torino, Italy,

Library of Congress Cataloging-in-Publication Data
Cellular and molecular methods in neuroscience research / editors, Adalberto Merighi, Giorgio Carmignoto.
    p. cm.
  Includes bibliographical references and index.
  ISBN 0-387-95386-8 (alk. paper)
    1. Molecular neurobiology—Laboratory manuals.   2. Neurons—Laboratory manuals.
  1. Merighi, Adalberto.   II. Carmignoto, Giorgio.
QP356.2 .C45 2002
573.8′48—dc21   2001054917

ISBN 0-387-95386-8   Printed on acid-free paper

© 2002 Springer-Verlag New York, Inc.

All rights reserved. This work may not be translated or copied in whole or in part without the written permission of the publisher (Springer-Verlag New York, Inc., 175 Fifth Avenue, New York, NY 10010, USA), except for brief excerpts in connection with reviews or scholarly analysis. Use in connection with any form of information storage and retrieval, electronic adaptation, computer software, or by similar or dissimilar methodology now known or hereafter developed is forbidden.
The use in this publication of trade names, trademarks, service marks, and similar terms, even if they are not identified as such, is not to be taken as an expression of opinion as to whether or not they are subject to proprietary rights.

Printed in the United States of America.

9 8 7 6 5 4 3 2 1   SPIN 10856186

www.springer-ny.com

Springer-Verlag New York Berlin Heidelberg
A member of BertelsmannSpringer Science+Business Media GmbH

*This book is dedicated to the memory of Giovanni Godina,
Emeritus Professor of Veterinary Anatomy at the University of Torino, Italy.*

Foreword

Of all the fields of medical research, Neuroscience is perhaps the most interdisciplinary. The intrinsic complexity of the nervous system demands it. Traditionally, the nervous system was explored in a unidisciplinary fashion, typically with neurophysiological, neurochemical and neuroanatomical approaches. In recent decades - and thanks to the development of a number of new and powerful technologies -it has been easier, and therefore compelling, to combine disciplines and methodologies in order to answer a single question. The revolution provoked by the progress in molecular biology has compounded and greatly enriched these possibilities, as demonstrated by the high quality and originality of present day publications.

The present book on " Cellular and Molecular Methods in Neuroscience Research" edited by Adalberto Merighi and Giorgio Carmignoto is an excellent representation of this integrated, multidisciplinary approach. The Editors have selected very relevant and current topics for each chapter and they have been able to attract very credible specialists to write them. I anticipate that the book will be an important reference publication for many years to come, as the choice of subjects makes it very attractive. The protocols cover a broad range of fields from downstream cellular signaling, transfections of neurons and glia, single cell mRNA analysis to integrated systems. This preface is not the place to analyze the individual merits of each chapter. However, it would be appropriate to highlight the fact that, contrary to analogous books, this particular publication has the merit of dwelling in some detail on the drawbacks and advantages of the procedures and on their best perceived applications. The book is written largely on the experimental evidence gathered by the authors, who provide a frank and clear explanation of the known limitations of the procedures and, in many cases, interesting accounts of the difficulties they personally encountered until optimal procedures were established.

The readers will find in "Cellular and Molecular Methods in Neuroscience Research" a most useful companion to their experimental work. Its bibliography is extensive and will prove valuable when searching for key methodological papers. Many of the chapters contain procedure flow charts showing experimental alternatives and tables with the key applications of the protocols described. These aspects, along with the description of the specific reagents, their applications and limitations and the name of suppliers, will greatly facilitate the transition from reading the chapter to the actual application of the protocols.

In closing, Drs. Merighi and Carmignoto should be congratulated for their vision in putting together such a valuable collection of chapters and for their ability in persuading very busy colleagues to set aside time and effort to describe in such detail experimental procedures. Springer-Verlag should also be congratulated for supporting the Editors in this enterprise. But it is the Neuroscience community as a whole that owes the authors the greatest debt of thanks for providing in such a clear fashion the best up-to-date "recipes" in their experimental "cookbook". I believe that a large cohort of contemporary neuroscientists will enjoy the reading and practice of this book. I wish them all new and exciting results!

A. Claudio Cuello
Research Chair in Pharmacology
McGill University
June 2002

PREFACE

Analysis of the nervous tissue presents unique and peculiar technical problems that are encountered in everyday bench work. While numerous books dealing with cellular and molecular protocols for general use in cell biology are available, very few are specifically devoted to neurobiology. Moreover, the "cross-talk" between researchers with different backgrounds, i.e., histologists, cell and molecular biologists, and physiologists, is still quite difficult, and very often one remains somehow confined to his own specific field of expertise, never daring to explore "mysterious" lands without the support of a big laboratory. The motivation behind this project was to put together the contributions from a number of well-known neuroscientists to produce a book that offers a survey of the most updated techniques for the study of nerve cells. We have chosen to cover a number of different topics, and therefore, each chapter should not be considered exhaustive of the matter but rather a guide to those who are willing to exploit a series of techniques that are not regularly used in their laboratories. This book is written by researchers who routinely perform their studies in different areas of neuroscience and have contributed to the development of new methodologies. It is designed as a method book to be routinely used by laboratory personnel, and each chapter also encompasses a background section, in which authors have delineated the rationale at the basis of the different approaches described, and a clear and accurate discussion of the advantages and disadvantages that are inherent to each technique. We asked the authors to put particular emphasis on the advantages of a multidisciplinary approach, which combines different techniques to obtain an in-depth structural and functional analysis of neural cells. Thus, it has been rather difficult to define a table of contents according to the classical way in which these books are organized. Nonetheless, we have tried to group together the chapters dealing with similar matters.

As a general indication to the readers, Chapters 1 and 2 describe a series of techniques that are suitable for analysis of signal transduction mechanisms in cell systems. Chapters 3 through 6 are devoted to the description of transfection methods and their applications in cultured cells and organotypic slices. Chapter 7 describes a sophisticated novel approach to the analysis of gene expression in single cells. It also contains a wide and punctual survey on the rationale for the choice of different approaches to the dissection of the complexity of CNS organization. The remaining chapters describe a series of techniques in situ to be used in analysis of gene expression (Chapters 8 and 9), neurochemical characterization of nerve cells and analysis of connectivity (Chapters 10 and 15), combined electrophysiological and morphological analysis (Chapter 11), tracing of neural connections (Chapters 12 and 13), apoptosis detection (Chapter 14) and calcium imaging (Chapter 16).

At the end of the book, we put particular care in the preparation of the Index, trying to make numerous cross-references to different indexing words, thus rendering easier the search for specific subjects.

We are grateful to Paula Challaghan, Life Science Editor at Springer-Verlag, New York for her confidence in our project. We also wish to mention the careful and patient work of Allan Abrams in assembling the book. Finally our deepest and sincere thanks go to all the scientists whose contributions made this manual possible.

If this book encourages even a few people to use one of the protocols described as a part of their regular techniques rather than leaving them to the aficionados, we will have more than satisfied our aim.

Adalberto Merighi
Giorgio Carmignoto
June 2002

Contents

	Foreword	vii
	Preface	ix
	Contributors	xiii
1	Analyses of Intracellular Signal Transduction Pathways in CNS Progenitor Cells	1
	Elena Cattaneo and Luciano Conti	
2	Confocal and Electron Microscopic Tracking of Internalized Neuropeptides and/or Their Receptors	15
	Alain Beaudet, Alexander C. Jackson, and Franck Vandenbulcke	
3	Transfection Methods for Neurons in Primary Culture	29
	Christoph Kaether, Martin Köhrmann, Carlos G. Dotti, and Francesca Ruberti	
4	Polyethylenimine: a Versatile Cationic Polymer for Plasmid-Based Gene Delivery in the CNS	37
	Barbara A. Demeneix, Gregory F. Lemkine, and Hajer Guissouma	
5	Transfection of $GABA_A$ Receptor with GFP-Tagged Subunits in Neurons and HEK 293 Cells	53
	Stefano Vicini, Jin Hong Li, Wei Jian Zhu, Karl Krueger, and Jian Feng Wang	
6	Neuronal Transfection Using Particle-Mediated Gene Transfer ...	67
	Harold Gainer, Raymond L. Fields, and Shirley B. House	
7	Analysis of Gene Expression in Genetically Labeled Single Cells ...	85
	Stefano Gustincich, Andreas Feigenspan, and Elio Raviola	
8	Immunocytochemistry and In Situ Hybridization: Their Combinations for Cytofunctional Approaches of Central and Peripheral Neurons	119
	Marc Landry and André Calas	
9	In Situ Reverse Transcription PCR for Detection of mRNA in the CNS	145
	Helle Broholm and Steen Gammeltoft	
10	Immunocytochemical Labeling Methods and Related Techniques for Ultrastructural Analysis of Neuronal Connectivity	161
	Patrizia Aimar, Laura Lossi, and Adalberto Merighi	

Contents

11 Combined Electrophysiological and Morphological Analyses of
    CNS Neurons 181
    Alfredo Ribeiro-da-Silva and Yves De Koninck

12 Tract Tracing Methods at the Light Microscopic Level 203
    Marina Bentivoglio and Giuseppe Bertini

13 Tract Tracing Methods at the Ultrastructural Level 221
    Isaura Tavares, Armando Almeida, and Deolinda Lima

14 In Vivo Analysis of Cell Proliferation and Apoptosis in the CNS 235
    Laura Lossi, Silvia Mioletti, Patrizia Aimar, Renato Bruno,
    and Adalberto Merighi

15 Confocal Imaging of Nerve Cells and Their Connections 259
    Andrew J. Todd

16 Confocal Imaging of Calcium Signaling in Cells
    from Acute Brain Slices 273
    Wim Scheenen and Giorgio Carmignoto

Index 285

Contributors

Patrizia Aimar
Department of Veterinary Morphophysiology
Neuroscience Research Group
University of Torino
Torino, Italy, EU

Armando Almeida
Institute of Histology and Embryology
Faculty of Medicine and IBMC
University of Oporto
Oporto, Portugal, EU

Alain Beaudet
Department of Neurology and
   Neurosurgery
Montreal Neurological Institute
Montreal, Quebec, Canada

Marina Bentivoglio
Department of Morphological and
   Biomedical Sciences - Section of
   Anatomy and Histology
Faculty of Medicine
University of Verona
Verona, Italy, EU

Giuseppe Bertini
Department of Morphological and
   Biomedical Sciences - Section of
   Anatomy and Histology
Faculty of Medicine
University of Verona
Verona, Italy, EU

Helle Broholm
Rigshospitalet
Copenhagen University Hospital
Department of Neuropathology
Copenhagen, Denmark, EU

Renato Bruno
Department of Veterinary
   Morphophysiology
University of Torino
Torino, Italy, EU

André Calas
Laboratorie de Cytologie, Institut des
   Neurosciences, UMR CNRS 7624
Université Pierre et Marie Curie, Paris VI
Paris Cedex, France, EU

Giorgio Carmignoto
Department of Experimental and
   Biomedical Sciences
CNR Center for the Study of
   Biomembranes
University of Padova
Padova, Italy, EU

Elena Cattaneo
Institute of Pharmacological Sciences
University of Milano
Milano, Italy, EU

Luciano Conti
Institute of Pharmacological Sciences
University of Milano
Milano, Italy, EU

Yves De Koninck
Department of Pharmacology and
   Therapeutics
McGill University
Montreal, Quebec, Canada

Barbara A. Demeneix
Laboratorie de Physiologie Générale et
   Comparée, UMR CNRS 8572
Muséum National d'Historie Naturelle
Paris Cedex, France, EU

Carlos G. Dotti
Cavalieri Ottolenghi Scientific Institute
Università degli Studi di Torino
Orbassano, Italy, EU

Andreas Feigenspan
Department of Neurobiology
Harvard Medical School
Boston, MA, USA

Raymond L. Fields
Laboratory of Neurochemistry
National Insitute of Health, NINDS
Bethesda, MD, USA

Harold Gainer
Laboratory of Neurochemistry
National Insitute of Health, NINDS
Bethesda, MD, USA

Steen Gammeltoft
Department of Clinical Biochemistry
Glostrup Hospital
Copenhagen University
Glostrup, Denmark, EU

Hajer Guissouma
Laboratorie de Physiologie Générale et
   Comparée, UMR CNRS 8572
Muséum National d'Historie Naturelle
Paris Cedex, France, EU

Stefano Gustincich
Department of Neurobiology
Harvard Medical School
Boston, MA, USA

Shirley B. House
Laboratory of Neurochemistry
National Insitute of Health, NINDS
Bethesda, MD, USA

Contributors

Alexander C. Jackson
Department of Neurology and Neurosurgery
Montreal Neurological Institute
Montreal, Quebec, Canada

Christoph Kaether
EMBL Heildelberg, Cell Biology Programme
Heildelberg, Germany, EU

Martin Köhrmann
EMBL Heildelberg, Cell Biology Programme
Heildelberg, Germany, EU

Karl Krueger
Department of Physiology and Biophysics
Georgetown University Medical School
Washington DC, USA

Marc Landry
INSERM EPI 9914
Instutut François Magendie,
1 Rue Camille Saint-Saëns,
33077 cedex, France, EU

Gregory F. Lemkine
Laboratorie de Physiologie Générale et Comparée, UMR CNRS 8572
Muséum National d'Historie Naturelle
Paris, France, EU

JinHong Li
Department of Physiology and Biophysics
Georgetown University Medical School
Washington DC, USA

Deolinda Lima
Institute of Histology and Embryology
Faculty of Medicine and IBMC
University of Oporto
Oporto, Portugal, EU

Laura Lossi
Department of Veterinary Morphophysiology
Neuroscience Research Group
University of Torino
Torino, Italy, EU

Adalberto Merighi
Department of Veterinary Morphophysiology
Neuroscience Research Group
Rita Levi Montalcini Center for Brain Repair
University of Torino
Torino, Italy, EU

Silvia Mioletti
Department of Veterinary Morphophysiology
University of Torino
Torino, Italy, EU

Elio Raviola
Department of Neurobiology
Harvard Medical School
Boston, MA, USA

Alfredo Ribeiro-Da-Silva
Departments of Pharmacology and Therapeutics and Anatomy and Cell Biology
McGill University
Montreal, Quebec, Canada

Francesca Ruberti
EMBL Heildelberg, Cell Biology Programme
Heildelberg, Germany, EU

Wim Scheenen
Department of Cellular Animal Physiology
University of Nijmegen
Nijmegen, The Netherlands, EU

Isaura Tavares
Institute of Histology and Embryology
Faculty of Medicine and IBMC
University of Oporto
Oporto, Portugal, EU

Andrew J. Todd
Laboratory of Human Anatomy
Institute of Biomedical and Life Sciences
University of Glasgow
Glasgow, UK, EU

Franck Vandenbulcke
Laboratoire de Biologie Animale
Université de Lille I, CNRS Unit 8017
Villeneuve d'Ascq, Cedex, France, EU

Stefano Vicini
Department of Physiology and Biophysics
Georgetown University Medical School
Washington DC, USA

Jian Feng Wang
Department of Physiology and Biophysics
Georgetown University Medical School
Washington DC, USA

WeiJian Zhu
Department of Physiology and Biophysics
Georgetown University Medical School
Washington DC, USA

1

Analyses of Intracellular Signal Transduction Pathways in CNS Progenitor Cells

Elena Cattaneo and Luciano Conti
Institute of Pharmacological Sciences, University of Milano, Milano, Italy, EU

OVERVIEW

Growth factors such as epidermal growth factor (EGF), fibroblast growth factor (FGF), platelet derived growth factor (PDGF), and the neurotrophins and cytokines, such as interleukines and interferons, have a profound influence on the proliferation, survival, and differentiation of central nervous system (CNS) cells. They exert their roles by binding to their respective membrane-bound receptors and stimulating phosphorylation cascades (4). These receptors have been classified into two major groups: *(i)* receptors that have an intrinsic tyrosine kinase domain. These are also known as receptor protein tyrosine kinases (RPTK) and are exemplified by the epidermal growth factor receptor (EGFR) and neurotrophin receptors; and *(ii)* receptors such as those for the interleukins, which lack a kinase domain and use cytoplasmic tyrosine kinases.

In both cases, common strategies are employed for intracellular propagation of the external stimulus. These include receptor dimerization, transphosphorylation of the receptor chains, as well as recruitment and phosphorylation of cytoplasmic signaling components. Phosphotyrosine residues function as binding sites for intracellular signaling proteins containing SH2 (src homology 2) or phosphotyrosyl binding (PTB) domains (10), thus allowing specific protein–protein interactions. The kinase domain of activated RPTKs, for example, undergoes transphosphorylation of the dimerized receptors and then phosphorylates adaptor proteins like Shc and Grb2, which then activate Ras. Subsequent events involve three kinases steps: a MAPKKK-like Raf1, which phosphorylates a MAPKK-like MEK, which, finally, phosphorylates the MAPKs. MAPKs ultimately translocate to the nucleus where they phosphorylate transcription factors to generate both immediate (*c-fos* gene expression) and delayed gene transcription responses (11).

In recent years, there has been much progress in the identification and characterization of the intracellular signaling pathways that mediate responses by CNS cells to growth factors. The first evidence of protein kinase presence and activity in the

CNS dates back to the early 1980s, when a coincidental increase in the activity of pp60^{src} with active neurogenesis in the striatum and hippocampus indicated that changes in protein tyrosine phosphorylation occurred during maturation. More recently, it has been demonstrated that regulation of phosphorylation events on specific signaling proteins may affect the behavior of CNS cells. We have found that marked changes occur in the availability of the Shc(s) molecules during neuronal maturation. In particular, levels of ShcA adaptor decrease sharply in coincidence with neurogenesis in the brain (7). We have suggested that changes in Shc levels at different stages of development may affect the activity of downstream components of signaling pathways (for example Ras-MAPK) and thereby cause either proliferation or differentiation (4).

Other pathways have been identified which affect the survival of neural cells. For example, once the high affinity nerve growth factor (NGF) receptor TrkA is activated in PC12 cells, it stimulates cell survival through a Ras independent mechanism that utilizes the phosphatidyl inositol 3-kinase (PI3-K) pathway. PI3-K is an SH2-containing enzyme associated with a variety of receptor and nonreceptor protein tyrosine kinases. The enzyme is a heterodimer that phosphorylates the 3′ position on a variety of inositol lipids and serines on protein substrates. It has been shown that exposure of various cell types (including cerebellar neurons) to survival factors induces activation of the PI3-K and of its crucial mediator, a serine–threonine protein kinase named protein kinase B (PKB) or Akt. PKB promotes cell survival via three mechanisms: *(i)* phosphorylation and inactivation of the pro-apoptotic BAD (Bcl2-associated death promoter); *(ii)* phosphorylation of FKHRL1, a member of the Forkhead family of transcription factors, thus inhibiting its nuclear translocation and transcriptional activation of death genes; and *(iii)* inhibition of caspase-9 activation, that normally leads to cell death.

Other signaling pathways are also known to exert important roles in CNS cells. Among them, the JAK/STAT pathway is critical for the transduction of signals from activated cytokine receptors (3). The JAKs (for janus kinases) are cytoplasmic tyrosine kinases that, once activated by the stimulated receptors, can phosphorylate the STAT (for signal transducers and activators of transcription) transcription factors. These translocate into the nucleus where they bind to specific DNA elements (DNA response elements) situated upstream of genes induced by cytokines (6,8). For example, STAT3 phosphorylation and activation has been demonstrated to be crucial for astrocytes differentiation from CNS progenitor cells (1).

This chapter will describe methods employed to study signaling pathways in CNS cells.

BACKGROUND

Phosphospecific Antibodies

Old techniques for the study of tyrosine phosphorylated molecules require biosynthetic labeling with ^{32}P-labeled inorganic phosphate. This is intrinsically quite simple, but requires the use of radiolabeled compounds that involve the risks of radioactive manipulation. Thus, an important advance in the analysis of protein tyrosine phosphorylation, and the regulation of signaling by such phosphorylation, was the development of antibody technology to generate phosphospecific antibodies. These recognize a phosphorylated epitope in a given protein, thereby avoiding cross-reaction with other phosphoproteins or with the unphosphorylated form of the protein. In fact, once the primary sequence around a phosphorylation site is known, it becomes possible to generate antibodies against any synthetic polypeptides modeled on these phosphorylation sites. Thus, unlike conventional and general antiphosphoamino acid (i.e., antiphosphotyrosine, antiphosphoserine, antiphosphothreonine) antibodies, which have broad reactivity, antiphosphospecific antibodies have unique specificity toward the cognate proteins. These reagents have provided new insights into the phosphorylation processes that control protein function. Thus, for example, using antiphosphopeptide antibodies and immunoblotting analyses, it is possible to identify and isolate distinct phosphorylated species of a phosphoprotein that contains multiple phosphorylation sites. Such reagents not only facilitate conventional in vitro analyses of phosphoproteins, but also permit the in situ analysis of the abundance and phosphorylation (and activation) state of individual proteins in preparations of cells and tissue. These antibodies can therefore be used with immunofluorescence on fixed cultured cells, with immunohistochemistry on formalin-fixed paraffin-embedded tissue sections, as well as for immunoprecipitation and immunoblotting analyses.

Antiphosphoamino Antibodies and Immunoprecipitation Assays

The list of antibodies that recognize phosphoproteins is growing rapidly, but is still limited, while the methods for production of phosphospecific antibodies is time-consuming and very expensive. As a result, other more classical techniques must be used to study protein phosphorylation. Historically, the advance in analyses of protein tyrosine phosphorylation, and the regulation of signal transduction pathways by such phosphorylation, occurred with the production of polyclonal and monoclonal antiphosphotyrosine antibodies. These antibodies proved capable of recognizing phosphorylated tyrosine residues in the context of virtually any flanking peptide sequence. Antiphosphotyrosine antibodies have been most useful in the analyses of tyrosine phosphorylation of proteins with a technique that combines immunoprecipitation and immunoblotting. Typically in this procedure, a protein is immunoprecipitated either with conventional antiprotein antibody or with antiphosphotyrosine antibody, then immunoblotted with whichever of these two antibodies was not used for the immunoprecipitation. Immunoprecipitation is a procedure by which peptides or proteins that react specifically with an antibody are removed from solution. As usually practiced, the name of the procedure derives from the removal of antibody–antigen complexes by the addition of an insoluble form of an antibody binding protein such as protein A or protein G (Figure 1). The choice of immobilized antibody binding protein depends upon the species that the antibody was raised in. Protein A binds well to rabbit, cat, human, pig, and guinea pig IgG as well as mouse IgG_{2a} and IgG_{2b}. Protein G binds strongly to IgG from cow, goat, sheep, horse, rabbit, and guinea pig, as well as to mouse IgG_1 and IgG_3. Protein G can also bind bovine serum albumin (BSA). Thus, BSA should be added to buffers used with protein G. Alternatively, recombinant protein G without BSA binding sites can be used. Second, antibody coupled to Sepharose® or Protein G-Sepharose (both from Amersham Pharmacia Biotech, Little Chalfont, Bucks, England, UK) can also be used instead. It is not crucial that Sepharose be used as a matrix, because other polymerized agaroses or even fixed strains of Staphylococcus cells expressing high amounts of surface protein A can also be used. Analysis of the immunoprecipitate is usually done by electrophoresis and western blot, although other techniques can be employed.

Kinase Assays

Another assay commonly used in studies on the regulation by reversible phosphorylation of specific biochemical events is the analysis of the kinase activity. To assay the phosphotransfer reactions catalyzed by protein kinases, it is necessary first to identify a target substrate for the transfer reactions. Essentially, this means a substrate that is quite specific. The basic strategy for protein kinase assays is based, therefore, on the use of a labeled donor substrate so that when phosphotransferase activity is present in the enzyme sample, accumulation of the label in the protein or peptide acceptor substrate can be easily detected. The most frequently used protocol requires [γ-^{32}P]ATP as the donor substrate and a specific protein or peptide as the acceptor substrate. Phosphotransfer is detected as the accumulation of ^{32}P-labeled protein or substrate. Clearly, the source of enzyme activity is critical to obtain acceptable results. In fact, the primary requirement is that the kinase activity be stable both under the conditions used to prepare the enzyme and under those used in the assay.

Immunocytochemical Assays

Another method to investigate the state of phosphorylation and activation of a protein is to analyze its subcellular localization by immunocytochemistry. For several kinases and transcription factors, the phosphorylation event is associated with their activation and nuclear translocation, because this is the zone where they will exert their roles. This is the case of a family of transcription factors, the STATs. In latent cells, STAT proteins are found in the cytoplasm in a monomeric form, while in stimulated cells, STAT proteins are subjected to tyrosine phosphorylation by the activated JAK(s) or other tyrosine kinases. Tyrosine phosphorylation of STAT proteins is known to be associated with nuclear translocation and activation of latent DNA binding activity, leading to transcriptional activation of target genes. Nuclear translocation is therefore indicative of STATs activation. Thus, description of subcellular STAT localization will also reveal their phosphorylation state.

PROTOCOLS

Protocol for Immunoprecipitation

Materials and Reagents

All chemicals are from Sigma (St. Louis, MO, USA) unless otherwise stated.

<u>Lysis buffer</u>

- 25 mM Tris-HCl, pH 7.5
- 150 mM NaCl
- 1 mM Sodium orthovanadate
- 20 mM NaF
- 5 mM EDTA
- 1 mM EGTA
- 2% Triton® X-100
- 10% Glycerol
- 1 mM $ZnCl_2$

Stored and stable at 4°C. Before use, add:

- 1 mM Phenylmethylsulfonyl fluoride (PMSF)

PMSF is very labile in water. A more expensive but more stable alternative is 4-(2-aminoethyl)-benzenesulfonyl fluoride hydrochloride (AEBSF) (Pefabloc®; Roche Molecular Biochemicals, Mannheim, Germany).

- 10 µg/mL Pepstatin
- 10 µg/mL Leupeptin
- 10 µg/mL Aprotinin

Washing Buffer

- 50 mM Tris-HCl, pH 7.5
- 0.3 M NaCl
- 0.5% Triton X-100
- 0.02% NaN_3

The washing buffer can be stored at 25°C for a number of months.

- 0.1 M Sodium orthovanadate (phosphatase inhibitor). Dissolve powder in water at pH 10.0. Boil. Keep at room temperature for up to 1 week in the dark.
- 10 mg/mL of aprotinin and leupeptin in water. Keep frozen at -20°C in aliquots.
- 10 mg/mL of pepstatin A in ethanol. Keep frozen at -20°C in aliquots.
- Phosphate-buffered saline (PBS)

50% Protein A-Sepharose Solution

Resuspend protein A-sepharose slurry in PBS (100 mg of protein A-sepharose powder once hydrated corresponds to 400 µL of volume). Wait until all crystals are dissolved. Pellet by centrifugation (14 000× g) for 10 minutes at full speed in a centrifuge at 4°C. Discard supernatant and resuspend in PBS containing 1% Triton X-100. Repeat centrifugation. Discard the supernatant and add to the pellet an equal volume of immunoprecipitation buffer containing 100 µg/mL BSA and 0.1% sodium azide. Store at 4°C for up to 6 months.

2× Electrophoresis Buffer

- 250 mM Tris, pH 6.8
- 100 mM Glycine
- 4% Sodium dodecyl sulfate (SDS)
- 10% Glycerol

Procedure

The procedure here assumes that a concentration step is required to obtain enough protein material for the immunoprecipitation analysis. It is possible to lyse 2 to 4 100-mm diameter dishes using 1 mL of lysis buffer in order to concentrate the protein content.

1. Decant the medium and follow with a rapid rinse in PBS. After the cells are washed, drain and aspirate the excess PBS.
2. Add 250 to 350 µL of lysis buffer to the plate. Scrape the cells from the dish and transfer them to a microcentrifuge tube. The viscosity of the sample can be reduced by a brief sonication or by several passages through a 26 gauge needle.
3. Leave the sample on ice for 30 minutes.
4. Centrifuge (14 000× g) at 4°C for 10 minutes at 14 k rpm in a microcentrifuge.
5. Collect the supernatant and measure the protein concentration using a BCA kit and protein standards (Pierce Chemical, Rockford, IL, USA). The cell lysates can be stored at -80°C. In many protocols, a preclearing step is performed to remove molecules that bind nonspecifically to the insoluble protein A or protein G (steps 7–8).
6. Use 1 to 2 mg (in a 1000 µL volume) of total proteins for immunoprecipitation.
7. Add 25 µL of protein-A-sepharose solution (shake to suspend slurry before pipetting) and incubate on a tube turner for 1 hour at room temperature.
8. Centrifuge (14 000× g) for 1 minute at 14 k rpm in a microfuge and transfer the supernatant to another tube.
9. Add 1 to 5 µg of antibody to each tube and incubate for 4 hours at room temperature or overnight at 4°C.

10. Add 100 µL of protein A-sepharose solution (shake to suspend slurry before pipetting) and incubate on a tube turner for 3 hours at room temperature.
11. Centrifuge (14 000× g) for 1 minute at 14 k rpm in a microcentrifuge and retain the pellet.
12. Add 500 µL of cold washing buffer, vortex mix, spin for 1 minute at 14 k rpm in a microcentrifuge, and discard the supernatant.
13. Add 1 mL washing buffer, vortex mix, spin for 1 minute, discard the supernatant, and repeat 2 times.
14. Add 1 mL 10 mM Tris-acetate, pH 7.5, vortex mix, spin for 1 minute, and discard the supernatant.
15. Solubilize all samples in 30 to 50 µL of the SDS sample buffer, vortex mix, boil for 5 minutes, and centrifuge.
16. Save the supernatant. Immunoprecipitates in sample buffer can be stored almost indefinitely at -80°C. Storage for more than 7 to 10 days at -20°C can lead to deterioration.
17. Electrophorese the sample on a SDS-polyacrylamide gel and transfer the proteins to a polyvinylidene fluoride (PVDF) membrane (Cat. No. 1722026; Roche Molecular Biochemicals). To prevent tyrosine dephosphorylation during the transfer procedure, we recommend adding 100 µM sodium orthovanadate to the transfer buffer.
18. Incubate the membrane in 50 mL of blocking buffer for 1 hour at room temperature. We recommend not using a blocking buffer that contains dry milk because antiphosphotyrosine antibodies bind to a number of the milk proteins. The following blocking buffer can be used: 5% (wt/vol) BSA, 10 mM Tris-HCl, pH 7.4, 0.15 M NaCl.
19. Incubate the membrane in antiphosphotyrosine antibodies in blocking buffer for 2 hours at room temperature or overnight at 4°C. Several antiphosphotyrosine antibodies are commercially available. We recommend monoclonal antibodies (PY20; Transduction Laboratories, Lexington, KY, USA; or 4G10; Upstate Biotechnology, Lake Placid, NY, USA). It is possible to make a mixture of the two antibodies (PY20 1:1000 dilution, 4G10 1:2500 dilution).
20. Wash the membrane for 1 hour with TBST (Tris-buffered saline with Tween®) at room temperature with agitation. Replace the washing solution every 10 to 15 minutes.
21. Incubate the membrane with horseradish peroxidase-conjugated secondary antibody for 1 hour at room temperature.
22. Wash the membrane as described in step 20.
23. Detect by using ECL™plus reagent (Amersham Pharmacia Biotech) following the manufacturer's instructions.

Protocol for In Vitro Kinase Assay

Materials and Reagents

Immunoprecipitation Buffer

- 10 mM Tris, pH 7.4
- 1.0% Triton X-100
- 0.5% Nonidet® P-40
- 150 mM NaCl
- 20 mM Sodium fluoride
- 0.5 mM Sodium orthovanadate
- 1 mM EDTA
- 1 mM EGTA
- 10 mM Sodium pyrophosphate
- 0.2 mM PMSF

- 10 μg/mL Aprotinin
- 10 μg/mL Leupeptin
- 10 μg/mL Pepstatin A

Kinase Buffer

- 10 mM Tris, pH 7.4
- 150 mM NaCl
- 10 mM $MgCl_2$
- 0.5 mM Dithiothreitol (DTT)

Staining Solution

- 0.25% Coomassie® blue
- 45% Methanol
- 10% Acetic acid

Destaining Solution

- 40% Methanol
- 10% Acetic acid

5× Electrophoresis Buffer

- 625 mM Tris, pH 6.8
- 20% SDS
- 50% Glycerol
- 0.05% Bromophenol blue
- 10% β-Mercaptoethanol
- 0.1 M Sodium orthovanadate (phosphatase inhibitor). Dissolve powder in water, pH 10.0. Boil. Keep at room temperature for up to 1 week in the dark.
- 10 mg/mL of aprotinin and leupeptin in water. Keep frozen at -20°C in aliquots.
- 10 mg/mL of pepstatin A in ethanol. Keep at -20°C.
- [γ-^{32}P]ATP
- Cold ATP
- X-ray film (Amersham Pharmacia Biotech)

Procedure

Preparation of Cell Lysate

1. Wash cells on a confluent 100-mm culture dish with 10 mL of PBS.
2. Lyse the cells by addition of 1 mL cold immunoprecipitation buffer.
3. Scrape the cells off the dish and pass 5 to 10 times through a 26 gauge needle to disperse large aggregates, then incubate for 20 minutes on ice.
4. Centrifuge (14 000× *g*) for 30 minutes in a microcentrifuge at 14 k rpm at 4°C. Retain the supernatant.
5. Repeat centrifugation in a new tube.
6. The supernatant is the total cell lysate. Measure the protein concentration using the BCA kit and protein standards.

Immunoprecipitation of the Protein Kinase

7. Incubate the cell lysate (0.5–1.0 mg protein) with 2 to 5 μg soluble antibody.
8. Immunoprecipitate for 1 hour at 4°C on a tube turner.
9. Add 30 μL of 50% protein A-sepharose suspension and incubate for 2 hours at 4°C on a tube turner.
10. Wash the complexes by resuspension in immunoprecipitation buffer, followed by a 3-minute centrifugation in a microcentrifuge 14 k rpm at 4°C. Repeat the wash twice.
11. Collect the complexes by centrifugation for 3 minutes in a microcentrifuge at 14 k rpm at 4°C.

Kinase Assay

12. Wash the immunocomplexes three times at 4°C with kinase buffer.
13. Remove the supernatant by aspiration

and, with the pellet on ice, add 40 µL of kinase buffer containing the appropriate protein substrate at 1.0 mg/mL (e.g., 5 µg of acid denatured enolase), 25 µM cold ATP, 2.5 µCi [γ-^{32}P]ATP.

14. Mix carefully by pipetting up and down.
15. Transfer the tubes to 30°C in a circulating water bath and incubate for 15 minutes.
16. Add 15 µL of boiling 5× concentrated electrophoresis sample buffer to stop the reaction. Boil for 5 minutes.
17. Centrifuge the samples and electrophorese the soluble fractions.
18. Fix and stain the gel in staining solution for 45 minutes at room temperature.
19. Destain the gel in destaining solution for 2 hours. Change the destaining solution 4 to 5 times.
20. Dry the gel and expose it to X-ray film at -80°C. Kinase activity will be indicated by a band of phosphorylated protein substrate.

RESULTS

The use of antiphosphospecific antibodies allows rapid detection of phosphorylated proteins by using the western blot assay or immunocytochemistry. There is an extensive list of phosphospecific antibodies that are commercially available. This includes tyrosine kinase receptors (such as EGFR, Trks), cytoplasmic tyrosine kinases (Src, JAKs), Ser-Thr kinases (Raf, PKB, PKC [protein kinase C]), MAPKs (Erks, p38, JNKs [c-Jun N-terminal kinase]), and transcription factors (c-Jun, STATs, CREB [cAMP response-element binding protein]). In our laboratory, these antibodies are primarily used to analyze the phosphorylation and activation state of various signaling proteins that are involved in the responsiveness of CNS progenitor cells to growth factors (4,6). Figure 2A carries an example of western blot assay performed using a phospho-STAT3 antibody on primary CNS progenitor cells stimulated with ciliary neurotrophic factor (CNTF). This

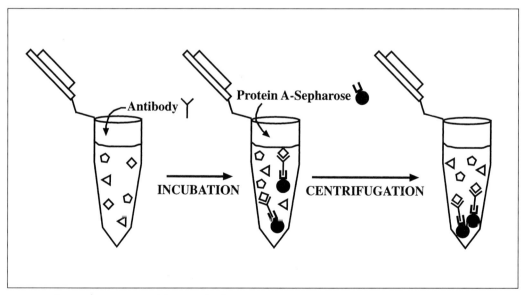

Figure 1. Schematic representation of the principle of immunoprecipitation. An antibody added to a mixture of proteins binds specifically to its antigen (◇). The antibody–antigen complex is absorbed from solution by addition of an immobilized antibody binding protein such as protein A-sepharose. Upon centrifugation, the antibody–antigen complex is collected in the pellet. Subsequent liberation of the antigen is achieved by boiling the sample in the presence of SDS.

antibody specifically recognizes the tyrosine 705 of phosphorylated STAT3 protein. This residue is normally phosphorylated by the JAKs and is important for STAT homo- or heterodimerization. In such assays, it is essential to evaluate the total content of the protein analyzed, in order to quantify the degree of activation. For this purpose, the same membrane is stripped and reacted with a STAT3 antibody recognizing the phosphorylated and nonphosphorylated STAT3 species (Figure 2A, lower panel).

Tyrosine phosphorylation of STAT proteins is known to be associated with nuclear translocation and activation of latent DNA binding activity, leading to transcriptional activation of target genes. Nuclear translocation is therefore indicative of STATs activation. In Figure 2B, we analyzed the occurrence of nuclear translocation of STAT3 after cytokine stimulation

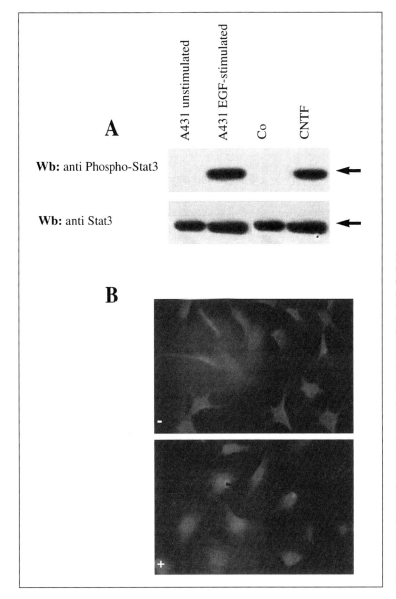

Figure 2. STAT3 activation in CNS progenitor cells following CNTF treatment. (A) Cell lysates obtained from untreated and CNTF-treated primary neuronal cultures generated from the E14 rat striatum primordia were immunoblotted with antiphospho-STAT3 antibody (upper panel). A tyrosine phosphorylated STAT3 band is visible in response to CNTF. The same membrane was stripped and reacted with anti-STAT3 antibody (lower panel). (B) STAT3 translocates into the nucleus of ST14A cells upon cytokine stimulation. The cells were incubated in the absence or presence of ligand for 15 minutes. The cellular distribution of STAT3 was examined by immunofluorescence. Untreated ST14A cells show a diffuse STAT3 distribution. On the other hand, STAT3 is detected exclusively in the nuclei of the treated cells, where a strong immunofluorescence signal is clearly visible (arrows). (See color plate A1.)

of ST14A CNS progenitor cells (5). STAT3 was found to vary its cellular distribution upon cytokine stimulation. In untreated ST14A cells (Figure 2B, upper panel), STAT3 antigene is visualized as a diffuse immunofluorescence both in the cytoplasm and, to a lesser extent, in the nucleus. Within 15 minutes following cytokine treatment (Figure 2B, lower panel), the nuclei of ST14A cells become brightly stained, indicative of nuclear translocation of STAT3.

The immunoprecipitation assay can be very informative for a number of situations. For example, it permits evaluation of direct and physical interaction between proteins by co-immunoprecipitation of one protein with another. We have applied this technique to identify signaling pathways activated by growth factors in embryonic CNS progenitor cells in vivo. We have exploited the fact that CNS progenitor cells in the embryonic brain are localized within the germinal zone which faces the ventricles. With this approach, we evaluated the extent of induced phosphorylation after injection of growth factors into the cerebral ventricular system of rat embryos (Figure 3, upper panel). In particular, we investigated whether ShcA adaptors, which are specifically expressed in CNS progenitor cells, were subjected to phosphorylation (7). For this purpose, lysates from EGF-treated and untreated (control) animals were subjected to immunoprecipitation with anti-ShcA antibodies, followed by western blot with antiphosphotyrosine antibodies. Figure 3 (lower panel) shows a basal level of $p52^{shcA}$ phosphorylation in lysates from control animals. On the other hand, when an equal amount of lysed telencephalic material obtained from EGF-injected embryos was subjected to the same immunoprecipitation procedure, $p52^{shcA}$ phosphorylation was markedly induced (Figure 3, lower panel A). In Figure 3, (lower panel B) the same membrane filter as in panel A was stripped and reacted with anti-ShcA monoclonal antibodies. As shown (Figure 3, lower panel, arrows), ShcA proteins were immunoprecipitated to the same extent in control and treated groups. The finding of in vivo EGF-induced $p52^{shcA}$ phosphorylation in CNS progenitor cells is further substantiated by the presence of a 170 kDa phosphorylated band in the EGF-treated lane (Figure 3, lower panel, arrow in A), which specifically reacts with a monoclonal antibody against the EGFR that is known to coprecipitate with phosphorylated ShcA proteins (not shown). Furthermore, since the activated ShcA proteins normally interact with the Grb2 adaptor protein, we evaluated the presence of Grb2 in the ShcA immunoprecipitates from control and EGF-stimulated groups. As shown in Figure 3 (lower panel C), the 23 kDa Grb2 protein coprecipitates with anti-ShcA antibodies in lysates from treated embryos, indicative of a functional activation of the Ras-MAPK pathway. We conclude that the use of immunoprecipitation assays not only allows evaluation of the phosphorylation state of a protein, but can also be used to dissect out the protein–protein interactions that occur in vitro and in vivo.

TROUBLESHOOTING

Lyses of the Cells

A critical step during cell lysis is the preservation of the phosphorylation state. This is accomplished by inhibition of protein phosphatases and other kinases by the addition to the lysis or homogenization buffer of inhibitors for serine–threonine kinases (such as sodium fluoride and okadaic acid) or for tyrosine kinases (such as sodium orthovanadate). The inclusion of chelating agents and protease inhibitors in the lysis buffer is also important. In fact,

1. Analyses of Intracellular Signal Transduction Pathways

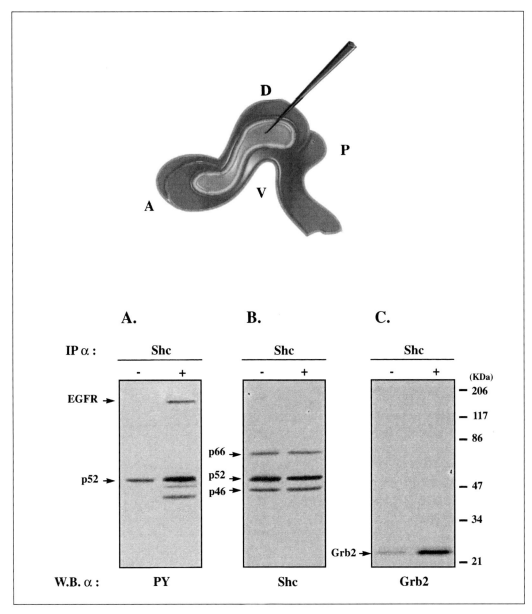

Figure 3. p52^{ShcA} phosphorylation and interaction with Grb2 in embryonic telencephalic vesicles following intraventricular injection of EGF. (Upper panel) Injection of growth factors into the telencephalic vesicles of E15 (embryonic day 15) embryos. Injection was performed by a procedure we have developed for embryonic transplantation of CNS progenitor cells (2,9). Ten microliters of EGF (10 ng/μL) were placed intraventricularly into E15 embryos. The figure shows a schematic drawing of the rat embryonic neural tube and of the ventricular system where the growth factors were delivered. At E15, the cells lining the neural tube are still immature and proliferating actively. Intraventricular injection of growth factors at this early stage of brain maturation can therefore target this particular population of CNS progenitor cells. (A, anterior; P, posterior; D, dorsal; V, ventral). (Lower panel) (A) Phosphotyrosine immunoblot of ShcA immunoprecipitates after in vivo EGF treatment. Ten minutes after injection, the embryos were removed, and the telencephalic vesicles were isolated and subjected to immunoprecipitation with anti-ShcA antiserum followed by immunodecoration with 4G10 antiphosphotyrosine antibodies. Phosphorylated p52^{ShcA} is indicated. Control embryos were injected with vehicle. A 170-kDa phosphorylated band (arrow) corresponding to the coprecipitated EGFR is also visible in the treated group. (B) Anti-ShcA immunoblot of the filter in panel A. The membrane was stripped and reacted with ShcA monoclonal antibody. As shown, ShcA proteins were immunoprecipitated to the same extent in control and EGF-treated groups. (C) Anti-Grb2 immunoblot of the same immunoprecipitates as in panel A. The arrow indicates the 23 kDa Grb2 protein, which coprecipitates with ShcA more abundantly in the treated group. (See color plate A2.)

many proteins such as kinases can be sensitive to limited proteolysis during the lysis. Compounds such as EDTA or EGTA chelate calcium and reduce the activity of calcium-activated proteases. The most commonly used protease inhibitors include PMSF, leupeptin, pepstatin A, antipain, and benzamidine. It is always a good idea to add these inhibitors to the lysis buffer just before it is used to lyse the cells.

Immunoprecipitation Assays

Like all immunochemical procedures, attention must be given to antibody cross-reactivity with other antigens. Nonspecific binding can be a particular problem if proteins that are immunologically distinct from the antigen are trapped in the pellets formed during immunoprecipitation. To reduce nonspecific binding, immunoprecipitation buffers usually contain a detergent that reduces hydrophobic interactions, a protein to block nonspecific binding sites, and high salt to reduce ionic interactions. Despite these precautions, nonspecific binding can occur. It is crucial, therefore, always to perform a control reaction where the antibody is replaced by a nonrelevant immunoglobulin (i.e., normal serum for polyclonal antibodies, control mouse ascytes fluid for ascytes, and isotype controls for purified mouse monoclonal antibodies).

Only use high quality siliconized tubes for immunoprecipitation. Many brands of Eppendorf® tube adsorb proteins which are released during the final boiling in SDS and contribute to the background. Some people transfer the sepharose pellet to the new tube immediately before addition of 2× electrophoresis buffer.

Only use ultra pure BSA in the buffer to resuspend protein A-sepharose.

Sensitivity can be a problem, especially when the antigen is a minor component of the protein pool. Effort should be made to use as much protein in the immunoprecipitation reaction as possible. Start with 1 or 2 mg of total protein extract.

Kinase Assays

The oxidation state of cysteines and the state of disulfide linkages may influence kinase activity. Inclusion of a reducing agent (2-β-mercaptoethanol or DTT) can therefore be essential to preserve enzyme activity.

It is important to choose the correct substrate for the kinase you are assaying; specific substrate information should be collected from previously published works. Synthetic peptide substrates are also commercially available.

For many kinases, it is important to identify the more appropriate Mg^{2+} and Mn^{2+} concentrations.

ACKNOWLEDGMENTS

The work of the authors is supported by Telethon Italy to E.C. (No. E840) and L.C. (No. E1025).

REFERENCES

1. **Bonni, A.**, Y. Sun, M. Nadal-Vicens, A. Bhatt, D.A. Frank, I. Rozovsky, N. Stahl, G.D. Yancopoulos, and M.E. Greenberg. 1997. Regulation of gliogenesis in the central nervous system by the JAK-STAT signaling pathway. Science *278*:477-483.
2. **Cattaneo, E.**, L. Magrassi, G. Butti, L. Santi, A. Giavazzi, and S. Pezzotta. 1994. A short term analysis of the behaviour of conditionally immortalized neuronal progenitors and primary neuroepithelial cells implanted into the fetal rat brain. Brain Res. Dev. Brain Res. *83*:197-208.
3. **Cattaneo, E.**, C. De-Fraja, L. Conti, B. Reinach, L. Bolis, S. Govoni, and E. Liboi. 1996. Activation of the JAK/STAT pathway leads to proliferation of ST14A central nervous system progenitor cells. J. Biol. Chem. *271*:23374-23379.
4. **Cattaneo, E. and P.G. Pelicci.** 1998. Emerging roles for SH2/PTB-containing Shc adaptor proteins in the developing mammalian brain. Trends Neurosci. *21*:476-481.
5. **Cattaneo, E. and L. Conti.** 1998. Generation and characterization of embryonic striatal conditionally

immortalized ST14A cells. J. Neurosci. Res. 53:223-234.
6. **Cattaneo, E., L. Conti, and C. De-Fraja.** 1999. Signalling through the JAK-STAT pathway in the developing brain. Trends Neurosci. 22:365-369.
7. **Conti, L., C. De-Fraja, M. Gulisano, E. Migliaccio, S. Govoni, and E. Cattaneo.** 1997. Expression and activation of SH2/PTB-containing ShcA adaptor protein reflects the pattern of neurogenesis in the mammalian brain. Proc. Natl. Acad. Sci. USA 94:8185-8190.
8. **De-Fraja, C., L. Conti, L. Magrassi, S. Govoni, and E. Cattaneo.** 1998. Members of the JAK/STAT proteins are expressed and regulated during development in the mammalian forebrain. J. Neurosci. Res. 54:320-330.
9. **Magrassi, L., M.E. Ehrlich, G. Butti, S. Pezzotta, S. Govoni, and E. Cattaneo.** 1998. Basal ganglia precursors found in aggregates following embryonic transplantation adopt a striatal phenotype in heterotopic locations. Development 125:2847-2855.
10. **Pawson, T.** 1995. Protein modules and signalling networks. Nature 373:573-580.
11. **Segal, R.A. and M.E. Greenberg.** 1996. Intracellular signaling pathways activated by neurotrophic factors. Annu. Rev. Neurosci. 19:463-489.

2

Confocal and Electron Microscopic Tracking of Internalized Neuropeptides and/or Their Receptors

Alain Beaudet, Alexander C. Jackson, and Franck Vandenbulcke
Montreal Neurological Institute, McGill University, Montreal, QC, Canada

OVERVIEW

This chapter describes confocal and electron microscopic methods for tracking peptide ligands and/or their receptors following their internalization in cell cultures or brain slices. The confocal microscopic techniques are based upon the use of high affinity fluorescent ligands that were originally developed in our laboratory to study the fate of internalized neurotensin (NT), somatostatin (SRIF), and opioid peptides (9,13,17). The electron microscopic techniques are adapted from the pre-embedding immunogold method developed by Virginia Pickel and her team (5) as applied by us to study the effect of ligand exposure on the subcellular distribution of various subtypes of neuropeptide receptors.

As with any recipe, the success of these different methods lies with the quality of the underlying ingredients. In other words, selective high affinity ligands and specific, sensitive antibodies are sine qua non. We have recently reviewed the factors to be considered when selecting a fluorescent ligand for confocal imaging studies (2). A vast selection of fluorescent peptides is currently available from Advanced Bioconcept (Montreal, QC, Canada), a subsidiary of NEN Life Science Products (Boston, MA, USA). However, not all of these peptides are applicable to the type of study described below, and it is strongly recommended that the ligand of choice be first tested in a heterologous transfection system. Neuropeptide receptor antibodies have also become widely available through a variety of commercial sources. Here again, however, these antibodies may not all be ideal for the type of double labeling or electron microscopic work detailed below. Furthermore, some of these antibodies may not recognize an important subset of receptors, because the sequence against which they are directed is either glycosylated or otherwise conformationally modified. Care should be taken, therefore, to first test the selected antibodies thoroughly in a model system and, preferably, to identify recognized molecules by Western blot.

Cellular and Molecular Methods in Neuroscience Research
Edited by A. Merighi and G. Carmignoto

BACKGROUND

The interaction between neuropeptides and some of their complimentary G protein-coupled receptors (GPCR) has been shown to result in the endocytosis of the receptor–ligand complex. This mechanism, referred to as ligand-induced internalization, has long been known to play a key role in receptor sequestration and resensitization (12) and was recently proposed to be involved in cell signaling (6,20).

Most of our knowledge concerning the fate of internalized ligands and/or receptors is derived from studies of single transmembrane domain receptors. For instance, the transferrin receptor has long been known to be constitutively internalized via clathrin-coated pits into early endosomes (19). In the acidic environment of endosomes, iron dissociates from transferrin, and both transferrin and its receptor return to the cell surface in recycling endosomes (7). Much less is known, however, concerning the fate of internalized GPCRs or their ligands. With regard to receptors, most of the available evidence is derived from studies of the prototypical GPCR, the β_2-adrenergic receptor, which was documented to recycle back to the plasma membrane following ligand dissociation in the acidic environment of endosomes (23). Other GPCRs, however, such as the luteinizing hormone receptor, are degraded in lysosomes (14). As for GPCR ligands, virtually nothing is known of their postinternalization trafficking, with the exception of some neuropeptides that were shown to be targeted to lysosomes for degradation (14,15).

The techniques described below were developed by us to monitor the fate of internalized neuropeptide receptor–ligand complexes by confocal and electron microscopy. Visualization of bound and internalized ligand molecules proved the most challenging, since peptide ligands are prone to dissociate from their receptors or to leak out from intracellular compartments during histological processing. We tackled this problem by resorting to fluorescent ligands and minimizing histological steps prior to their visualization by confocal microscopy. However, internalized ligand molecules may also be detected by other techniques, such as autoradiography, as described elsewhere (4). Receptors are easier than their ligands to track at cellular and subcellular levels, since they are membrane-bound and are therefore preserved in situ during histological processing. Furthermore, they are strongly antigenic and are therefore amenable to immunocytochemical detection. For studies in heterologous transfection systems, this detection may be facilitated by tagging the receptors with immunogenic or fluorescent sequences (1,15). The techniques that we describe below were developed for studying GPCR trafficking in cell cultures and brain slices. However, similar approaches have been used equally effectively by others to study the effect of agonist exposure on receptor trafficking in vivo (8,16).

PROTOCOLS

Protocol for Tracking Internalized Ligands by Confocal Microscopy

Principle of Technique

The fate of internalized ligands is monitored by confocal microscopy following the labeling of cells in culture or of brain slices ex vivo with nanomolar concentrations of fluorescent derivatives of either native or metabolically stable analogs of peptide ligands. The distribution of the label may be analyzed either immediately after ligand exposure, as described below for studies in cell cultures, or after varying periods of chasing with physiological buffer, as described below for studies in brain slices.

Studies in Cell Culture

The protocol described here was used to monitor the fate of internalized NT (18), somatostatin (17,21), and opioid peptides (13) in transfected COS-7 cells or in primary neuronal cultures.

Materials and Reagents

- COS-7 cells transfected with cDNA encoding the appropriate receptors (for details on transfection procedure see Reference 17) or:
- Neuronal cultures from embryonic or neonatal rat brain prepared as previously described (18,22).
- α-Bodipy-neurotensin 2-13 (fluo-NT), α-Bodipy-[D-Trp^8]somatostatin (fluo-SRIF), α-Bodipy-dermorphin, α-Bodipy-deltorphin. These fluorescent compounds were originally synthesized and purified for us by Dr. J.-P. Vincent (University of Nice-Sophia Antipolis, France). They are currently available from NEN Life Science Products.
- 12-mm polylysine-treated glass coverslips (25 µg/mL polylysine, 15 min at room temperature) (Sigma, St. Louis, MO, USA).
- Earle's buffer: 50 mM 4-(2-hydroxyethyl)-1-piperazineethanesulfonic acid (HEPES) buffer, pH 7.4, containing 140 mM NaCl, 5 mM KCl, 1.8 mM $CaCl_2$, 3.6 mM $MgCl_2$ (all salts are from Sigma).
- Supplemented Earle's buffer: Earle's buffer containing 0.1% bovine serum albumin (BSA), 0.01% glucose, and 0.8 mM 1,10-phenanthroline (peptidase inhibitor), pH 7.4.
- Hypertonic acid buffer: 0.2 M acetic acid and 0.5 M NaCl in Earle's buffer, pH 4.0.
- Aquamount (Polysciences, Warrington, PA, USA).

Procedure

1. For experiments on transfected epithelial cells, plate the cells as a monolayer on 12-mm polylysine-coated glass coverslips and let them adhere for 1 to 2 hours at 37°C. For experiments in primary cultures, plate the cells onto polylysine-coated glass coverslips and allow them to grow in a humidified atmosphere at 37°C and 5% CO_2 until fully differentiated (6–10 days).
2. Preincubate the cells for 10 minutes at 37°C in supplemented Earle's buffer.
3. Incubate the cells for various periods of time (5, 10, 15, 30, 45, and 60 min) with 10 to 20 nM of the appropriate fluorescent ligand in supplemented Earle's buffer. For determination of nonspecific labeling, add a hundredfold concentration of nonfluorescent peptide or antagonist to the incubation medium.
4. At the end of the incubation, rinse the cells 3 times with ice-cold Earle's buffer or with hypertonic acid buffer to dissociate surface-bound ligand. At this point, cells may be fixed with 4% paraformaldehyde in 0.1 M phosphate buffer, pH 7.4. The latter procedure offers the advantage of allowing for coimmunolocalization of cellular antigens (see below).
5. Air-dry the cells rapidly and mount them up on glass slides with Aquamount. It is imperative that the cells themselves not be exposed to an aqueous medium (unless they were fixed) as this would promote dissociation of receptor–ligand complexes.
6. Examine by confocal microscopy. Images may be acquired as single midcellular optical sections or through multiple serial Z levels at 32 scans per frame.

Studies in Brain Slices

The protocol described below was used to monitor the fate of internalized NT in slices from rat ventral midbrain tegmentum (11) and basal forebrain (10).

Materials and Reagents

- Adult male Sprague-Dawley rats.
- Fluorescent ligand (same as above).
- Ringer buffer: 130 mM NaCl, 20 mM $NaHCO_3$, 1.25 mM KH_2PO_4, 1.3 mM $MgSO_4$, 5 mM KCl, 10 mM glucose, and 2.4 mM $CaCl_2$.
- 4% Paraformaldehyde (PFA) (Electron Microscopy Science, Fort Washington, PA, USA): 4% PFA in 0.1 M phosphate buffer, pH 7.4.
- Cryoprotectant solution: 30% sucrose in 0.1 M phosphate buffer, pH 7.4.
- Cryoprotectant solution: 30% sucrose in 0.1 M phosphate buffer, pH 7.4.
- Aquamount (Polyscience)

Procedure

1. Following decapitation of the rat, rapidly remove and immerse the brain in a cold oxygenated (95% O_2, 5% CO_2) Ringer buffer.
2. Cut 300 to 400-μm-thick slices through the regions of interest with a Vibratome.
3. Equilibrate the slices for 45 minutes in oxygenated Ringer buffer at room temperature.
4. Superfuse the slices for 3 minutes at 37°C with 20 to 40 nM fluorescent ligand in Ringer buffer.
5. Rinse with oxygenated Ringer buffer for 5, 10, 15, 30, 45, or 60 minutes at 37°C. To control for nonspecific labeling, incubate additional slices in the presence of 100- to 1000-fold excess of nonfluorescent probe or antagonist.
6. After rinsing, fix the slices for 30 minutes at room temperature with 4% PFA.
7. Immerse the slices overnight in the cryoprotectant solution, flatten on tissue chuck, snap freeze in isopentane at -60°C, and resection at 45 μm thickness in the plane of the slice on a freezing microtome.
8. Mount frozen sections on gelatin-coated glass slides with Aquamount and examine by confocal microscopy.

Protocol for Simultaneous Detection of Internalized Ligand and of either Receptors or Cell Compartment Markers

Principles of Technique

Combination of fluorescent ligand binding and immunohistochemistry makes it possible to simultaneously track down ligand and receptor following neuropeptide binding and internalization. Intracellular trafficking of ligand can also be monitored through combined visualization of the fluorescent ligand and of specific markers of intracellular compartments. Because significant amounts of ligand are lost in the course of immunohistochemical processing, this type of study is best performed in transfected cells as these express high concentrations of receptors and thus bind and internalize commensurately large amounts of fluorescent ligand. Presumably, the same type of approach should be applicable to cells or tissue slices expressing endogenous receptors, provided that the ligand is cross-linked to the receptor prior to immunohistochemical processing.

Materials and Reagents

- COS-7 cells transfected with cDNA encoding either native or epitope-tagged receptors.

- Appropriate α-Bodipy-labeled fluorescent ligand.
- Antibodies directed against either the receptor itself or against the immunogenic epitope in the case of epitope-tagged receptors; or antibodies directed against compartment-specific cellular antigens (e.g., rab proteins, lamp proteins, etc.).
- Fluorescein isothiocyanate (FITC)-tagged secondary antibodies.
- Normal serum from the same species as the secondary antiserum.
- Phosphate-buffered saline (PBS): 0.9% NaCl in 0.1 M phosphate buffer, pH 7.4.
- Earle's buffer.
- Polylysine (Sigma).
- Aquamount (Polyscience).

Procedure

1. Plate the transfected cells on 12-mm polylysine-coated glass for 1 to 2 hours at 37°C.
2. Incubate the transfected cells with the fluorescent ligand (20 nM) for various periods of time at 37°C as described above and rinse 3 times in ice-cold Earle's buffer.
3. Fix the cells with 4% PFA for 20 minutes at room temperature.
4. Rinse twice with PBS.
5. Preincubate the cells for 20 minutes in PBS containing 3% normal serum.
6. Incubate for 60 minutes at room temperature with appropriate dilution of primary antibody in PBS containing 1% normal serum and 0.02% Triton® X-100.
7. Rinse 3 × 5 minutes with PBS.
8. Incubate with the FITC-tagged secondary antibody diluted 1:100 to 1:500 in PBS for 30 minutes at room temperature.
9. Rinse 3 × 5 minutes in PBS.
10. Mount the coverslips, cell-side down on glass slides with Aquamount and examine by confocal microscopy. FITC signal is imaged by exciting samples with 488 nm and Bodipy red signal by exciting samples with 568 nm.

Protocol for Monitoring the Effect of Ligand Exposure on the Subcellular Distribution of Neuropeptide Receptors

Principles of Technique

The present technique applies standard pre-embedding immunogold procedures, as originally developed in V. Pickel's laboratory (5), to the electron microscopic detection of neuropeptide receptors in transfected epithelial cells, primary neuronal cultures, or brain slices, following stimulation by an unlabeled agonist. As for confocal microscopic studies, labeling may be carried out either by pulse chase as described below for studies in cell culture, or immediately after stimulation with the agonist, as described below for studies in brain slices.

Studies in Cell Cultures

Materials and Reagents

- COS-7 cells transfected with cDNA encoding the appropriate receptors (for details on transfection procedure see Reference 17) or:
- Neuronal cultures from embryonic or neonatal rat brain prepared as previously described (18,22).

To Be Prepared Fresh on Day 1

- 0.2 M Sörensen's phosphate buffer (SPB), pH 7.4: 154 mM Na_2HPO_4, 23 mM $NaH_2PO_4H_2O$. Do not adjust pH.

- 0.1 M Tris-buffered saline (TBS), pH 7.4: 1.2% (wt/vol) Trizma base, 0.9% NaCl. Adjust to pH 7.4 with HCl.
- 2% PFA in 0.1 M SPB, pH 7.4.
- 2% Acrolein (Electron Microscopy Science) and 2% PFA in 0.1 M SPB.
- Blocking buffer: 1.5% normal serum (Sigma) from the same species as secondary antibody diluted in 0.1 M TBS.
- Antibody dilution buffer: 0.05% Triton X-100 and 0.5% normal serum in 0.1 M TBS.
- Primary antibodies directed against either the receptor itself or an epitope tag and diluted in antibody dilution buffer (dilution must be worked out for each antibody).
- Earle's buffer: 140 mM NaCl, 5 mM KCl, 1.8 mM $CaCl_2$, 0.9 mM $MgCl_2 \cdot 6 H_2O$, 25 mM HEPES.
- Appropriate receptor agonist diluted in concentrations ranging from 10 nM to 10 µM in binding buffer consisting of 0.8 mM 1,10-phenanthroline, 0.1% D-glucose, 1% BSA in Earles' buffer.

To be Prepared Fresh on Day 2

- 0.01 M PBS: 0.01M SPB in double-distilled water and 0.9% NaCl. Adjust to pH 7.4.
- Washing incubation buffer: 0.5% gelatin stock and 8.0% (wt/vol) BSA in 0.01 M PBS.
- 2% glutaraldehyde (Electron Microscopy Science) in 0.01 M PBS.
- 1 nm gold particle-tagged secondary antibodies directed against species in which primary antibody was raised (IgG-gold conjugate; Amersham Pharmacia Biotech, Little Chalfont, Bucks, England, UK), diluted 1:20 in washing incubation buffer.

- 0.2 M citrate buffer: 5.95% (wt/vol) sodium citrate (trisodium citrate, dehydrated) in double-distilled water; adjust to pH 7.4 with 0.2 M citric acid (2.1 g in 50 mL distilled water).
- 2% osmium tetroxide (OsO_4) in 0.2 M SPB (prepare immediately before use and keep in the dark at all times).
- Silver intensification kit (Amersham Pharmacia Biotech).

Procedure

These experiments are carried out on cells that have been cultured directly into the bottom of plastic culture dishes (21).

Day 1

Incubate cells with desired concentration of agonist. For pulse-chase labeling over multiple time points: *(i)* preincubate in binding buffer for 5 minutes at 4°C; *(ii)* pulse with agonist dissolved in binding buffer for 30 minutes at 4°C; and *(iii)* chase with binding buffer for various time points at 37°C.

1. Fix with 2% acrolein in 2% PFA for 20 minutes at room temperature.
2. Post-fix with 2% PFA for 20 minutes at room temperature.
3. Rinse 2 × 10 minutes in 0.1 M TBS.
4. Incubate in blocking buffer for 30 minutes.
5. Incubate overnight at 4°C with appropriate dilution of primary antibody directed against the receptor in antibody dilution buffer.

Day 2

1. Rinse 3 × 10 minutes in 0.01 M PBS.
2. Rinse for 10 minutes in washing incubation buffer.
3. Incubate for 2 hours at room tempera-

ture with the IgG-gold conjugate.
4. Rinse for 5 minutes in washing incubation buffer.
5. Rinse 3 × 5 minutes in 0.01 M PBS.
6. Fix 10 minutes with 2% glutaraldehyde.
7. Rinse for 5 minutes in 0.01 M PBS.
8. Rinse twice in 0.2 M citrate buffer.
9. Silver intensification: mix solutions A and B in each well and develop for 7 minutes.
10. Rinse twice in 0.2 M citrate buffer.
11. Rinse for 10 minutes in 0.1 M SPB.
12. Postfix for 10 minutes in 2% osmium tetroxide (in the dark).
13. Dehydrate in graded ethanols:
50% EtOH for 2 × 5 minutes.
70% EtOH for 2 × 5 minutes.
80% EtOH for 5 minutes.
90% EtOH for 5 minutes.
95% EtOH for 10 minutes.
100% EtOH for 2 × 15 minutes.
14. Embed in Epon as follows:
   a. Apply 1:1 Epon: propylene oxide solution for 1 minute and aspirate.
   b. Apply 1:3 Epon: propylene oxide solution for 3 minutes and aspirate.
   c. Apply one drop of 100% Epon to the surface of the cells.
15. Place the flush surface of the cylindrical plastic mold onto the bottom of the well. Ensure a good seal between the mold and the bottom of the well, and that the mold is perpendicular to the bottom.
16. Carefully fill in the area between the inside surface of the culture plate and the plastic mold with plasticine modeling clay.
17. Fill the plastic mold with 100% epon.
18. Replace the 4-well plate cover, add lead weights to the lid, and cure in a 60°C oven for 13 to 16 hours.
19. Remove the plasticine modeling clay and crack off the polymerized Epon blocks from the surface of the culture plate.
20. Examine the bottom surface of the polymerized Epon block under the dissecting microscope for labeled cells.
21. Trim the block around the labeled cells.
22. Incubate in a 60°C oven for at least another 24 to 48 hours before cutting with the ultramicrotome.

Studies in Brain Slices

Materials and Reagents

- Adult male Sprague-Dawley rats (200–250 g).
- Ringer buffer: 124 mM NaCl, 5 mM KCl, 1.2 mM NaH_2PO_4, 2.4 mM $CaCl_2$, 1.5 mM $MgSO_4$, 26 mM $NaHCO_3$, 10 mM glucose, pH 7.4.
- Fixative: 4% PFA and 0.3% glutaraldehyde in 0.1 M SPB, pH 7.4.
- Cryoprotectant: 0.1 M SPB, 25% sucrose, 3% glycerol.
- Isopentane at -70°C.
- Liquid nitrogen.
- Same complement of immunohistochemical reagents and buffers as listed for studies in cell cultures.

Procedure

Day 1

1. Decapitate the rats and rapidly remove the brains.
2. Block and section the region(s) of interest on a vibratome and collect slices (100 μm) in ice-cold Ringer buffer, continuously oxygenated by a mixture of 95% O_2 and 5% CO_2.
3. Equilibrate slices in Ringer buffer for 40 minutes at room temperature.
4. Preincubate slices in Ringer buffer for 15 minutes at 37°C.

5. Incubate slices in Ringer buffer containing various concentrations of agonist for 10 to 60 minutes at 37°C.
6. Fix slices with fixative.
7. Rinse twice in 0.1 M SPB.
8. To permeabilize the tissue, incubate sections in cryoprotectant solution for 30 minutes, freeze in isopentane at -70°C, dip in liquid nitrogen, thaw in 0.1 M SPB at room temperature.
9. Immerse in blocking buffer for 30 minutes.
10. Incubate overnight at 4°C with appropriate dilution of receptor antibody in antibody dilution buffer.

Day 2

1. Carry out immunolabeling with secondary antibody, postfixation in 2% OsO_4 for 40 minutes, and dehydration of slices as described for cell cultures above (day 2, steps 1 through 13).
2. Then, embed slices as follows:
   a. Immerse in 1:1 Epon: propylene oxide solution for 30 minutes and aspirate.
   b. Immerse in 1:3 Epon: propylene oxide solution for 30 minutes and aspirate.
   c. Immerse in 100% Epon overnight at 4°C.
3. Flat-embed the sections in between two sheets of acetate film; lay on a flat, smooth surface, and add lead weights to the top.
4. Cure in a 60°C oven for 24 to 48 hours.
5. Gently remove one of the acetate sheets from the surface of the embedded tissue.
6. Add a thin film of cyanoacrilate glue to the surface of the embedded tissue and quickly affix to the flush bottom surface of a polymerized Epon block.
7. Trim the block around the labeled cells.
8. Cure at 60°C for at least another 24 to 48 hours before cutting with the diamond knife.

RESULTS AND DISCUSSION

Confocal Microscopic Studies

Studies in Cell Cultures

The result of an experiment in which the fate of fluo-NT, specifically bound to high affinity NT1 receptors, was monitored over time in COS-7 cells transfected with cDNA encoding the NT1 receptor is illustrated in Figure 1. Cells were exposed to 20 nM fluo-NT for periods ranging between 5 and 60 minutes, and the intracellular distribution of the ligand was visualized by confocal microscopy following hypertonic acid stripping of surface-bound ligand. At short time intervals (0–30 min), the label formed small "hot spots" distributed throughout the cytoplasm of the cells, but sparing the nucleus (Figure 1A). At later time points (>30 min), these endosome-like particles decreased in number and progressively clustered towards the center of the cells next to the nucleus (Figure 1B). This fluorescent labeling was specific in that it was not observed in nontransfected parent cells or in transfected cells incubated in the presence of a hundredfold concentration of nonfluorescent NT (Figure 1C).

Figure 2 illustrates the results of an experiment in which we monitored in parallel the fate of a fluorescent analog of somatostatin, fluo-SRIF, with that of one of its receptors, sst_{2A}, in COS-7 cells transfected with cDNA encoding the sst_{2A} receptor subtype. sst_{2A} receptors were labeled by immunocytochemistry using an antibody directed against its C terminal sequence (kindly provided by Dr. Agnes Schonbrunn, University of Texas). To obtain a good compromise between immunocytochemical detection of the receptor and preservation of internalized ligand, all immunolabeling steps were shortened as much as they could, and double-labeled cells were examined immediately after mounting. After short incubations (0–20 min), receptor and ligand molecules were extensively colocalized at the outskirts of the cells (Figure 2, A and A'). By contrast, after longer incubation periods, receptor and ligand were almost entirely dissociated (Figure 2, B and B'). Indeed, while the ligand was confined to the core of the cells, the receptors had largely recycled back to the cell surface.

To characterize the intracellular routing of fluo-SRIF, detection of the fluorescent label was combined with the immunocytochemical detection of marker proteins for different endocytic compartments. Thus, following short (0–20 min) incubations of the transfected cells with fluo-SRIF, most of the intracellular ligand was detected in compartments that immunostained positively for rab 5A, a small GTP-binding protein known to be associated with early endosomes (not shown). By contrast, at longer time intervals (45 min) it was extensively colocalized with syntaxin 6, a marker of the Trans-Golgi Network (TGN) and of the pericentriolar recycling endosome (Figure 2, C and C').

Studies in Brain Slices

Examination of rat brain slices pulse-labeled with fluo-NT and chased for short periods of time (5–10 min) revealed a fairly diffuse distribution of the fluorescent label over nerve cell bodies and neuronal processes within regions documented to express high concentrations of NT1 receptors, such as the basal forebrain (10) and the ventral midbrain (11) (Figure 3A). This distribution was specific in that it was totally prevented by the addition of a hundredfold concentration of nonfluorescent ligand. It resulted from clathrin-mediated internalization, as it was entirely blocked by the endocytosis inhibitor, phenylarsine oxide. At high magnification, the label was found to be accumulated within small endosome-like structures distributed throughout perikarya and dendrites. After longer periods of chase (20–60 min), the label had totally disappeared from neuronal processes and was densely concentrated within perikarya (Figure 3B), suggesting that it had been transported retrogradely from the dendrites to the cell bodies. Furthermore, at the level of neuronal perikarya themselves, a decrease in number but an increase in size of labeled endosomes was observed, suggesting that the original endocytic vesicles had coalesced over time. In addition, there was a decrease in their mean distance from the nuclear envelope, suggesting a progressive migration of the late endosomes towards a juxtanuclear recycling compartment.

Electron Microscopic Studies

Studies in Cell Cultures

As can be seen in Figure 4, SRIF sst_5 receptors, immunolabeled in COS-7 cells transfected with cDNA encoding the sst_5 receptor subtype, were detected in the form of small rounded silver-intensified gold particles. In the absence of stimulation by the agonist, approximately 25% of the gold particles were associated with plasma membranes versus 75% with intracellular vesicular organelles (21) (Figure 4A). By contrast, after 10 minutes of exposure to 20 nM D-Trp^8-SRIF, the proportion of gold particles associated with the membrane increased significantly (to 47%), while the mean distance of intracellular particles from the plasma membrane decreased significantly (from 2–3 to 0.25–1 μm), indicating a SRIF-induced mobilization of receptors from the cell center to the periphery (Figure 4B). Furthermore, in cells exposed to SRIF, gold particles were detected more frequently in

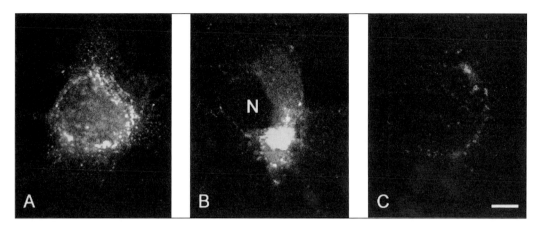

Figure 1. Confocal microscopic imaging of internalized fluo-NT in COS-7 cells transfected with cDNA encoding the NT1 receptor. Images were acquired as single midcellular optical sections at 32 scans per frame. (A) After a 20-minute incubation at 37°C, acid-wash resistant fluo-NT labeling is segregated within small endosome-like particles distributed throughout the cytoplasm of the cell. (B) At 45 minutes, intracellular fluorescent particles are less numerous and clustered next to the nucleus. (C) This internalization is receptor-mediated, as it is no longer detectable when the incubation is carried out in the presence of an excess of nonfluorescent NT. Abbreviation: N, nucleus. Scale bar: 10 μm.

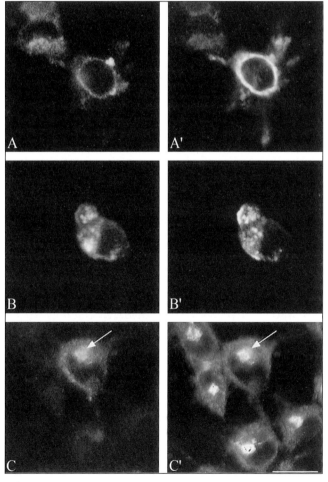

Figure 2. Dual localization of internalized fluo-SRIF (A, B, and C) and of sst_{2A} immunoreactivity (A' and B') or of the TGN marker syntaxin 6 (C') in COS-7 cells transfected with cDNA encoding the sst_{2A} receptor. Cells were incubated with 20 nM fluo-SRIF at 37°C for 5 to 45 minutes, fixed, and immunocytochemically reacted with either sst_{2A} or syntaxin 6 antibodies. (A and A') After 5 minutes of incubation, there is complete overlap between the ligand (A) and sst_{2A} immunoreactivity (A') at the periphery of the cell. (B and B') After 30 minutes of incubation, fluo-SRIF (B) and sst_{2A} immunoreactivity (B') are both distributed more centrally within the cells and are partially dissociated. (C and C') At 45 minutes, fluo-NT (C) is concentrated next to the nucleus, where it colocalizes extensively (arrows) with the TGN marker syntaxin 6 (C'). Abbreviation: N, nucleus. Scale bar: 10 μm. (See color plate A3.)

clathrin-coated pits at the cell membrane (Figure 4C) as well as in coated vesicles in the subplasmalemmal zone (Figure 4D) than in nonexposed cells, indicating that sst_5 receptor internalization proceeded through clathrin-mediated mechanisms.

Studies in Brain Slices

By contrast, in brain slices immunolabeled with antibodies directed against sst_{2A} receptors, exposure to the agonist (40 min) resulted in a significant decrease in membrane to intracellular receptor ratios (3) (Figure 5). The same decrease was apparent in the two brain regions sampled (based on their high sst_{2A} receptor content), namely the basolateral amygdala and the claustrum (Figure 5). It was entirely prevented by the endocytosis inhibitor phenylarsine oxide, indicating that it resulted from clathrin-mediated receptor internalization (Figure 5).

Taken together, these two sets of data indicate that depending on the receptor subtype involved, receptor-mediated internalization may either increase or decrease cell surface receptor density and, hence, play differential roles in cellular desensitization.

Technical Notes

There are a variety of fixative agents suitable for the investigation of receptor localization at the ultrastructural level. We have been equally successful with mixtures of glutaraldehyde (0.3%) and paraformaldehyde (4%) or of acrolein (3.75%) and paraformaldehyde (2%). The use of acrolein confers a greater degree of ultrastructural preservation but, at high concentrations, can adversely affect tissue antigenicity. In either case, primary fixation should be followed by two postfixation steps, one with a higher concentration of glutaraldehyde (2%) for cross-linking the secondary antibody to the primary antibody, and another with osmium tetroxide to aid ultrastructural preservation and the staining of lipid bilayers.

The silver-intensification step is perhaps the most delicate of the entire immunogold staining protocol. Care must be taken not only to find the optimal incubation time

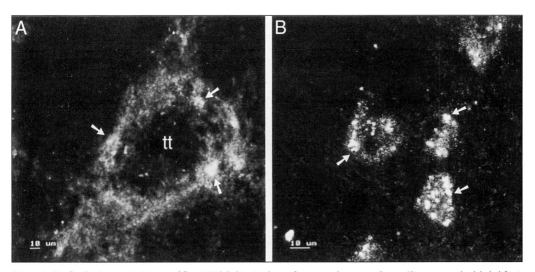

Figure 3. **Confocal microscopic images of fluo-NT labeling in slices of rat ventral tegmental area.** Slices were pulse-labeled for 3 minutes with 10 nM fluo-NT, and sections were scanned 10 minutes (A) and 30 minutes (B) after washout with Ringer buffer. At 10 minutes (A), labeling is evident over both perikarya (arrows) and neuropil. At 30 minutes (B), nerve cell bodies are still intensely labeled, but the neuropil labeling is markedly reduced. Note that at 30 minutes, the labeling is detected in the form of small puntate fluorescent granules that pervade the perikaryal cytoplasm (arrows). Images were reconstructed from a stack of 25 serial optical sections separated by 0.12 µm steps and scanned at 32 scans per frame. Scale bars: 10 µm. (**See color plate A4**.)

for one's specific preparation, but also to reduce any variability in time between conditions (e.g., between agonist-free and agonist-exposed cells or tissue). Overshooting the incubation time will result in a supersaturation of specific signal and the generation of extensive background. Within a certain range, however, varying the incubation time will merely modulate the diameter of the silver grains: longer incubations will result in larger particles making it easier to scan the specimen at lower magnification and making the analysis more rapid. Nonetheless, silver-intensified gold particles should not be so large as to obscure some of the fine detail of the preparation and thereby make it impossible, for example, to appraise the association of the label with small intracellular compartments.

A major advantage of immunogold cytochemistry is its amenability to quantitative analysis. As demonstrated above, this technique may be put to advantage to monitor receptor trafficking events following ligand-induced endocytosis with a spatial resolution impossible with any other method. The distribution of immunolabeled receptors may be assessed in a number of ways. Gold particles may be classified according to their association with specific subcellular structures and the number of labeled structures expressed as a proportion of total

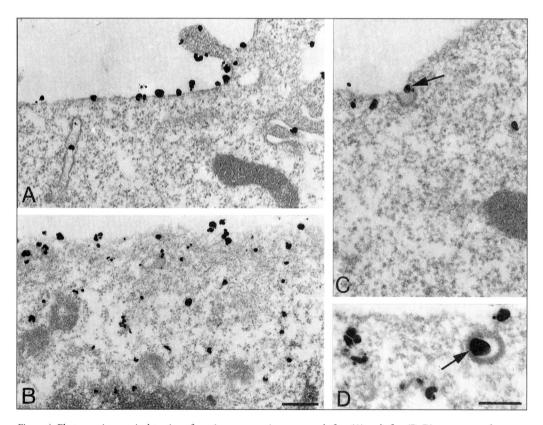

Figure 4. Electron microscopic detection of sst$_5$ immunoreactive receptors before (A) and after (B–D) exposure to the metabolically stable SRIF analog D-Trp^8-SRIF. Transfected COS-7 cells were immunolabeled before or after 30 minutes of pulse-labeling with 20 nM D-Trp^8-SRIF and 10 minutes of chase at 37°C. (A) In cells unexposed to SRIF, the bulk of labeled receptors is associated with the plasma membrane. (B–D) In cells exposed to D-Trp^8-SRIF, immunoreactive sst$_5$ receptors are much more abundant than in controls in the subplasmalemmal zone of the cytoplasm (B). Furthermore, several immunogold particles are observed in association with invaginating coated pits (C, arrows) and within clathrin-coated vesicles (D, arrow). Scale bars: A and B, 400 nm; C, 200 nm; D, 185 nm.

(e.g., labeled coated pits expressed as a percentage of coated pits evident at the cell surface). Alternatively, the number of labeled receptors associated with a particular structure may be expressed as a proportion of the total number of labeled receptors in the cell profile (e.g., proportion of labeled receptors associated with plasma membranes), or as densities of receptors associated with that profile (e.g., number of labeled receptors per unit length of membrane or per unit area of profile). In the context of receptor-mediated endocytosis, labeled receptors may also be classified according to their distance from the plasmalemma and then sorted into bins of increasing distance from the cell surface (21).

Readers are referred to Chapter 10 for further discussion on the immunogold labeling procedures.

ACKNOWLEDGMENTS

We thank Catherine M. Cahill, Anne Morinville, and Pierre Villeneuve for their help in the preparation of the manuscript and Naomi Takeda for secretarial assistance.

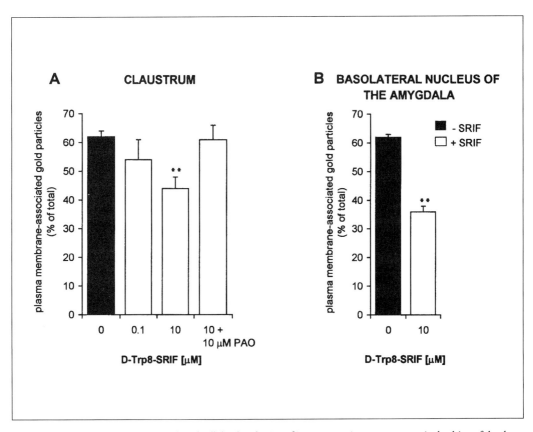

Figure 5. Effect of SRIF stimulation on the subcellular distribution of immunoreactive sst_{2A} receptors in dendrites of the claustrum and basolateral amygdala. Slices were incubated for 45 minutes at 37°C with or without D-Trp^8-SRIF (100 nm–10 μm). Dendrite-associated immunogold particles were counted and classified as either membrane-associated or intracellular, and the results expressed as membrane-associated per total. In the absence of agonist, approximately 60% of gold particles were associated with dendritic plasma membranes in both cerebral regions. Addition of agonist induced a decrease in the percentage of receptors associated with plasma membranes. The agonist-induced decrease in surface receptors was prevented by adding 10 μM phenylarsine oxyde to the incubation medium. Values are the mean ± standard error of the mean (SEM) from three animals. **, $P < 0.01$.

REFERENCES

1. Barak, L.S., S.S.G. Ferguson, J.I.E. Zhang, C. Martenson, T. Meyer, and M.G. Caron. 1997. Internal trafficking and surface mobility of a functionally intact beta2-adrenergic receptor-green fluorescent protein conjugate. Mol. Pharmacol. *51*:177-184.
2. Beaudet, A., D. Nouel, T. Stroh, F. Vandenbulcke, C. Dal Farra, and J.-P. Vincent. 1998. Fluorescent ligands for studying neuropeptide receptors by confocal microscopy. Braz. J. Med. Biol. Res. *31*:1479-1489.
3. Boudin, H., P. Sarret, J. Mazella, A. Schonbrunn, and A. Beaudet. 2000. Somatostatin-induced regulation of sst_{2A} receptor expression and cell surface availability in central neurons: role of receptor internalization. J. Neurosci. *20*:5932-5939.
4. Castel, M.-N., J. Woulfe, X. Wang, P.M. Laduron, and A. Beaudet. 1992. Light and electron microscopic localization of retrogradely transported neurotensin in rat nigrostriatal dopaminergic neurons. Neuroscience *50*:269-282.
5. Chan, J., C. Aoki, and V.M. Pickel. 1990. Optimization of differential immunogold-silver and peroxidase labeling with maintenance of ultra-structure in brain sections before plastic embedding. J. Neurosci. *15*:113-127.
6. Daaka, Y., L.M. Luttrell, S. Ahn, G.J. Della Rocca, S.S.G. Ferguson, M.G. Caron, and R.J. Lefkowitz. 1998. Essential role for G protein-coupled receptor endocytosis in the activation of mitogen-activated protein kinase. J. Biol. Chem. *273*:685-688.
7. Dautry-Varsat, A., A. Ciechanover, and H.F. Lodish. 1983. pH and the recycling of transferrin during receptor-mediated endocytosis. Proc. Natl. Acad. Sci. USA *80*:2258-2262.
8. Dumartin, B., I. Caillé, F. Gonon, and B. Bloch. 1998. Internalization of D1 dopamine receptor in striatal neurons in vivo as evidence of action by dopamine agonists. J. Neurosci. *18*:1650-1661.
9. Faure, M.-P., P. Gaudreau, I. Shaw, N.R. Cashman, and A. Beaudet. 1994. Synthesis of a biologically active fluorescent probe for labeling neurotensin receptors. J. Histochem. Cytochem. *42*:755-763.
10. Faure, M.-P., A. Alonso, D. Nouel, G. Gaudriault, M. Dennis, J.-P. Vincent, and A. Beaudet. 1995. Somatodendritic internalization and perinuclear targeting of neurotensin in the mammalian brain. J. Neurosci. *15*:4140-4147.
11. Faure, M.-P., D. Nouel, and A. Beaudet. 1995. Axonal and dendritic transport of internalized neurotensin in rat mesostriatal dopaminergic neurons. Neuroscience *68*:519-529.
12. Ferguson, S.S.G., J. Zhang, L.S. Barak, and M.G. Caron. 1998. Role of β-arrestins in the intracellular trafficking of G-protein-coupled receptors. Adv. Pharmacol. *42*:420-424.
13. Gaudriault, G., D. Nouel, C. Dal Farra, A. Beaudet, and J.-P. Vincent. 1997. Receptor-induced internalization of selective peptidic μ and δ opioid ligands. J. Biol. Chem. *272*:2880-2888.
14. Ghinea, N., M.T. Vu Hai, M.T. Groyer-Picard, A. Houllier, D. Schoevaert, and E. Milgrom. 1992. Pathways of internalization of the hCG/LH receptor: immunoelectron microscopic studies in Leydig cells and transfected L-cells. J. Cell Biol. *118*:1347-1358.
15. Grady, E.F., A.M. Garland, P.D. Gamp, M. Lovett, D.G. Payan, and N.W. Bunnett. 1995. Delineation of the endocytic pathway of substance P and its seven-transmembrane domain NK1 receptor. Mol. Biol. Cell *6*:509-524.
16. Mantyh, P.W., C.J. Allen, J.R. Ghilardi, S.D. Rogers, C.R. Mantyh, H. Liu, A.I. Basbaum, S.R. Vigna, and J.E. Maggio. 1995. Rapid endocytosis of a G protein-coupled receptor: substance P-evoked internalization of its receptor in the rat striatum in vivo. Proc. Natl. Acad. USA *92*:2622-2626.
17. Nouel, D., G. Gaudriault, M. Houle, T. Reisine, J.-P. Vincent, J. Mazella, and A. Beaudet. 1997. Differential internalization of somatostatin in COS-7 cells transfected with sst_1 and sst_2 receptor subtypes: a confocal microscopic study using novel fluorescent somatostatin derivatives. Endocrinology *138*:296-306.
18. Nouel, D., M.-P. Faure, J.-A. St. Pierre, R. Alonso, R. Quirion, and A. Beaudet. 1997b. Differential binding profile and internalization process of neurotensin via neuronal and glial receptors. J. Neurosci. *17*:1795-1803.
19. Pearse, B.M. 1982. Coated vesicles from human placenta carry ferritin, transferrin, and immunoglobulin G. Proc. Natl. Acad. Sci. USA *79*:451-455.
20. Sarret, P., D. Nouel, C. Dal Farra, J.-P. Vincent, A. Beaudet, and J. Mazella. 1999. Receptor-mediated internalization is critical for the inhibition of the expression of growth-hormone by somatostatin in the pituitary cell line AtT-20. J. Biol. Chem. *274*:19294-19300.
21. Stroh, T., A.C. Jackson, P. Sarret, C. Dal Farra, J.-P. Vincent, H.J. Kreienkamp, J. Mazella, and A. Beaudet. 2000. Intracellular dynamics of sst5 receptors in transfected COS-7 cells: maintenance of cell surface receptors during ligand-induced endocytosis. Endocrinology *141*:354-365.
22. Stroh, T., A.C. Jackson, C. Dal Farra, A. Schonbrun, J.P. Vincent, and A. Beaudet. Receptor-mediated internalization of somatostatin in rat cortical and hippocampal neurons. Synapse (In press).
23. von Zastrow, M. and B.K. Kobilka. 1992. Ligand-regulated internalization and recycling of human $β_2$-adrenergic receptors between the plasma membrane and endosomes containing transferrin receptors. J. Biol. Chem. *267*:3530-3538.

3

Transfection Methods for Neurons in Primary Culture

Christoph Kaether[1], Martin Köhrmann[1], Carlos G. Dotti[1,2], and Francesca Ruberti[1]
[1]EMBL Heidelberg, Cell Biology Programme, Heidelberg, Germany, EU;
[2]Cavalieri Ottolenghi Scientific Institute, Università degli Studi di Torino, Orbassano, Italy, EU

OVERVIEW

We describe two simple, quick, and reproducible methods to transfect primary hippocampal neurons. The first method consists of a modified calcium phosphate (CP) transfection. We have developed this method to analyze localization of proteins in fully mature hippocampal neurons, however it can also be used for others purposes, such as the analysis of gene regulation or function of proteins in neurons. This method offers the advantage of transfecting small numbers of neurons at different developmental stages. The second method is an efficient electroporation (EP) protocol to perform large-scale biochemistry experiments. Examples demonstrate the suitability of both techniques. These protocols could be a useful framework to transfect primary neurons cultured from other brain regions.

BACKGROUND

Many neuroscientists have successfully developed viral methods to transfect primary neurons (reviewed in Craig, 1998; see Reference 1). However, production of recombinant virus remains complex and time-consuming. Moreover, potential toxicity to the neurons and probable safety hazard to laboratory personnel cannot be excluded. In the last few years, several laboratories have made an effort to transfect postmitotic primary neurons in dissociated culture adapting classical transfection methods (e.g., CP precipitation and cationic lipid transfection) used for dividing cells. Transfection by CP has been successfully applied for introducing cloned eukaryotic DNAs in cultured mammalian cells. Although the mechanisms remain obscure, it is believed that the transfected DNA enters the cytoplasm by endocytosis and is transferred to the nucleus (4).

In the last few years, a series of papers have been published where this DNA/CP transfection method was adapted to transfect a primary culture of rat cortical, hippocampal, spinal cord, and dorsal root ganglion neurons (6,9,10). However, the precise conditions for successful CP-mediated DNA transfection have not been fully investigated, and the efficiency reported is relatively low, at most 1% to 2% (for review see Reference 1). Furthermore, in previous reports neurons are mostly transfected before plating or at an immature stage.

In our laboratory, we are interested in the mechanisms that guard intracellular transport of membrane proteins and their sorting to the axon or the dendrites in neurons. We use as a cellular system primary neurons from rat embryo hippocampi developed by Goslin and Banker (3). Hippocampal neurons are grown at low density on glass coverslips, which are then inverted in order to face a monolayer of glia grown on a plastic dish. The use of the hippocampal–glia coculture system allow us to reproduce in vitro neuronal polarization, which leads to differentiated neurons with one axon and several dendrites. Cells can be maintained in culture more than three weeks after plating.

For our work it would be convenient to have a simple protocol that allows: (*i*) to transfect embryonic hippocampal neurons at different developmental stages including fully mature neurons; (*ii*) to transfect several cDNAs at the same time in parallel experiments instead of producing a recombinant virus for each construct; and (*iii*) to cotransfect two plasmids into the same neurons. Thus, we adapted a CP protocol for efficient transfection of hippocampal neurons (5). This method has been successfully applied to transfect adult hippocampal neurons, but the cells were grown in a different manner. We describe a protocol and emphasize the crucial factors for obtaining efficient transfection by CP in "our" hippocampal neurons.

We also describe a second protocol that is based on EP (8). There are very few examples of applications of this method to primary neurons, and transfection efficiencies are below 2% (7). We present a protocol to transfect hippocampal neurons at a rate of up to 20% efficiency. Other transfection protocols are described in Chapters 4 to 6 of this book.

PROTOCOLS

Protocol for Calcium Phosphate Precipitation

We use CP for transfecting exogenous proteins into cultured hippocampal neurons and subsequent localization studies using immunofluorescence or video microscopy. CP can also be used for small-scale biochemical studies like western blot to characterize an antibody or to verify the full-length expression of a transfected protein.

The preparation and culturing of hippocampal neurons is described in detail in Reference 2 and 3.

Materials and Reagents

- Neurons 3 to 14 days in vitro (DIV)
- Waterbath at 37°C
- 3.5-cm Plastic dishes
- Sterile forceps
- 2-mL Eppendorf® tubes
- Vortex mixer
- 2.5% CO_2 incubator
- 250 mM $CaCl_2$ solution, sterile and warmed to room temperature
- 2× BBS, (280 mM NaCl, 1.5 mM Na_2HPO_4, 50 mM BES [N,N-bis(2-hydroxyethyl)-2-aminoethanesulfonic acid]; Sigma, St. Louis, MO, USA), pH 7.1
- HBS (135 mM NaCl, 20 mM HEPES [4-(2-hydroxyethyl)-1-piperazineethanesufonic acid], pH 7.1, 4 mM KCl, 1 mM Na_2HPO_4, 2 mM $CaCl_2$, 2 mM $MgCl_2$, 10 mM glucose)
- Dishes with conditioned N2 medium
- DNA (2–5 mg/mL)

Preparation of DNA

For optimal results, prepare endotoxin-free DNA with a plasmid purification kit (EndoFree™ Plasmid Maxi Kit; Qiagen, Hilden, Germany). After the final precipitation, dissolve DNA in endotoxin-free water or Tris-EDTA (TE) at a concentration of 1 to 3 mg/mL. Prepare 50 to 100-μL aliquots and store at -20°C. Store thawed DNA at 4°C and avoid frequent freezing and thawing.

Preparation of Conditioned N2

Older, polarized neurons survive better in medium enriched in factors secreted by astrocytes. To keep neurons in conditioned medium throughout the transfection, it is necessary to have preconditioned medium at hand. Therefore, we grow astrocytes on 6-cm plastic dishes as described in Reference 2. The N2 medium in which astrocytes are grown can be used for transfection. Check that the pH is not too acidic, as this will prevent the precipitate to form.

Procedure

1. Warm 2× BBS and $CaCl_2$ to room temperature. Place HBS in a 37°C waterbath.
2. Add 2 mL of conditioned N2 into a 3.5-cm dish and transfer coverslips into the dishes such that the neurons are facing up. Immediately place them in the 5% incubator for 30 minutes.
3. In a 2-mL tube mix 2 to 5 μg of DNA with the $CaCl_2$ solution to a final volume of 60 μL. Mix briefly by vortex mixing.
4. Slowly add 60 μL of 2× BBS while stirring with the tip.
   Vortex mix briefly. If several dishes are transfected with the same DNA, prepare a master mixture. This ensures equal transfection for all dishes.
5. After 1 minute, take the cells out of the incubator and check the pH of the medium. If necessary, add drops of diluted hydrochloric acid to reach the desired pH. Color must be red, not yellowish and not purple.
6. Add DNA–phosphate precipitate and stir gently. The medium should now become yellowish. Transfer dishes to a 2.5% CO_2 incubator.
7. Let precipitate form and settle onto cells for 30 to 120 minutes. Check the medium after 30 to 60 min. If the medium is very purple and turbid, the precipitate is probably already very big. Check on a microscope for size of the precipitate. Precipitate should have the size and appearance of bacterial contamination. Big precipitates are toxic, and the cells will die.
8. After precipitate has formed, gently wash the cells 2 times with prewarmed HBS, then add the original medium in which they were grown.
9. Flip coverslips so that the neurons are facing the bottom of the dish and incubate until analysis. Depending on the construct, we analyze cells 8 hours to 3 days after transfection.

Pitfalls

Several parameters are critical for an efficient transfection.

1. We found that the use of endotoxin-free DNA significantly increases the number of transfected cells.
2. The use of a 2.5% CO_2 incubator during the transfection procedure is essential in our hands.
3. Successful transfection is dependent on the pH of the medium to which the precipitate is added. A too acidic medium will prevent the precipitate to form, and a too alkaline medium will lead to huge precipitates that are neurotoxic.

Drawbacks

Even by carefully following the protocol transfection, efficiencies vary from transfection to transfection. Moreover different constructs generally give different efficiencies, even if the same expression vector is used (C.K. and F.R., unpublished observation). Also, transfected cells on a dish or coverslip are not evenly distributed, but tend to cluster.

Protocol for Transfection of Neurons for Biochemical Experiments

For analysis of exogenous proteins by western blot or radioactive labeling, neurons can be grown on 3.5-cm plastic dishes and transfected in principle as described above.

Preparation of Polylysine-Coated Plastic Dishes: Prepare 2 Days Before Plating

1. Cover bottom of dish with 1 mg/mL polylysine and leave overnight at room temperature.
2. Remove polylysine solution and wash 2 times for 1 hour in distilled water.
3. Add 1 mL medium (minimum essential medium [MEM] + 10% horse serum) and keep overnight in the incubator.
4. Replace medium with N2 supplemented with insulin (5 µg/mL).
5. Seed cells and allow to grow for desired time. For better survival neurons can be covered with a big coverslip with paraffin dots.

Procedure

1. Collect the original medium of the cells and store in the incubator.
2. Perform the transfection as described above. Afterwards, add the original medium back and incubate the cells until needed.

For analysis, prepare a cell lysate (see below) and pool three dishes per lane on a sodium dodecyl sulfate (SDS) gel.

Protocol for Electroporation

For large-scale biochemistry, we electroporate neurons before plating and grow them in 75-cm^2 flasks.

Materials and Reagents

- Gene Pulser® (Bio-Rad Laboratories, Hercules, CA, USA)
- 0.4-cm Gene Pulser cuvettes (Bio-Rad)
- Horse-MEM
- Hank's balanced salt solution (HBSS)
- 75-cm^2 flasks coated with polylysine
- DNA, endotoxin-free, concentration of 1 to 4 mg/mL

Procedures

Coating of Flasks: Prepare at Least 1 Day Before Preparing the Neurons

1. Add 10 mL of a 10 µg/mL polylysine solution to the flask and incubate for 1 hour at room temperature.
2. Wash 3 times with water and add 10 mL horse-MEM. Incubate overnight in a 37°C, 5% CO_2 incubator.
3. Replace the medium with N2 medium and equilibrate in the incubator for at least 3 hours.
4. Dissect hippocampi, triturate, and count.
5. Add 2.5 to 3 × 10^6 cells to an electroporation cuvette.
6. Add 50 µg DNA and HBSS if necessary, to a final volume of 0.8 mL and mix with a pasteur pipet. Leave at room temperature for 3 to 5 minutes with occasional mixing.
7. Electroporate at 850 V, 25 µF, and 200 Ohm. Time constant should be 0.8.
8. Carefully resuspend cells by pipetting

twice with the pasteur and transfer cells into flask. To check efficiency, a small aliquot can be plated in a 3.5-cm coated dish. Neurons can be fixed and processed for immunofluorescence directly in the dish. For microscopy, a coverslip can be mounted with Mowiol (Merck, West Point, PA, USA), and the neurons can be analyzed for transfection efficiency. We analyze neurons typically 6 to 7 days after EP, but probably later time points are also possible.

Expression of Cytomegalovirus (CMV)-Promoter Driven Genes Can be Enhanced by Induction with Sodium Butyrate

Add sodium butyrate to a final concentration of 2 mM and incubate for 17 to 24 hours. A stock solution of butyrate is prepared by dissolving butyrate in water to a final concentration of 0.5 M. Stock solution is stored at 4°C.

Protocol for Preparation of Cell Lysates for Western Blotting

Materials and Reagents

- Lysis buffer 1 (L1): 0.1% SDS in water
- Lysis buffer 2 (L2): 150 mM NaCl, 50 mM Tris, 2 mM EDTA, 1% Nonidet® P-40 (NP40), 1% Triton® X-100
- Methanol
- Chloroform

Procedure

1. Lyse cells grown in flasks by incubating in L1 or L2 for 30 minutes on ice. Lyse and pool cells grown on 3.5-cm dishes by adding 0.3 mL of L1 or L2, scraping cells using a cell scraper, and transferring the lysate to a second dish and subsequently to a third dish. Leave for 30 minutes on ice.
2. Transfer lysates into 15-mL plastic tubes (when flasks were used) or 2-mL centrifuge tubes (when 3.5-cm dishes were used).
3. Add 3.2 volumes of MeOH and vortex mix.
4. Add 0.8 volumes of $ClCH_3$ and vortex mix.
5. Add 2.4 volumes water, vortex mix vigorously for 1 minute, and spin for 2 minutes.
6. Remove and discard upper phase.
7. Add 2.4 volumes MeOH, vortex mix, and spin for 5 minutes.
8. Remove supernatant and air-dry pellet.
9. Resuspend in 1× SDS running buffer and boil for 2 minutes.

SDS gel electrophoresis and western blotting can then be performed.

Immunocytochemistry

Materials and Reagents

The following primary antibodies were used: mouse anti-HA (12CA5 from Roche Moleculer Biochemicals, Mannheim, Germany), polyclonal antibody 514 anti-MAP2 (a gift from C. Sanchez, Centro de Biologia Molecular, Madrid), and polyclonal anti-human APP (kindly provided by C. Haass, München).

Procedure

1. Fix cells grown on glass coverslips with 4% paraformaldehyde in phosphate-buffered saline (PBS) for 15 minutes.
2. Quench paraformaldehyde with 50 mM ammonium chloride in PBS for 10 minutes.
3. Permeabilize with 0.2% Triton X-100 in PBS for 5 minutes.

4. Then incubate with blocking solution (2% bovine serum albumin [BSA], 2% fetal calf serum [FCS], 0.2% fish skin gelatin) in PBS for 30 minutes.
5. For double labeling, incubate the cells with the primary antibodies diluted in 10% blocking solution in PBS for 1 hour at room temperature or overnight at 4°C.
6. Wash with PBS 3 times.
7. Incubate with appropriate fluorochrome-conjugated secondary antibodies diluted in 10% blocking solution in PBS for 20 to 30 minutes at room temperature.
8. Mount coverslips with mowiol containing 100 mg/mL DABCO (Sigma) as an antifading agent.

RESULTS AND DISCUSSION

Calcium Phosphate Transfection

The CP protocol was applied to study the sorting of exogenous neuronal transmembrane proteins. Figure 1 shows the result of an experiment in which we have transfected 9 days in vitro (DIV) hippocampal neurons with the cDNA for the metabotropic glutamate receptor (mGluR2, kindly provided by S. Nakanishi) in which an HA epitope tag has been inserted at the N terminus of the protein. Twenty-four hours after transfection, cells were analyzed by immunofluorescence using antibodies against the dendritic cytoskeletal protein MAP2 and antibodies against the HA epitope tag to detect the recombinant mGluR2. As shown in Figure 1, the recombinant HA-mGluR2 is distributed to the somatodendritic domain like the protein MAP2.

The CP method also allows the analysis of the distribution of two different exogenous proteins in the same cell. Almost 100% cotransfection can be obtained by mixing plasmids prior to precipitate formation. Figure 2 shows the results of a cotransfection of the cDNA for glutamate receptor 1 (GluR1) and amyloid precursor protein (APP). These two proteins are dendritic and axonal–dendritic, respectively.

This CP transfection method can also

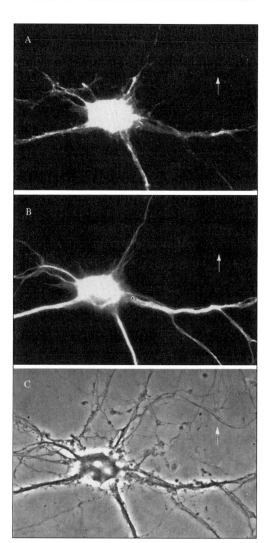

Figure 1. Immunofluorescence showing the dendritic distribution of N-HA-mGluR2 in 10 DIV hippocampal neurons. Nine DIV neurons were transfected with the cDNA for HA-mGluR2 and fixed 24 hours later and processed for immunofluorescence using antibodies against HA and MAP-2. Panels A and B show the distribution of the HA-mGluR2 protein and the endogenous dendritic marker MAP2, respectively. Panel C shows a phase contrast micrograph. HAmGluR2 labeling is restricted to cell body and dendrites (identified by MAP2 labeling). The axon is indicated by the arrow.

be used for localization studies by time-lapse video microscopy using green fluorescent protein (GFP) fusion protein. Transport of GFP-tagged transmembrane proteins such as synapthophysin and APP (C. Kaether and C.G. Dotti, unpublished data) and the RNA binding protein Staufen (6a) have been successfully analyzed in living hippocampal neurons. The rate of transfection using this protocol allowed us to analyze by western blotting that the GFP fusion proteins are accurately processed (see protocol section).

Calcium Phosphate Transfection Efficiency

In order to determine transfection efficiency, we have transfected cDNAs coding for different neuronal transmembrane proteins under the control of the CMV promoter. We can detect protein expression by immunofluorescence 8 hours after transfection. However, in most cases, we have analyzed protein distribution 24 hours after CP. We can transfect cells after 1 to 11 DIV, but we found that transfection efficiency decreases when cells after 8 DIV were used. In optimal conditions, we were able to obtain a transfection efficiency of 10% to 20% for three different constructs. Yet we want to stress, as mentioned already in the protocol section, that the efficiency can vary a lot between transfections and between constructs.

Several other applications of this CP method can be envisaged, for example functional studies of wild-type and mutated proteins in single cells, using electrophysiology or time-lapse microscopy. Although we established optimal CP transfection conditions for embryonic hippocampal neurons, it will be also possible to adapt this protocol to other types of primary neuronal cultures.

Electroporation

To demonstrate the efficient transfection of hippocampal neurons by EP, we dissected the hippocampi of 13 to 15 embryos. Dissociated cells were electroporated without or with pEGFP-N1 (CLONTECH Laboratories, Palo Alto, CA, USA), which is an expression vector coding for enhanced GFP, and cultured in 75-cm^2

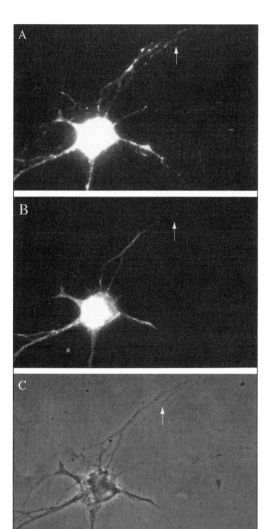

Figure 2. Cotransfection of 9 DIV hippocampal neurons with HA-GluR1 and human APP-cDNAs. Twenty-four hours after cotransfection, the cells were analyzed by immunofluorescence using antibodies against HA and human APP. Human APP (panel A) extends to the axonal domain (arrow) as well as to the dendritic compartment. HA-GluR1 distribution is restricted to the dendrites (panel B). Panel C shows a phase contrast micrograph. Scale bar is 10 μm.

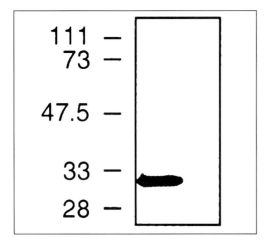

Figure 3. Western blot of electroporated neurons cultured for 7 days. Neurons were electroporated with (lane 1) or without (lane 2) pEGFP-N1 cDNA, cultured for 7 days, and subjected to Western blotting. Western blotting using GFP-antiserum detected a 30 kDa band in the cell lysates of transfected neurons (lane 1). The positions of molecular weight markers are indicated (in kDa).

flasks. After 7 days, EP neurons were lysed in lysis buffer and processed for immunoblotting. Blotting was performed using an antiserum against GFP (CLONTECH) (Figure 3). A single band at 30 kDa corresponding to GFP is detected in cell lysates of transfected cells (Figure 3, lane 1) but not in control cells (Figure 3, lane 2). A 15th of the total cell lysate of a 75-cm^2 flask could be readily detected on a western blot (data not shown). This corresponds to approximately 70 000 neurons, out of which 10% to 20% are transfected.

In summary, the described protocols provide tools for a variety of approaches where the expression of exogenous proteins in primary neurons is required. In the future, new approaches using this technique might include screening of an expression library to pick up relevant genes for specific neuronal function.

ACKNOWLEDGMENTS

We thank B. Hellias for technical assistance and Jose Abad for providing the protocol on growing hippocampal neurons in 75-cm^2 flasks. C.K. was a recipient of a fellowship of the Fritz-Thyssen-Stiftung, F.R. was a recipient of a European Community long-term fellowship.

REFERENCES

1. Craig, A.M. 1998. Transfecting cultured neurons, p. 79-111. *In* G. Banker and K. Goslin. (Eds.), Culturing Nerve Cells. MIT Press, Boston.
2. de Hoop, M.J., L. Meyn, and C.G. Dotti. 1998. Culturing hippocampal neurons and astrocytes from fetal rodent brain, p. 154-163. *In* J.E. Celis (Ed.), Cell Biology: A Laboratory Handbook, Vol. 1. Academic Press, San Diego.
3. Goslin, K. and G. Banker. 1991. Rat hippocampal neurons in low-density culture, p. 251-282. *In* G. Banker and K. Goslin. (Eds.), Culturing Nerve Cells. MIT Press, Cambridge, MA.
4. Graham, F.L. and A.J. van der Eb. 1973. Transformation of rat cells by DNA of human adenovirus 5. Virology *52*:456-467.
5. Haubensak, W., F. Narz, R. Heumann, and V. Lessmann. 1998. BDNF-GFP containing secretory granules are localized in the vicinity of synaptic junctions of cultured cortical neurons. J. Cell Sci. *111*:1483-1493.
6. Kanai, Y. and N. Hirokawa. 1995. Sorting mechanisms of tau and MAP2 in neurons: suppressed axonal transit of MAP2 and locally regulated microtubule binding. Neuron *14*:421-432.
6a. Kohrmann, M., M. Luo, C. Kaether, L. DesGroseillers, C.G. Dotti, and M.A. Kiebler. 1999. Microtubule-dependent recruitment of Staufen-green fluorescent protein into large RNA-containing granules and subsequent dendritic transport in living hippocampal neurons. Mol. Biol. Cell. *10*:2945-2953.
7. Li, H., S.T. Chan, and F. Tang. 1997. Transfection of rat brain cells by electroporation. J. Neurosci. Methods *75*:29-32.
8. Potter, H., L. Weir, and P. Leder. 1984. Enhancer-dependent expression of human kappa immunoglobulin genes introduced into mouse pre-B lymphocytes by electroporation. Proc. Natl. Acad. Sci. USA *81*:7161-7165.
9. Watson, A. and D. Latchman. 1996. Gene delivery into neuronal cells by calcium phosphate-mediated transfection. Methods *10*:289-291.
10. Xia, Z., H. Dudek, C.K. Miranti, and M.E. Greenberg. 1996. Calcium influx via the NMDA receptor induces immediate early gene transcription by a MAP kinase/ERK-dependent mechanism. J. Neurosci. *16*:5425-5436.

4 Polyethylenimine: A Versatile Cationic Polymer for Plasmid-Based Gene Delivery in the CNS

Barbara A. Demeneix, Gregory F. Lemkine, and Hajer Guissouma
Laboratoire de Physiologie Générale et Comparée, UMR CNRS 8572
Muséum National d'Histoire Naturelle, Paris, France, EU

OVERVIEW

We have developed and optimized a method for delivering plasmid-based genes into the vertebrate brain. The method involves condensation of DNA into stable, small (<100 nm ø), and diffusible complexes using a cationic polymer, polyethylenimine (PEI). The approach is extremely versatile as it relies on use of plasmid, not viral-based constructions. This means that construction preparation and amplification is easy to carry out, risk free, and rapidly verified. The method is extremely efficient, giving very high yields for small (nanogram) quantities of plasmid. An overriding advantage is that it provides a convenient technique for studying physiological regulation of neuronal gene function within defined brain areas at defined developmental stages or in specific physiological states. One can thus obtain data on integrated transcriptional responses that, obviously, could not be obtained by an in vitro approach without resorting to germinal transgenesis. We have applied it to the analysis of protein function and promoter regulation in the central nervous systems (CNS) of mouse, rat, and Xenopus.

BACKGROUND

Gene transfer into cells of the CNS, whether in vivo or ex vivo, is a delicate and difficult endeavor. Its uses are, however, so promising that it is not surprising to note that the field has drawn much attention over the last decade. Two broad classes of uses can be distinguished. On the one hand, there is gene therapy where the ultimate target will be the modification of an endogenous gene by homologous recombination or the remedial addition of a gene coding for a deficient protein. On the other hand, we have analytical approaches where the aim may be either to determine the physiological relevance of a given protein by blocking or by bolstering its expression or to dissect the regulatory mechanisms impinging on promoter function in an integrated setting. Furthermore, analysis of promoter regulation will be a prerequisite for creating constructs with optimized regulatory sequences for expressing therapeutic proteins in physiologically appropriate conditions.

In the field of basic research, stereotaxic microinjection of different permutations of a specific promoter into defined brain regions is a rapid and inexpensive method

Cellular and Molecular Methods in Neuroscience Research
Edited by A. Merighi and G. Carmignoto

for assessing function and for mapping transcriptional regulatory elements. Indeed, using somatic gene transfer can provide results on spatial and temporal regulation that otherwise could only be obtained by labor-intensive germinal transgenesis, an approach that also requires a great deal of organizational prowess and expense for maintenance of the numerous lines created. Gene transfer into the CNS can be based on cell grafting or direct delivery. For direct delivery, a variety of methods, viral or nonviral, are available. As to the viral methods, in mammalian systems, reports have appeared describing the use of adenoviruses, lentiviruses, herpes virus, and adenoassociated virus-derived vectors. In amphibians, vaccinia virus has also been applied. However, besides their inherent safety problems in therapeutic settings, viral constructs are laborious to construct and verify. Moreover, their production in large quantities is often problematic. For these reasons many groups have turned to synthetic, or nonviral, vectors to achieve gene transfer in the CNS. Two main classes of synthetic vectors have been tested in the intact CNS: cationic lipids and cationic polymers. Here, we describe the use of a cationic polymer PEI. Indeed, of all the synthetic vectors so far tested in the mammalian CNS, low molecular weight PEIs (1) provide the most efficient gene delivery.

One of the most important features of PEI is its lack of toxicity in vivo. In the CNS, the lesions inevitably created by microinjection into the brain tissue are no different following injection of carrier or injection of PEI/DNA complexes in carrier. This lack of toxicity with PEI is no doubt related to its high efficiency, which allows the use of very low amounts of DNA (in the nanogram range). The high efficiency of PEI is in turn related to its capacity for protonation (2). The fact that, in PEI, one in every third atom is an amino nitrogen that can be protonated, makes PEI the cationic polymer having the highest charge density potential available. Moreover, the overall protonation level of PEI increases from 20% to 45% between pH 7.0 and 5.0 (12).

PEI can be used for delivering plasmid DNA or oligonucleotides to brains of adult and newborn mice (1,3) and rats (9). Moreover, it can be used in the CNS for intrathecal (1,9) or intraventricular delivery (6), the latter route being one that could be particularly useful for delivery of therapeutic proteins such as nerve growth factors. Work from our group and others is showing that the method can be used for up and down modulation of protein production in defined brain regions and for analysis of neuron-specific promoter regulation (5,7).

When starting up a gene transfer protocol in vivo, it is always preferable, whether one's aim is promoter analysis or production of protein, to optimize delivery by examining the quantitative aspects of transgene expression in the region targeted with a luciferase and then spatial aspects with a β-galactosidase (β-gal) construct. Indeed, we have found that the optimal ratio of PEI amine to DNA phosphate can vary according to species and brain region targeted. Such preliminary work will also enable one to test which promoter will perform best in a given cell population or developmental stage. For these reasons, we detail our methods for extracting and assaying firefly luciferase (from *Photinus pyralis*) in the brain, this luciferase being the best reporter gene available for setting up gene transfer protocols. It is three orders of magnitude more sensitive than β-gal, and the fact that it can be quantified with precision is an overriding factor for choosing it for optimization of PEI/DNA ratios, amounts of DNA to be used, time course evaluation, and for promoter analysis. Other reporter genes, such as chloramphenicol acetyltransferase (CAT) and β-gal can of course be quantified, but each has its drawbacks compared to luciferase. The main problem with colorimetric quantification of β-gal expression in the CNS is excessive interference from endogenous enzymatic activity. Suppliers of kits (such as Promega, Lyon, France) for such methodology recommend heat inactivation of endogenous enzymes. However, in our hands, such precautions have proven ineffective, and the extremely high background found throughout the mouse brain precludes precise quantification of expression of transgenes encoding β-gal. Similarly, when using histochemical procedures it is vital to use appropriate fixation conditions to avoid interference from endogenous activity that can be high, particularly in the hippocampus. CAT is a good alternative for quantification, but whichever method is chosen [usually enzyme-linked immunosorbent assay (ELISA) or the method of Seed and Sheen (11)], assay time is longer and the methodology laborious. Thus, we recommend starting off with luciferase, thereby determining first, optimal PEI/DNA ratios, and then kinetics and dose-response curves can be established (Figure 1). Such experiments will also reveal the inherent variability of transfer efficiency in the target examined and determine the need or not for normalization in experiments involving promoter regulation with luciferase.

However, it remains that β-gal is one of the best markers for following spatial aspects of expression. Green fluorescent protein (GFP) is equally sensitive, but the fluorescent imaging, although esthetically pleasing, is not as satisfactory as standard light microscopy for anatomical detail. For this reason, we provide the methodology for β-gal revelation in whole brains. Indeed, histochemical analysis of β-gal expression on whole brains allows for rapid assessment of transfer efficiency and transgene distribution in small sized samples (newborn mouse brains or regions of adult brains). We also provide a methodology for revealing β-gal expression on histological sections, a step that obviously permits more precise anatomical analysis, which is of course essential for determining brain regions and morphology of transfected neurons and glial cells. Further, it is on such sections (prepared by vibratome or cryostat sectioning) that double labeling by immunocytochemistry can be performed to identify the cells expressing the transgenes. For instance, to identify neurons, one can use a neurofilament (NF) antibody or a NeuN antibody, and to identify astrocytes, an antibody against glial acid fibrillary protein (GFAP) can be employed. Above all, one must remember that if one obtains just a few cells labeled with β-gal (a very insensitive method when dealing with transient expression in vivo), this level of efficiency will be more than sufficient to allow one to proceed with either promoter analysis using luciferase or to study the biological effects of a given gene of interest. Finally, we also suggest a very sensitive immunoautoradiographic method for measuring variation in expression of genes of interest at low levels. Readers should refer to Chapters 3, 5, and 6 of this book for descriptions of other transfection procedures.

PROTOCOLS

Protocol for PEI Preparation

Materials and Reagents

- Branched 25 kDa PEI (Sigma, St. Quentin Fallavier, France) in anhydrous form.
- Linear (0.1 M) 22 kDa PEI (Euromedex, Souffleweyersheim, France).

Both preparations should be stored at 4°C having adjusted the pH to ≤4.0.

Procedure

The stock solution (0.1 M) of PEI provided by Euromedex (22 kDa) is used as supplied. The 25 kDa obtained is prepared as follows: weigh 4.5 mg of the solution, mix with 800 µL of sterile water thus obtaining a 0.1 M solution. Adjust the pH to ≤4.0 with 0.1 N HCl and the volume to 1 mL. Solutions are kept as aliquots at 4°C or at -20°C for long term storage (>1 month).

Protocol for Plasmid Preparation

Materials and Reagents

- For setting up in vivo gene transfer with PEI, one can use commercially available plasmids [e.g., p cytomegalovirus (CMV)-luciferase from Promega; pCMV(nls)-Lac-Z from CLONTECH (Montigny-le-Bretonneux, France)]. For CAT, the most efficient

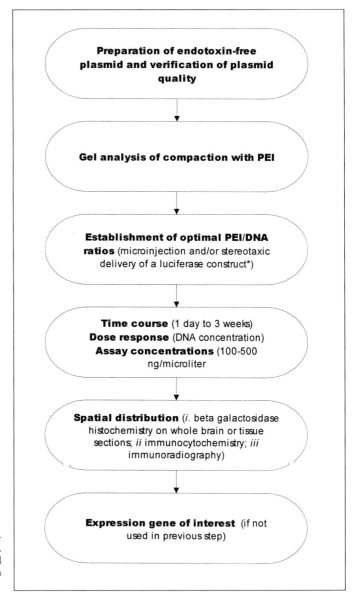

Figure 1. Flow chart for optimizing PEI-based gene delivery into the intact CNS. We recommend setting up this method with a luciferase reporter gene under a CMV promoter.

construct we have tested is pcis-CMV-CAT provided by R. Debs and co-workers (13).
- Endotoxin-free plasmid DNA.
- Appropriate restriction enzymes for verifying plasmids.
- Agarose gels [0.8% in 1× Tris-acetate EDTA (TAE) or Tris-borate EDTA (TBE), see Reference 10].
- Spectrophotometer for measurements of DNA concentrations (OD_{260}) and purity of DNA ($OD_{260}/OD_{280} > 1.8$).

Procedure

1. For plasmid DNA preparation and purification we recommend use of Jetstar columns (GENOMED, Raleigh, NC, USA). The system is based on anion exchange columns. According to the manufacturer's instructions and solutions supplied, bacteria resulting from maxiculture are lysed by alcali. Large membrane debris are eliminated using potassium acetate and centrifugation. The resulting supernatant is loaded on columns, and DNA is eluted with a solution containing approximately 2 M NaCl, then centrifuged after isopropanol precipitation. The pellet is washed with ethanol 70% (-20°C) and resuspended in water or Tris-EDTA (TE) at high DNA concentration (≥0.5 μg/μL in TE). This is to ensure that when diluting DNA to its working concentration (≤0.5 μg/μL in 5% glucose), the final TE concentration is not greater than 1 mM Tris/0.1 mM EDTA (standard TE/10). These plasmid preparations are endotoxin free.

2. One microliter of each DNA preparation is diluted in 1 mL of sterile water and analyzed at 260 and 280 nm. According to Sambrook et al. (10), 1 U OD_{260} correspond to 50 μg/mL of double-stranded DNA.

3. Agarose gel electrophoresis is used to verify that the plasmid DNA is not denatured, is free of RNA, and is mainly supercoiled. Restriction map analysis can be used to check constructions at this point. To this end, native and digested plasmids are analyzed using 0.8% agarose gel electrophoresis in TAE, with bromophenol-blue and a DNA molecular weight marker (10). The gel is observed on a UV transilluminator (312 nm) and photographed with a Sony video equipment from OSI (Maurepas, France).

Protocol for Condensation of DNA with PEI and Analysis

Materials and Reagents

- Filtered (0.22 μm) 5% glucose solution.
- Autoclaved 0.9% NaCl solution.
- DNA resuspended in water at a final concentration of <0.5 μg/μL.
- PEI solutions diluted extremporaneously to 10 mM.
- Sterile polypropylene tubes (1.5 mL).
- Vortex mixer.
- Electrophoresis equipment for checking complexation (not required each time, only for the first round of experiments).

Procedure

1. Plasmid DNA is diluted in sterile (0.22 μm filtered) 5% glucose to the chosen concentration (usually 0.5–2 μg/μL). After vortex mixing, the appropriate amount of a 0.1 M PEI solution is added, and the solution is vortex mixed again. The required amount of PEI, according to DNA concentration and number of equivalents needed, is calculated by taking into account that 1 μg

DNA is 3 nmol of phosphate, and that 1 µL of 0.1 M PEI is 100 nmol of amine nitrogen. So to complex 10 µg of DNA (30 nmol phosphate) with N/P (PEI amine/DNA phosphate) ratio of 5 eq PEI, one needs 150 nmol of PEI (1.5 µL of a 0.1 M solution). To minimize pipetting errors, it is best to dilute PEI down to 50 or 10 mM, but this should be done extemporaneously. Dilute DNA to 10 µg (final concentration 0.5 µg/µL) in 20 µL of final 5% glucose. Add the necessary volume of PEI to form the desired N/P ratio and, water to a final volume of 20 µL.

2. When using a plasmid preparation for the first time, we recommend analysis of complexes by agarose gel electrophoresis (Figure 2) and comparing their migration to that of naked DNA (plasmid alone, N/P ratio = 0) after adding 1 to 2 µL of bromophenol blue. This gives a gel retardation profile in which DNA/PEI complexes formed at N/P <1 migrate similarly to naked DNA. At N/P >6, complexes are so positively charged that they migrate to the negative pole.

Protocol for Microinjections and Animal Care

Materials and Reagents

- Animals: adult and newborn OF-1 mice or Sprague-Dawley rats, both supplied by Iffa Credo (L'Arbresle, France).
- Stereotaxic apparatus from David Kopf, Phymed, Paris, France. Stereotaxic coordinates for mice are determined according to Lehmann (8).
- Micromanipulator (Narishige; OSI) and microcapillaries (ext ø 1 mm; OSI). Capillaries are pulled to ext ø of 10 to 15 µm.
- Capillary puller (Narishige).
- Hamilton syringe (10 µL) with a 21 gauge needle (ext ø of 460 µm; OSI).
- Ice to anesthetize newborn mice (10 min on ice).

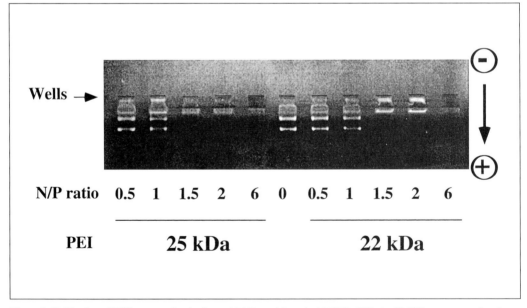

Figure 2. Verification of DNA compaction by linear 22 kDa and branched 25 kDa PEI. A pCMV-luc construct (1 µg) was mixed with PEI at various charge ratios, and DNA/PEI complexes were electrophoresed in 0.8% agarose gel stained with ethidium bromide. The position of the wells and the direction of the electrophoretic migration are indicated on the right.

- Sodium pentobarbital (Sanofi, Paris, France) diluted to 10% to anesthetize adult animals (70 mg/kg weight, i.p.).
- Recovery chamber. Animals should be kept under an infrared lamp or a specially constructed heated cage until fully recovered from anesthesia and surgery.

Procedure

1. Obviously, all the procedures described herein that involve animals and their care must be conducted in conformity with appropriate institutional guidelines that are in accordance with national and international laws and polices.
2. Adult (1–2 month old) mice are anesthetized using sodium pentobarbital diluted to 10% in 0.9% NaCl. Animals are anesthetized by an i.p. injection (70 mg/kg). Adult mice or rats are placed in the stereotaxic apparatus. An incision is made to expose the cranial skull and a hole made with a 21 gauge needle at chosen stereotaxic coordinates. Between 0.5 and up to 5 µL of complexes in 5% glucose are injected slowly (<5 min) with a Hamilton syringe adapted on a stereotaxic apparatus, small volumes are used for intrathecal injection and larger volumes for intraventricular injection. The needle is left in place for 5 minutes postinjection, to limit backflow from the injection site.
3. Newborn mice are anesthetized by hypothermia on ice, and 1 µL of complexes are injected with a microcapillary adapted on a micromanipulator. The head of the anesthetized pup is held manually for direct microinjection. Again, injections should be as slow as possible, and the capillary left in place for at least a minute to limit backflow.
4. Animals are left in a recovery chamber until active. When optimizing PEI delivery, newborns are returned to the dam for 24 hours before sacrifice, and adults are kept for 72 hours before sacrifice, as expression is usually maximal at these time points.

Protocol for Following Luciferase Activity in Brain Homogenates

Materials and Reagents

- Microdissection tools for dissecting out brain areas.
- Ultra-Turrax (OSI) equipped with a small plunger to homogenize tissue samples directly in a polyethylene tube.
- Refrigerated bench-top centrifuge for Eppendorf® tubes to pellet cell debris after homogenization so as to recuperate supernatant for luciferase assay.
- Luciferase assay kit from Promega. This system is based on the oxidation of luciferin by luciferase, in the presence of ATP and O_2, with photon production.
- Luminometer (model ILA-911; Tropix, Bedford, MA, USA) to quantify light emitted.

Procedure

To quantify luciferase expression, animals are sacrificed by decapitation after anesthesia. The dissected brains are separated in hemispheres and homogenized using an Ultra-Turrax, in 2-mL Eppendorf tubes containing 200 µL (for the newborn) or 500 µL (for the adults) of luciferase lysis buffer. Homogenates are centrifuged, and 20 µL of each supernatant with 100 µL of luciferase assay substrate are vortex mixed. The light emitted is quantified in relative light units (RLUs) using a luminometer.

Protocols for β-Galactosidase Revelation

Materials and Reagents

- 5-Bromo-4-chloro-3-indolyl-β-D-galactopyranoside (X-gal) (Eurogentec, Seraing, Belgium) sold in powder form. Both the powder and stock solution (40 mg/mL) prepared in dimethyl sulfoxide (DMSO; Sigma) must be kept at -20°C.
- Phosphate-buffered saline (PBS) 0.1 M.
- EGTA.
- Paraformaldehyde (PFA; Sigma). Prepare a stock solution of 20% in PBS and store at -20°C.
- $MgCl_2$ (1 M).
- Tween®20 (Sigma).
- Heparin (10 U/L in 0.9% NaCl) for perfusion.
- Peristaltic pump (Polylabo, Strasbourg, France) for perfusion of animals and fixation of tissues by perfusion.
- Vibratome (Leica, Rueil-Malmaison, France) to section the tissues (20–40 μm thickness).
- DMSO to make stock solutions of X-gal.
- Small paint brush to transfer tissue sections from one solution to another. Obtain from any art equipment supplier.
- 25-mL sterile plastic vials to collect sections.
- Potassium ferricyanide and potassium ferrocyanide (Sigma): prepare 0.2 M stock solutions of each.
- Alcohol series for dehydration (baths: 70%, 95%, and 100% of ethanol).
- Xylene, benzyl benzoate, and benzyl alcohol (all from Sigma) for delipidation of whole newborn brains and for small blocks of tissue from adult brains.
- Appropriate sized coverslips and glycerol (glycerol/PBS, 1/3 vol/vol) for mounting slides.
- Plastic gloves. Benzyl benzoate and benzyl alcohol are irritants.
- Glassware for delipidation solutions.
- Microscope equipped with activation and emission filters for fluorescein.

Procedure

β-gal can be revealed by several means:

1. Histochemical X-gal revelation, whether on sections or carried out in toto, requires intracardial perfusion of the anesthetized animals using a peristaltic pump. First, tissues are fixed by perfusing 2% PFA (in PBS), then fixation is continued by leaving tissues blocks overnight in the same solution. Tissues are then vibratome sectioned, and sections are incubated in a 0.4 mg/mL X-gal solution for 2 to 4 hours (30°C). It is important to precede the fixation by perfusion with saline containing 10 U of heparin to help remove blood and blood cells from the vessels. The fixative (2% PFA) can contain EGTA 1.25 mM and $MgCl_2$ 2 mM, which improves the X-gal reaction. To make the reaction mixture, the stock solution of X-gal (40 mg/mL in DMSO) is diluted to 0.8 to 1 mg/mL PBS containing: 0.1% Tween 20, 4 mM potassium ferricyanide, 4 mM potassium ferrocyanide, and 2 mM $MgCl_2$.

2. Immunocytochemical revelation also requires fixed vibratome sections mounted on cromallun–gelatin-coated slides. Polyclonal anti-β-gal monoclonal anti-GFAP and monoclonal anti-NeuN antibodies are used in our experiments. Monoclonal as well as polyclonal antibodies can be labeled using appropriate labeling kits from Amersham Pharmacia Biotech (Piscataway, NJ, USA). We used

Cy™3.5 (Fluorolinf-ab™) according to the manufacturer's instructions to label monoclonal antibodies. Primary antibodies are diluted to the concentrations recommended by each manufacturer in PBS 0.1 M (containing gelatin 0.2%, Triton®X-100 0.3%, 3% normal goat serum). Anti-β-gal is revealed using fluorescein coupled antirabbit antibody. Sections are protected from light to avoid fading of the fluorescence and mounted with glycerol/PBS (1/3 vol/vol) or Vectashield and examined under a fluorescence microscope.

Luciferase antibodies can also be used to follow transgene expression in a double labeling protocol. However, there are currently some problems with obtaining good luciferase antibodies for in vivo work (see Discussion).

3. Fluorescein digalactoside (FDG) is another substrate for β-gal. Hydrolysis of FDG by β-gal results in the liberation of both a monogalactoside and fluorescein. This second product is easily detectable and theoretically makes this method very sensitive. Its main limitation for us is that it is not suitable for fixed tissues, so it is difficult to obtain good morphology in brain preparations. Moreover, on unfixed tissue, as cells die, the fluorescein product diffuses out. Thus, the revelation procedure must be very fast. Also, it is not possible to perform double staining to identify cell types. So this method, which has the theoretical advantage of higher sensitivity than X-gal, is in fact rather limited for in vivo studies.

4. In toto X-gal revelation require intracardial perfusion of the anesthetized animals using a peristaltic pump. Fixation and postfixation are performed as for vibratome sections, but organs are treated as whole mounts. Incubate tissues in a 0.4 mg/mL X-gal solution for 2 to 4 hours (30°C), rinse in 0.1 M PBS (2 × 5 min) and transfer in series of ethyl alcohol under mild agitation: 70% (2 × 2 h), 95% (overnight and another bath of 1 h), 100% (2 × 2 h). Dehydration times can be adapted depending on the tissue size. To ensure thorough dehydration, one can leave the tissue in the second 100% alcohol bath overnight. Put dehydrated tissues into xylene (2 × 2 h in glass), then transfer them into benzyl benzoate–benzyl alcohol (2/1 vol/vol) in a glass container until clarification.

Note: Use gloves and glass containers at all steps involving benzyl benzoate and benzyl alcohol. These agents are irritants and also dissolve plastic.

Protocol for CAT Assay

Materials and Reagents

- [^{14}C]chloramphenicol (57 mCi/mmol; Amersham Pharmacia Biotech). Aliquots stored at -20°C.
- Butyryl-CoA 10 mM (100 mg; Cat. No. B1508; Sigma). Aliquots stored at -80°C.
- Tris-HCl buffer 250 mM, pH 7.5.
- 2,6,10,14-tetramethylpentadecane/xylene (2:1, TMPD; Sigma).
- Miniature polybrene vials for scintillation counter (Packard, Meriden, CT, USA).

Procedure

We use CAT activity to normalize for luciferase expression when the site of injection produces mean values with intra- and interassay variability of more than 15%. In such conditions, it is difficult to obtain statistically valid results without normalizing for transfection efficiency with a constitutively active construct (e.g., CMV-CAT).

1. Before using a co-injected ubiquitously

expressed gene (*CMV-CAT*) to normalize for expression from a physiologically regulated transgene, it is appropriate to validate this approach by quantifying the correlation between the expression of two constituitively expressed genes co-injected into the same brain area. For this, two plasmids (0.35 µg pCMV-luc and 0.15 µg pCMV-CAT in 2 µL 5% glucose) are complexed with PEI 22 kDa (4 eq) and co-injected into brain area targeted. After 18 hours, mice are anesthetized, decapitated, and brains removed for luciferase and CAT assays.

2. For CAT assay, transfer a 50-µL supernatant aliquot to a 1.5-mL polypropylene tube, keep the tube on ice before adding 40 µL of 0.25 M Tris-HCl buffer (pH 7.5). Start reaction by adding 10 µL of mix solution of butyryl-CoA (0.53 mM; Sigma) and [^{14}C] chloramphenicol (0.01 mM, 1.85 kBq per tube; Amersham Pharmacia Biotech). After 3 to 5 seconds of vortex mixing, incubate for 1 hour at 37°C. Then, stop the reaction by adding 200 µL of TMPD/xylene solution (2:1). Vortex mix for 20 seconds and place tube on ice. To separate products, centrifuge for 5 minutes at 4°C (11 000× *g*), remove 150 µL of supernatant, and quantify products in a scintillation counter (Amersham Pharmacia Biotech, Piscataway, NJ, USA).

3. Assay luciferase on other sample of supernatant (see above).

4. Plot luciferase against CAT values from the same sample. Correlation should be significant.

Protocol for Immunoradiography

Materials and Reagents

- Appropriate primary polyclonal antibody raised in rabbit against gene of interest.
- Cryostat (Leica).
- Cryostat sections (15–20 µm thick) cut at -20°C.
- Cromallun–gelatin-coated slides (see above).
- Dessicator.
- PFA (see above).
- PBS.
- BSA.
- Normal goat serum (Sigma).
- Donkey antirabbit [^{35}S]IgG (200–2000 Ci/mmol, 100 µCi/mL; Amersham Pharmacia Biotech).
- β max films (Amersham Pharmacia Biotech).
- Computerized image analysis system (Biocom, Les Ulis, France).

Procedure

This protocol is adapted from Reference 4.

1. The cryostat sections are dessicated at 4°C and frozen at -80°C until used. After a 3-minute fixation (4% paraformaldehyde in PBS) at 4°C sections are preincubated for 1 hour in PBS supplemented with 3% BSA and 1% goat serum, then incubated overnight in an appropriate concentration of polyclonal primary antibody raised in rabbit.

2. After extensive washes in PBS, sections are incubated for 2 hours at room temperature in donkey antirabbit [^{35}S]IgG.

3. After abundant washing, sections are air-dried and apposed to β max film for 1 to 2 days.

4. Optical density is measured by computerized image analysis.

RESULTS AND DISCUSSION

Choice of PEI

A number of PEIs with different mean

molecular weights are available commercially. For example, preparations of branched PEI synthetized to different degrees of polymerization are available (800, 50, or 25 kDa). Preparations of very low molecular weight (0.7 and 2 kDa) are also available from Sigma, but do not complex DNA efficiently. We have found that the branched 25 kDa and linear 22 kDa (Euromedex) polymers work best in the CNS (1,6).

It is most important to note that when using either the 22 or 25 kDa preparations to deliver oligonucleotides, one should only use phosphodiesters and not phosphorothioates. We have found (as has E. Saison Behmoaras, CNRS/MNHN, Paris) that complexes of PEI and phosphorothioates are of lower efficiency than PEI/phosphodiesters complexes in vitro and in vivo.

Choice of Plasmid DNA Preparation Kit

Even though our own (unpublished) results show that the presence of small amounts (≤1 endotoxin units [EU] endotoxin/µg DNA) of endotoxins (also referred to as lipopolysaccarides [LPS]) have no deleterious effects on short term (<4 days) expression in the CNS, we have no data on the possible effects of their presence in the longer term. For this reason, we always use endotoxin-free plasmid preparations. We recommend Jetstar columns. The resulting DNA has ≤0.1 EU endotoxins/µg DNA. This value is not statistically different from the values we find in plasmids prepared with Endotoxin-free columns (Qiagen, Valencia, CA, USA).

If other methods of plasmid preparation are used, we recommend measurement of endotoxin content by Limulus Amebocyte Lysate Assay (LAL; or Coatest™ Endotoxin) manufactured by Charles River Endosafe (Charleston, SC, USA) and distributed in Europe by Chromogenix (Mölndal, Sweden). If values are ≥4 EU/µg DNA, then endotoxins should be removed by chromatography through Affi-prep™ Polymysin Matrix (Bio-Rad Laboratories, Hercules, CA, USA) according to the manufacturer's instructions and endotoxin content reverified. After chromatography, values should be ≤0.05 EU/µg DNA.

Optimal PEI/DNA Ratios

We have found that the optimal ratio of PEI amine/DNA phosphate (N/P ratio) providing the best level of gene expression can vary according to species (and perhaps can also vary according to site of injection). In the mouse and Xenopus tadpole brains, we have consistently found the N/P ratio of 6 provides the best transfection conditions. In contrast, when injecting PEI/DNA complexes into the adult rat substantia nigra, we found that the optimal ratio was 3 (9). In this light, it is important to note that increasing the N/P ratio from 2 to 6 not only increases the overall charge of complexes but also decreases complex size, complexes excluding ethidium bromide (BET), and becoming less visible in the gel. This greater condensation has been confirmed by zeta-sizing (6). Condensation of DNA can be followed by use of BET and electrophoresis (Figure 2), but the optimal N/P ratio can only be ascertained by transfection.

Optimization of Amounts of DNA to be Used

We have found in a number of in vivo systems that lower amounts of DNA give better yields and more faithful physiological regulation than larger amounts. For instance, in a number of promoter studies in the mouse CNS, we use no more than 100 ng of expression vectors coding for the different transcription factors to be tested. Optimization of yield can be carried out by varying the amounts of DNA used either in a constant volume (and varying concentration) or by using a constant DNA concentration (and varying the volume). We have found in the newborn mouse brain that whichever method is used the optimal yield is around 100 ng per injection site (Figure 3). It is interesting to note also that very low amounts of DNA (20 ng) give measurable amounts of luciferase, an observation that is difficult to obtain in culture conditions.

Choice of Reporter Gene and Detection Method

When assessing efficiency of gene transfer, one must take into account the sensitivity of the method used. Histochemical β-gal assay with X-gal is rather insensitive but can give quite satisfactory results in some situations. We find that it gives roughly the same image of transgene distribution and number of cells labeled to an equivalent amount of GFP plasmid. It has been estimated that revealing a good GFP signal requires 10^5 to 10^6 molecules per cell to show up over background (14). For this reason, we have also tried immunocytochemistry with a polyclonal antibody against β-gal (Cappel [ICN, Orsay, France]) or against GFP (CLONTECH) (Figure 4), but this method is generally no more sensitive than histochemistry. However, it is important to note that one can obtain very good, and statistically significant, modulation of endogenous proteins with PEI-based gene transfer in systems where transferring an equivalent amount of β-gal reveals no apparent production of

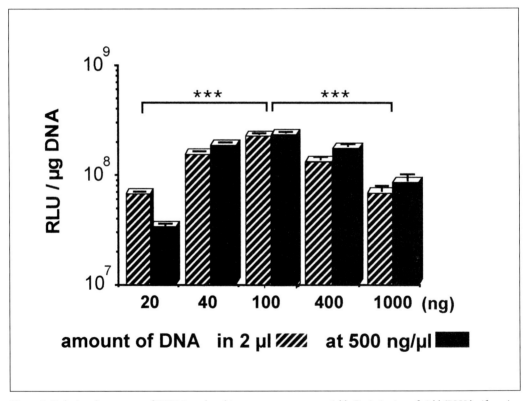

Figure 3. Reducing the amounts of DNA introduced improves reporter gene yield. Optimization of yield (RLU luciferase/µg DNA used) was carried out by varying amounts of DNA used either in a constant volume (2 µL) and varying concentration (striped bars) or by using a constant DNA concentration (500 ng/µL) and varying the final injection volume (black bars). Whichever method is used, the optimal yield is around 100 ng per injection site in this model (newborn mouse brain).

protein. This was the case for a recent series of experiments where we transfected plasmids expressing either sense or antisense sequences of the dopamine transporter (DAT) into the rat substantia nigra (9). In transfection conditions where no β-gal activity could be revealed (0.5 μg CMV-β-gal in 1 μL 5% glucose), use of plasmid encoding DAT significantly increased DAT content. This was shown by immunoautoradiography and by biological measurements of dopamine uptake (9). For this reason, we particularly recommend immunoautoradiography for following low levels of expression of genes of interest against which good antibodies have been raised.

Although luciferase is an excellent reporter gene for kinetic and quantitative studies, the lack of good antibodies against it, limits its use for precise examination of spatial aspects of gene expression. To our knowledge, the only reasonable luciferase antibody currently available is that from Cortex Biochem (San Leandro, CA, USA) distributed in Europe by Europa Research Products, Ely, UK. We have obtained variable results with this polyclonal antibody according to batch number. Promega commercialized a polyclonal antifirefly luciferase up until the end of 1997, but it

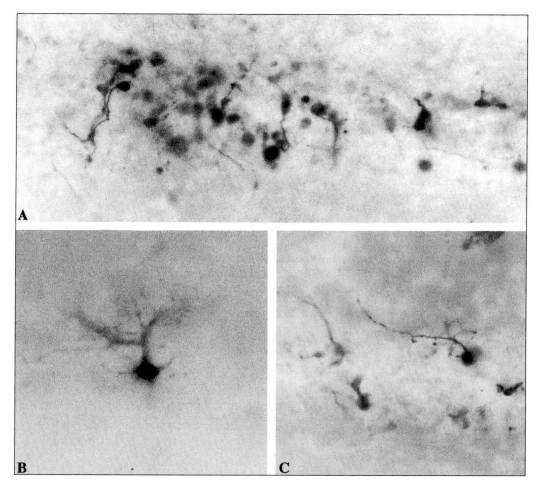

Figure 4. **Use of immunocytochemistry to reveal expression of GFP reporter gene in newborn mouse brains.** Vibratome sections were prepared and exposed to a rabbit polyclonal (CLONTECH) antibody against GFP. The second antibody was an antirabbit antibody linked to alkaline phosphatase (CLONTECH).

has been withdrawn from circulation due to problems of titer. However, we have been able to obtain reasonable results with some batch numbers. Santa Cruz Biotechnology (Santa Cruz, CA, USA) also produces a luciferase antibody, but we have not yet tested this antibody on section from brains transfected in vivo.

Using PEI-Based Gene Transfer to Follow Promoter Regulation in Defined Areas of the CNS

A major topic of discussion in the field at the moment is whether faithful physiological regulation can be obtained by transient transfection (whether in vitro or in vivo)

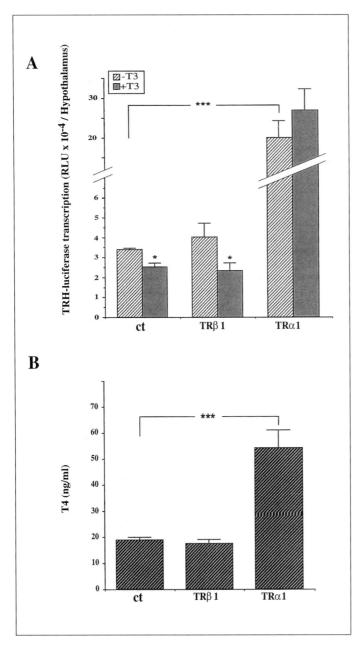

Figure 5. PEI-based gene transfer is a powerful method for carrying out analysis of promoter regulation in vivo. Example of thyroid hormone-dependent down regulation of TRH promoter activity in the hypothalamus of newborn mice. (A) TRH-luc is regulated in a physiologically faithful manner being significantly repressed in the presence of T_3 (controls, ct; left hand columns). Transcriptional regulation is isoform specific: overexpression of TRβ allows for T_3-dependent transcription (middle columns), whereas overexpression of TRα abrogates T_3-dependent transcription and increases basal expression 5-fold (right hand columns). (B) TR isoform effects on TRH-luc expression are played out down the hypothalamo–hypophyseal–thyroid axis. Overexpression of an empty (ct) plasmid or a plasmid expressing TRβ does not affect basal levels of circulating thyroid hormone (T_4), whereas expression of TRα increases T_4 secretion 5-fold. Note that effects of TR isoforms on TRH-luc transcription and T_4 secretion are equivalent. T_4 levels were measured by radioimmunoassay.

using plasmid-based delivery. This is because with plasmid vectors, as with adenoviral vectors, transcription results from episomally-located sequences. This contrasts with the situation where one follows transcription from integrated sequences, as is the case with retroviruses or adenoassociated viruses, or with stably transfected cell lines whatever the method initially chosen for gene delivery. However, we have found that, despite the lack of integration into the genome, we obtain remarkably tight physiologically appropriate regulation of cell-specific promoters in vivo. A particularly illustrative example is that of thyrotropin releasing hormone (TRH) promoter regulation in the hypothalamus. We have microinjected complexes containing 1 µg of construct containing 900 bp of the rat TRH promoter upstream of the luciferase coding sequence, complexed with 22 kDa PEI into the hypothalamus of newborn mice in different thyroid states. In hypothyroid animals, transcription is twice that in normal euthyroid animals, and in hyperthyroid animals, transcription is reduced to half that of the normal group (7). This transcriptional regulation faithfully reflects the negative feedback effect of circulating thyroid hormone on hypothalamic TRH production (Figure 5A). What is more, the effects of different thyroid hormone receptor isoforms (TRα and TRβ) can be assessed both on TRH-luc transcription and on endogenous TRH (Figure 5B). In effect, when animals are transfected with an expression vector for TRα, not only is TRH-luc transcription increased 4/5-fold, but the effect is played out down the hypothalamo–hypophyseal–thyroid axis, and T4 secretion is increased 4/5-fold. We deduce from this that the effects of the different TRs on TRH-luc transcription are echoed on endogenous TRH transcription. Another example of physiological regulation of transgenes introduced by somatic gene transfer includes the Krox-24 gene in the newborn mouse brain (5).

REFERENCES

1. Abdallah, B., A. Hassan, C. Benoist, D. Goula, J.P. Behr, and B.A. Demeneix. 1996. A powerful nonviral vector for in vivo gene transfer into the adult mammalian brain: polyethylenimine. Hum. Gene Ther. 7:1947-1954.
2. Behr, J.P. and B.A Demeneix. 1998. Gene delivery with polycationic amphiphiles and polymers. Curr. Res. Mol. Ther. 1:5-12.
3. Boussif, O., F. Lezoualc'h, M.A. Zanta, M. Mergny, D. Scherman, B. Demeneix, and J.P. Behr. 1995. A novel, versatile vector for gene and oligonucleotide transfer into cells in culture and in vivo: polyethylenimine. Proc. Natl. Acad. Sci. USA. 92:7297-7303.
4. Gérard, C., M.P. Martres, K. Lefèvre, M.C. Miquel, D. Vergé, L. Lanfumey, E. Doucet, M. Hamon, and S. El Mestikawy. 1997. Immuno-localization of serotonin 5-HT$_6$ receptor-like material in the rat central nervous system. Brain Res. 146:207-219.
5. Ghorbel, M.T., I. Seugnet, N. Hadj-Sahraoui, P. Topilko, G. Levi, and B.A. Demeneix. 1998. Thyroid hormone effects on Krox-24 transcription in the post-natal mouse brain are developmentally regulated but are not correlated with mitosis. Oncogene 18:917-924.
6. Goula, D., J. Remy, P. Erbacher, M. Wasowicz, G. Levi, B. Abdallah, and B. Demeneix. 1998a. Size, diffusibility and transfection performance of linear PEI/DNA complexes in the mouse central nervous system. Gene Ther. 5:712-717
7. Guissouma, H., M.T. Ghorbel, I. Seugnet, T. Ouatas, and B.A. Demeneix. 1998. Physiological regulation of hypothalamic TRH transcription in vivo is T3 receptor isoform-specific. FASEB J. 12:1755-1764.
8. Lehmann, A. 1974. Atlas Stéréotaxique du Cerveau de la Souris. CNRS Editions, Paris.
9. Martres, M.P., B.A. Demeneix, N. Hanoun, M. Hamon, and B. Giros. 1998. Up- and down-expression of dopamine transporter by plasmid DNA transfer in the rat brain. Eur. J. Neurosci. 10:3607-3616.
10. Sambrook, J., E.F. Fritsch, and T. Maniatis. 1989. In N. Ford, C. Nolan, and M. Ferguson (Eds.), Molecular Cloning: A Laboratory Manual. CSH Laboratory Press, Cold Spring Harbor, NY.
11. Seed, B. and J.Y. Sheen. 1988. A simple phase extraction assay for chloramphenicol acetyl transferase. Gene 67:271-277.
12. Suh, J., H.-J. Paik, and B.K. Hwang. 1994. Ionization of polyethylenimine and polyallylamine at various pHs. Bioorg. Chem. 22:318-327.
13. Zhu, N., D. Liggitt, Y. Liu, and R. Debs. 1993. Systemic gene expression after intravenous DNA delivery into adult mice. Science 261:209-211.
14. Zlokarnik, G., P.A. Negulescu, T.E. Knapp, L. Mere, N. Burres, L. Feng, M. Whitney, K. Roemer, and R.Y. Tsien. 1998. Quantitation of transcription and clonal selection of single living cells with β-lactamase as reporter. Science 279:84-88.

5. Transfection of $GABA_A$ Receptor with GFP-Tagged Subunits in Neurons and HEK 293 Cells

Stefano Vicini, Jin Hong Li, Wei Jian Zhu, Karl Krueger, and Jian Feng Wang

Department of Physiology and Biophysics, Georgetown University Medical School, Washington, DC, USA

OVERVIEW

We describe an approach that may allow to study changes in the γ-aminobutyric acid ($GABA_A$)receptor distribution with development and pharmacological treatments in living neurons. We produced expression vectors containing chimeras of the green fluorescent protein (GFP) linked to the C terminus of $GABA_A$ receptors α1, γ2, or the δ subunits. Human embryonic kidney (HEK) 293 cells were successfully transfected with α1-GFP cDNAs together with β3 subunit as indicated by the formation of green fluorescent clusters of receptor subunits that colocalized with immunospecific staining for the α1 subunits and by whole-cell recordings of GABA-activated Cl⁻ currents. Although the current density was lower in these cells, GABA, bicuculline, and $ZnCl_2$ actions were unaltered. Similarly, transfection with cDNAs encoding for the γ2-GFP chimera together with α1β3 subunit cDNAs produced clusters of subunits and GABA-activated chloride currents that were insensitive to blockade by $ZnCl_2$ and that were potentiated by zolpidem. Lastly, δ-GFP chimeras transfection in HEK cells produced receptor insensitive to the potentiation by the neurosteroid THDOC. We then successfully transfected primary cultures of neocortical and cerebellar neurons with these $GABA_A$ receptor subunits—GFP chimeras. We obtained evidence of elongated cluster formation in both cell types that matched well, although not completely, endogenous receptor clusters as indicated by β2/3 staining, and also partially corresponded to synaptophysin positive punctae indicating synaptic localization of transfected subunits. Electrophysiological recordings from transfected neurons indicated that functional GABAergic synapses were still maintained. This approach will allow to follow targeting and distribution of $GABA_A$ receptor clusters in living neurons during development in culture and in different experimental conditions.

BACKGROUND

Receptor protein dynamics and interactions in living cells can been studied with

Cellular and Molecular Methods in Neuroscience Research
Edited by A. Merighi and G. Carmignoto
©2001 Eaton Publishing, Natick, MA

fluorescent microscopy, photobleaching, and resonance energy transfer. It is therefore essential to be able to identify strong ligands or selective antibodies to link to fluorescent probes to allow similar studies. An alternative approach takes advantage of the possibility to tag receptor proteins with a fluorescent protein. To this aim, the GFP from the jellyfish *Aequorea victoria* is beginning to be extensively used for this purpose. Indeed, GFP-tagged proteins, both in the cytosol and in the plasma membrane, have been studied recently (See References 6 and 7 for review). A particularly useful application of GFP tagging of protein is to study the localization, distribution, and dynamics of neurotransmitter receptors. An essential condition for this study however is that tagging does not alter binding or functional properties of the receptor. A GFP-tagged version of the *N*-methyl-D-aspartate (NMDA) receptor subunit NR1 (12) has been transfected in mammalian HEK 293 cells and has been demonstrated to produce functional NMDA receptors when cotransfected with NR2A subunit. Functional integrity of GFP-conjugated glycine receptor channels has been recently reported (8). Similar findings have been shown for both α and β adrenergic receptors (2,3,10) and voltage-gated calcium channels (9).

Ligand gated receptor channels are heterooligomeric proteins made by multiple subunit. In particular, $GABA_A$ receptors are pentameric structures that comprise an ion channel for chloride ion. These receptors, localized in postsynaptic membranes throughout the central nervous system are responsible for inhibitory synaptic transmission (See Reference 11 for review). Tagging of one of the subunit may allow to observe the formation of functional receptor complexes as demonstrated for NMDA receptors in HEK cells (12). Indeed, Connor et al. (5) have recently demonstrated the formation of functional GABA channels tagged with GFP on the surface of Xenopus oocytes. These receptors have maintained both binding and electrophysiological properties of nontagged receptors. Additionally, the level of receptor expression was unaltered, and both Ec50 for GABA and benzodiazepine potentiation were also unaffected. Lastly, the interaction between distinct subunit that gives rise to specific pharmacological properties was preserved when GFP-tagged α1 subunit is expressed. Calcium phosphate-based transient transfection technique (4) has been classically applied to transfect eukaryotic expression vectors in mammalian tumoral cell lines. Recently however, successful transfection of primary neuronal culture has been reported, although with a limited efficiency (1,16). Here, we report the successful construction of cDNAs expressing $GABA_A$ receptor subunits tethered to the GFP protein and its expression and colocalization in primary cultures of cortical and cerebellar neurons.

PROTOCOLS

Materials and Reagents

- Basal Eagle's medium (Life Technologies, Gaithersburg, MD, USA).
- Coverslips were mounted on slides using Vectashield (Vector Laboratories, Burlingame, CA, USA) as mounting medium.
- Fluorescein isothiocyanate (FITC)-conjugated goat antimouse (Jackson ImmunoResearch Laboratories, West Grove, PA, USA).
- Gentamycin (Life Technologies).
- Glass capillaries (Wiretrol II; DRUMMOND Scientific, Broomall, PA, USA).
- Glass coverslips (Fisher Scientific,

- Pittsburgh, PA, USA).
- HEK 293 cells (No. CRL1573; ATCC, Rockville, MD, USA).
- Indocarbocyanine (Cy3™) (Jackson ImmunoResearch Laboratories).
- Minimal essential medium (MEM) (Life Technologies).
- MEM with Hanks' salts (Life Technologies).
- Monoclonal mouse antisynaptophysin (Roche Molecular Biochemicals, Mannheim, Germany).
- Penicillin (Life Technologies).
- Plasmid pGREENLANTERN™ (Life Technologies).
- Polyclonal rabbit anti-α1 and β2/3 antibodies (CHEMICON International, Temecula, CA, USA).
- Poly-L-lysine (Sigma, St. Louis, MO, USA).
- Rabbit IgG antibodies (Jackson ImmunoResearch Laboratories).
- Streptomycin (Life Technologies).
- Trypsin (Sigma).
- Vector pCDM8 (Invitrogen, Carlsbad, CA, USA).
- Vector pEGFP-N1 (CLONTECH Laboratories, Palo Alto, CA, USA).
- Patch amplifier (Axopatch 1D; Axon Instruments, Foster City, CA, USA).
- 8-pole low-pass Bessel filter (Frequency Devices, Haverhill, MA, USA).
- Digitization software (Clampex 8; Axon Instruments).
- Analysis software (Origin; MicroCal Software, Northampton, MA, USA and Clampfit 8; Axon Instruments).

Procedures

HEK 293 Cell Line

HEK 293 cells were grown in MEM, supplemented with 10% fetal bovine serum, 100 U/mL penicillin, and 100 U/mL streptomycin, in a 6% CO_2 incubator. Exponentially growing cells were dispersed with trypsin, seeded at 2×10^5 cells/35-mm dish in 1.5 mL of culture medium and plated on 12-mm glass coverslips.

Primary Cultures

Primary cultures of rat cerebellar granule neurons were prepared from 7 days old Sprague Dawley rat cerebella. Cortical neurons from newborn rat pups were prepared with similar procedure. Cells were dispersed with trypsin (0.25 µg/mL) and plated at a density of 0.8 to 1×10^6 on 35-mm Nunc dishes coated with poly-L-lysine (10 µg/mL). Cells were cultured in basal Eagle's medium supplemented with 10% bovine calf serum, 2 mM glutamine, 100 µg/mL gentamycin, and maintained at 37°C in 6% CO_2. Cytosine arabinoside (10 µM) was added to all cultures 18 to 24 hours after plating to inhibit glial proliferation. The final concentration of KCl in the culture medium was adjusted to 25 mM.

Construction of Plasmid DNA for GABA Subunit with GFP Tags

Rat α1, γ2, and δGABA$_A$ receptor subunit cDNAs were individually subcloned into the expression vector pCDM8. To fuse those cDNAs with GFP, we used pEGFP-N1 vector that contains multiple cloning sites before the EGFP coding sequence. The GABA$_A$ receptor subunit cDNAs was modified via polymerase chain reaction (PCR), and the stop codons were replaced by suitable cloning site. Then the target gene was cloned into pEGFP-N1 so that it is in frame with the EGFP coding sequences. The α1 subunit was cloned into pEGFP-N1 using *Hin*dIII and *Kpn*I sites. The PCR product of γ2 subunit was cut by

BamHI and SpeI, and ligated to the pEGFP-N1 cut by NheI and BamHI. The δ subunit was inserted into EcoRI and BamHI sites of pEGFP-N1.

cDNA Transient Transfection

HEK 293 cells were transfected using the calcium phosphate precipitation method (4) with various combinations of subunit cDNAs. The following plasmid combinations were mixed: α1:β3:γ2-GFP, α1-GFP:β3:γ2, and α1:β3:δGFP, (1 μg each construct) and the coprecipitates were added to culture dishes containing 1.5 mL MEM for 12 to 16 hours at 37°C under 3% CO_2. The media was removed, the cells were rinsed twice with culture media, and finally incubated in the same media for 24 hours at 37°C under 6% CO_2.

Cortical neurons and cerebellar granule cells at 4 days in vitro (DIV) were transfected with pEGFP-N1 expression vector, γ2-GFP, and α1-GFP chimera expression vector using a modified calcium phosphate precipitation method. Briefly, the culture medium was replaced and saved. After one time washing with transfection medium (MEM with Hanks' salts), 3 μg plasmids were added to cells of each dish containing 1.5 mL transfection medium and then incubated for 1 hour at room temperature in cell culture hood. After rinsing 2 times with transfection medium, the cells were put back to saved medium and then were maintained at 37°C under 6% CO_2. For α1-GFP chimera transfected cortical neurons, insulin was added to the medium at the concentration of 2 μm/mL in neurons at DIV 13 for 2 hours.

Electrophysiological Studies

Cultured granule cells or isolated HEK 293 transfected cells were voltage-clamped at -50 mV in the whole-cell configuration using the patch clamp technique on the stage of an inverted microscope at room temperature. The recording pipet contained (mM) 145 CsCl, 5 $MgCl_2$, 11 ethylene glycol bis (β-aminoethlether)-N, N, N′,N′-tetraacetic acid (EGTA), 5 Na-adenosine-5′-triphosphate (ATP), 0.2 guanosine-5′-triphosphate (GTP), and 10 mM 4-(2-hydroxyethyl)-1-piperazineethanesulfonic acid (HEPES) at pH 7.2 with CsOH. Cells were bathed in (mM) 145 NaCl, 5 KCl, 2 $CaCl_2$, and 5 HEPES at pH 7.2 with NaOH. Osmolarity was adjusted to 325 mosm with sucrose. The culture dish in the recording chamber (< 500 μL total volume) was continuously perfused (5 mL/min) to prevent accumulation of drugs. All drugs dissolved in bath solution contained, dimethyl sulfoxide (DMSO) at a maximal final concentration of 0.01%, which failed per se to modify GABA responses. GABA was applied directly by a gravity-fed Y tubing delivery system placed within 100 μm of the recorded cell. Bath perfusion for the 2 minutes preceding coapplication with GABA was required to observe full potentiation of GABA responses by flunitrazepam. GABA (0.5 M in water adjusted to pH 4.0 with HCl) was also applied by iontophoresis with 30-millisecond pulses of positive current. With GABA iontophoretic currents in the 25 to 50 nA range, outward currents were generated in transfected cells in such a way as to obtain a peak amplitude of 150 to 200 pA. Benzodiazepines (BZs) were a gift from Hoffman La Roche (Switzerland), DMCM and β-CCM were from Ferrosan (Denmark) and Zolpidem was from Syntelabo (France). All drugs were dissolved in bath solution containing DMSO at a maximal final concentration of 0.01%.

Currents were monitored with a patch amplifier (Axopatch 1D), filtered at 1.5 kHz (8-pole low-pass Bessel), digitized in an IBM-PC computer with the Clampex 8 software for off-line analysis. After normalization, fitting of the dose-response relationship was performed using the logistic equation:

% $I_{max} = 100/I_{max} * \{1+(EC_{50}/[GABA]^{n_h})\}$

where I_{max} is the maximal Cl⁻ current, elicited by GABA, EC_{50} is the GABA concentration eliciting the half-maximal response, and n_h is the Hill coefficient. Results are expressed as mean ± SEM. Origin and Clampfit 8 software were used for figure preparation and statistical analysis using ANOVA with a $P < 0.05$ and a paired t-test with $P < 0.01$. The Bonferroni correction was applied for multiple group comparison.

Antibodies

The monoclonal mouse anti-GAD65 antibody was a kind gift from Dr. Samuel Rabkin (Georgetown University, Washington, DC).

Immunocytochemistry

Cultured HEK 293 cells and were fixed with 4% paraformaldehyde, 4% sucrose in phosphate-buffered saline (PBS) for 15 minutes at room temperature and permeabilized with 0.25% Triton® X-100 for 3 to 5 minutes. Cells were preincubated in 10% goat serum for 30 minutes at room temperature and then incubated in primary antibody in PBS containing 5% goat serum overnight at 4°C. The concentrations of primary antibodies were: rabbit anti-β2/3 subunit (10 μg/mL), rabbit anti-α1 subunit (1:100), and mouse anti-GAD65 (1 μg/mL). After washing 3 times in PBS, cells were incubated with goat antirabbit and/or antimouse (Cy3-conjugated 1:200, FITC-conjugated 1:50) secondary antibodies for 1 hour at room temperature. Coverslips were visualized using a microscope equipped for the visualization of fluorescence (Vanox, Olympus, Japan). Spectral characteristics of the excitation–emission filters used were 490/530 nm (green fluorescence for FITC and GFP) and 545/610 nm (red fluorescence for Cy3), respectively. Bleed-through was minimal as seen looking at immunostaining with only single color labeling and by the lack of colocalization of some clusters in double staining experiments. Cells were visualized through 10×, 40×, and 100× objectives, and images were captured using a digital camera and transferred to a computer workstation. Controls were performed by omitting primary antibodies. Only a weak and nonspecific staining was observed under these experimental conditions.

Receptor cluster colocalization was quantified over a length of 100 μm in 3 dendrites evaluated in each of 5 selected neurons. Clusters were counted after background subtraction and manual fluorescent thresholding correspondent to twice the intensity of the diffuse fluorescence of the dendritic shaft (modified from Reference 13).

RESULTS AND DISCUSSION

Transfection of GABA$_A$ Receptor Subunit EGFP Constructs in HEK Cells

Using a commercially available plasmid kit, we made several constructs of GABA$_A$ receptor subunits with the EGFP protein tethered to the C terminus of the receptor subunit. When cotransfected in mammalian HEK 293 cells diffuse punctae staining could be observed throughout the entire cells including the plasma membrane in living cells. The nucleus was not stained. In Figure 1 are shown examples of cells transfected with α1-GFP and β3 cDNAs transfection and γ2-GFP with α1β3 cDNAs transfections. Similar distribution was observed with δ-GFP subunit transfection (not shown).

Immunocytochemical staining with antibodies against the α1 subunit of GABA receptor confirmed that GFP punctae corresponded to the α1 subunit protein (not shown). Similarly, staining with β2/3 subunit selective antibodies (not shown) demonstrated that α1-GFP clusters matched β3 subunit clusters, indicating that the majority of receptors comprising α1-GFP tandems also contained the β3 subunit.

We then performed electrophysiological studies on HEK 293 cells transfected with constructs of $GABA_A$ receptor subunits tethered to the EGFP protein. Applications of increasing GABA concentrations to voltage-clamped cells elicited currents of increasing peak characterized by faster desensitization (Figure 2, A and B). Dose-response indicated that the presence of the GFP tether did not alter the half maximal concentration nor the maximal response recorded from cells transfected with the α1 and β3 cDNAs.

The presence of specific subunit in the $GABA_A$ receptor complex determines selective pharmacological regulation by distinct pharmacological agents (11). Zinc has been widely used as a selective agent to determine the presence of the γ2 subunit in the receptor channel complex (11). Figure 2B illustrates that currents elicited by ionophoretical application of GABA were blocked by $ZnCl_2$. Among the distinct pharmacological agents that regulate $GABA_A$ receptor, the benzodiazepines and neurosteroids have a better characterized structural requirement in term of subunit composition (11). In Figure 2C, we illustrate GABA-activated currents in a HEK cell transfected with α1β3 and γ2-GFP cDNAs. The allosteric modulator

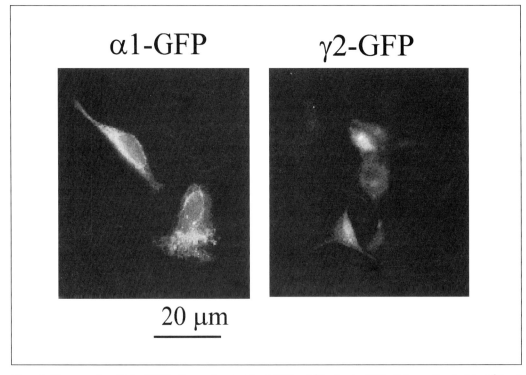

Figure 1. Two examples of living HEK cells transfected with α1-GFP and β3 subunit cDNAs (left panel) and with α1β3 and γ2-GFP subunit cDNA (right panel). Diffuse punctate staining could be observed throughout the cells including the plasma membrane. The nucleus was not stained.

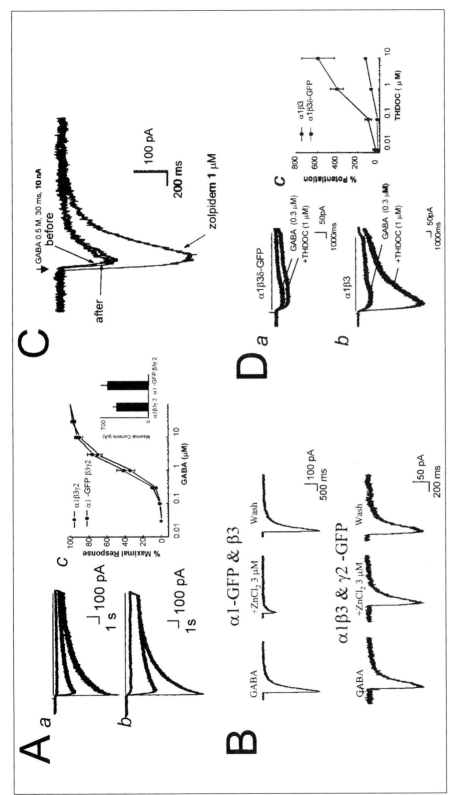

Figure 2. GABAR-GFP receptors with GFP tethers at the C terminus preserve pharmacological properties. (A) Half maximal concentration and maximum peak of GABA-activated currents from HEK 293 cells expressing GABA receptors with α1-GFP subunit together with β3 and γ2 subunits (a) does not change as compared to cells expressing with α1β3 and γ2 subunits (b). A summary of the results obtained is shown in (c). (B) γ2-GFP tandem preserve sensitivity to $ZnCl_2$ inhibition of GABA currents produced by ionophoretic applications in a HEK cell transfected with α1-GFP/β3 subunit cotransfected with α1β3 subunit cDNA inhibits $ZnCl_2$ sensitivity. (C) γ2-GFP tandem preserve zolpidem potentiation. Ionophoretic application of GABA produced currents in a HEK cell transfected with α1-GFP/β3/γ2 subunit cDNAs that were potentiated by the imidazopyridine Zolpidem. (D) GABA current potentiation by neurosteroid THDOC in cells transfected with α1β3 (a) was abolished when δ-GFP cDNA was cotransfected with α1β3δ-GFP (b). A summary of the results obtained is shown in (c).

Zolpidem potentiated considerably the response to ionophoretically applied GABA, demonstrating that the benzodiazepine recognition site was not affected by the GFP tether on the γ2 subunit. Coexpression of δ-GFP tandem strongly reduced the potentiation by the neurosteroid THDOC (Figure 2D), as observed with wild-type δ subunit (18). These results taken together indicate that the expression of GABA$_A$ receptor subunit, modified to include a fluorescent marker at their C terminus, does not affect the formation of functional chloride channels, agonist sensitivity, or allosteric regulation of channel activity.

Transfection of Neurons in Primary Culture

Using a modification of the CaPO$_4$ precipitation technique (4), we transfected primary culture of rat cortical and cerebellar neurons. GFP transfected neurons were observed as early as 6 hours after transfection. In Figure 3 are shown examples of cortical neurons expressing GFP protein and one example of GFP positive cerebellar granule neuron where dendrite and axon are clearly distinguishable. The percent success of transfection was highly variable in distinct experiments but never exceeded 0.5% of the neurons. Cells remained transfected for the reminder of the culture lives (up to 4 weeks), but the number of fluorescent cells decrease with time along with the total cell number. Fluorescent intensity of transfected neurons was extremely bright and allowed us to distinguish many details of anatomical features such as axonal branches and growth cones and dendritic spines (not shown). Functional properties such as action potential generation, occurrence of

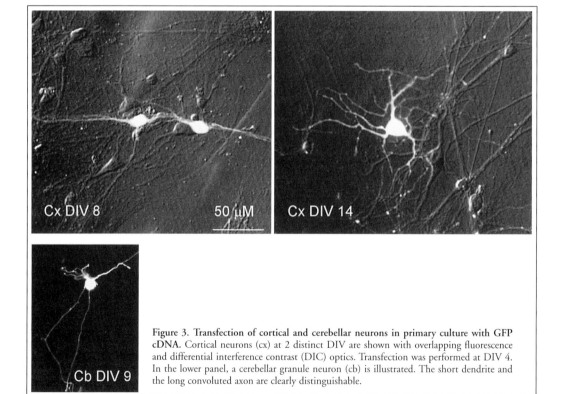

Figure 3. **Transfection of cortical and cerebellar neurons in primary culture with GFP cDNA.** Cortical neurons (cx) at 2 distinct DIV are shown with overlapping fluorescence and differential interference contrast (DIC) optics. Transfection was performed at DIV 4. In the lower panel, a cerebellar granule neuron (cb) is illustrated. The short dendrite and the long convoluted axon are clearly distinguishable.

spontaneous synaptic currents, and the capability to produce functional synapses were not affected as observed in parallel electrophysiological recordings (not shown). For comparison, GFP expression was attempted with viral transfection using a defective herpes simplex virus vector (courtesy of Dr. S. Rabkin). Virus infection produced GFP expression and labeling of neuronal cells, but the intensity of staining was 100-fold lower than with $CaPO_4$ transfection, and the degree of toxicity was greatly enhanced.

Transfection of GABA$_A$ Receptor Subunit EGFP Constructs in Neurons

After demonstrating successful transfection of the pEGFP construct in neurons, we attempted to transfect α1-GFP and γ2-GFP constructs. Cells were successfully transfected with both cDNAs, and the percent cell transfected was not considerably different for the distinct plasmids. However, the intensity of fluorescence and the pattern of protein distribution differed considerably. First, with α1-GFP cDNA, the formation of clear punctae of GFP fluorescence was observed on dendrites and occasionally on axons of transfected neurons (Figure 4). Second, a diffuse and weak fluorescence staining was observed throughout the cells allowing the identification of dendritic branches of the transfected cell. With γ2-GFP cDNA, we observed only the diffuse fluorescence, and we failed to observe clusters similar to those seen with α1-GFP transfection. It is possible that for both γ2 and α1 subunit tagging with GFP, the C terminus may alter the subunit assembly allowing release of cytoplasmic-free subunit-GFP tandems. It is also possible that cleavage of the GFP portion of the construct produced a considerable amount of free GFP. Lastly, diffuse staining may correspond to the cytoplasmic assembly of the receptor that precedes membrane insertion. Whatever mechanism underlie the diffuse staining, it is clear that with α1-GFP cDNA we also observed clear formation of subunit clusters that may indicate the correct assembly and targeting of labeled GABA$_A$ receptors in living neurons. The reason for the failure of γ2-GFP construct to perform equally well remain to be investigated further. This raises the concerns that different cDNA constructs may or may not have the capability for proper processing. Thus, to achieve successful expression, more than one construct per protein should be made.

Electrophysiological recordings demonstrate the occurrence of spontaneous inhibitory postsynaptic currents (s.i.p.s.cs) in α1-GFP transfected cells (Figure 4B). This indicates that the formation of inhibitory synapses and the function of postsynaptic receptors was unaffected by the tagging of the α1 subunit.

Once we observed the successful formation of subunit clusters in neurons, we wanted to study the colocalization of α1-subunit constructs and the presence of the β subunit with selective antibodies. As illustrated in Figure 5 (top panel), there was clear matching between α1-GFP clusters and the presence of clusters of β2/3 subunit, indicating, although not totally proving, that coassembly of native and transfected subunits may occur together with appropriate targeting to the dendrites of receptor subunits. The most suggestive evidence that α1-GFP clusters reveal functional GABA$_A$ receptor comes from staining of α1-GFP transfected neurons with antibodies against glutamic acid decarboxylase 65, a marker of GABAergic presynaptic terminals. As it can be observed in Figure 5B, many α1-GFP clusters were facing presynaptic GABAergic terminals, suggesting that they are part of postsynaptic densities at those sites. At the same time, a few extrasynaptic clusters were also identified consistently with the demon-

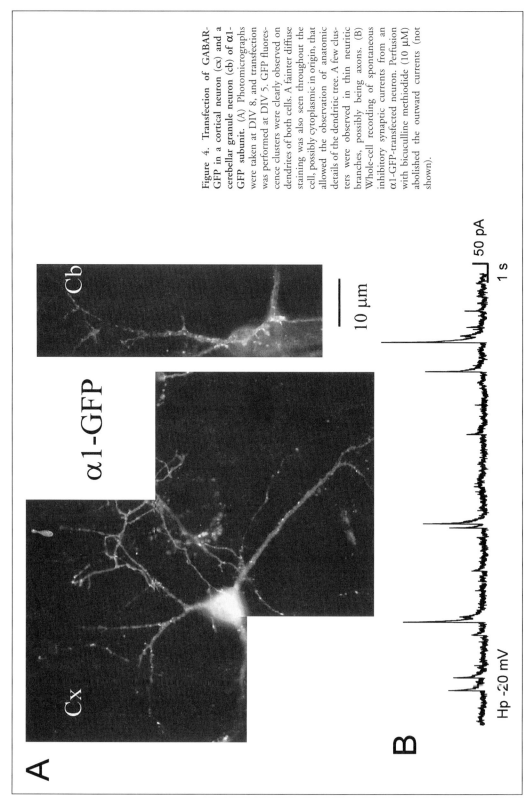

Figure 4. Transfection of GABAR-GFP in a cortical neuron (cx) and a cerebellar granule neuron (cb) of α1-GFP subunit. (A) Photomicrographs were taken at DIV 8, and transfection was performed at DIV 5. GFP fluorescence clusters were clearly observed on dendrites of both cells. A fainter diffuse staining was also seen throughout the cell, possibly cytoplasmic in origin, that allowed the observation of anatomic details of the dendritic tree. A few clusters were observed in thin neuritic branches, possibly being axons. (B) Whole-cell recording of spontaneous inhibitory synaptic currents from an α1-GFP-transfected neuron. Perfusion with bicuculline methiodide (10 μM) abolished the outward currents (not shown).

5. Transfection of GABA$_A$ Receptor

strated evidence for extrasynaptic receptor clusters.

Insulin Treatment

A report demonstrated the relatively rapid regulation of GABA$_A$ receptor subunit by insulin (17). To verify that the transfected constructs expressing α1-GFP subunit in cortical neurons were capable to undergo the same regulation as demonstrated by insulin treatments, we incubated cortical neurons for 1 or 2 hours with insulin and measured the number of α1-GFP clusters per 100 μm of dendritic length. As seen in Figure 6, insulin treatment produced a time-dependent increase of fluorescence subunit clusters, indicating that the tagging of the subunit protein with the fluorescent tether did not affect

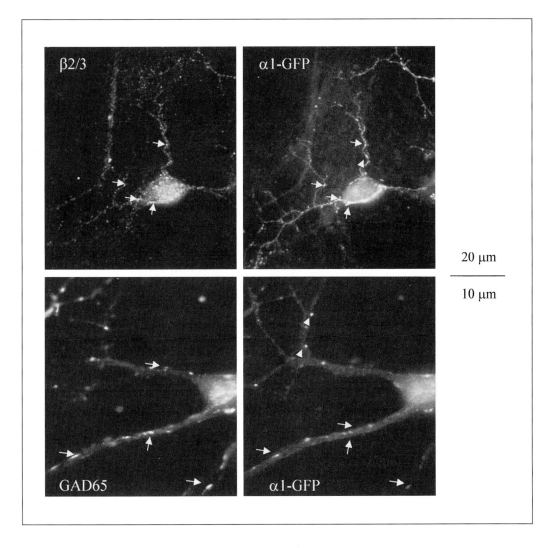

Figure 5. Colocalization of GABAR-GFP in cortical neurons with β2/3 subunit and GAD 65. In the two top panels α1-GFP clusters in fixed cortical neuron (DIV 10) is compared to Cy3-β2/3 subunit immunostaining. In the lower panels, α1-GFP clusters are compared in another cortical neuron to immunostaining for the selective presynaptic marker of GABAergic terminals GAD 65, shown at higher magnification. In both panels, arrows indicate matching clusters while arrowheads indicate nonmatching clusters.

63

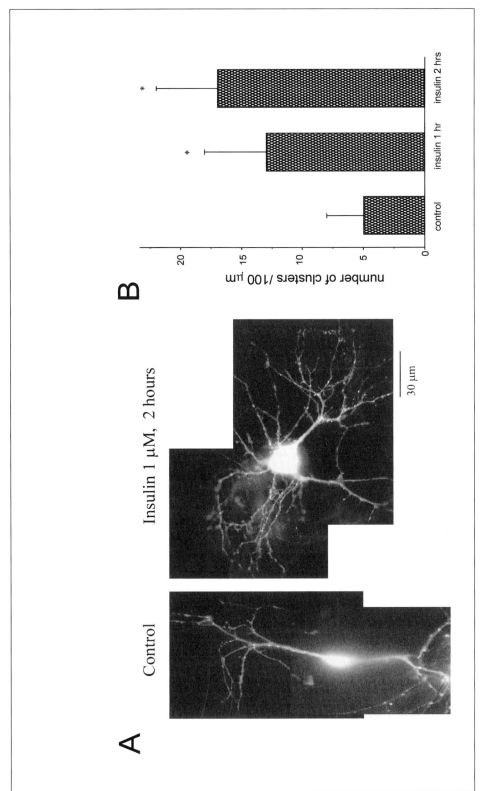

Figure 6. Insulin increase the clusters of α1-GFP subunit. (A) α1-GFP clusters are compared between a control cortical neuron in the same culture after 2 hours of addition to the culture medium of 1 μM insulin. (B) Summary of the results on the number of α1-GFP clusters per 100 μm dendritic length obtained with 2 insulin incubation times. * indicates significant difference (independent student t-test, $P < 0.05$).

regulation of expression.

CONCLUSIONS

As previously demonstrated, tagging $GABA_A$ receptor subunit with GFP does not alter functional expression and pharmacological properties. In addition, our data present evidence that transfection of GFP-tagged subunits is possible in neocortical and cerebellar neurons in primary culture. Clusters of fluorescent proteins that matched well with endogenous $GABA_A$ receptor clusters are formed. Functional GABAergic synapses are still maintained in transfected neurons that possibly utilize $\alpha 1$-GFP subunit as seen by correspondence with glutamic acid decarboxylase (GAD) positive punctae. Insulin treatment increased the expression of $\alpha 1$ clusters indicating that the receptor can undergo a similar regulation as the native receptor. Our data also highlight possible problems of transfecting exogenous subunit into neurons, $\gamma 2$-GPF transfection did not give rise to clusters and axonal localization may be an artifact of the transfection procedure. However, the results presented here indicate that if the transfected construct can form clusters in neurons in culture, assembly targeting and regulation can proceed as for native receptors. This raised the hope that this approach will allow us to follow the targeting and distribution of $GABA_A$ receptor clusters in living neurons during development in culture and in various experimental conditions. Furthermore, the possibility to simultaneously visualize two GFP variants in living neurons may allow, in the future, to tag two distinct proteins (14) with the goal of studying the coassembly and independent regulation of subunit expression. This will be essential to the understanding of the significance of the large heterogeneity of subunits for ligand-gated channels, which constitute postsynaptic receptors in the central nervous system (CNS). Indeed, studies on AMPA receptors GFP-tagged have demonstrated rapid spine delivery and redistribution after synaptic activation (15).

ACKNOWLEDGMENTS

Supported by National Institute of Neurological Disorders and Stroke (NINDS) Grant Nos. NS32759 and MH01680.

REFERENCES

1. **Ango, F., S. Albani-Torregrossa, C. Joly, D. Robbe, J.-M. Michel, J.-P. Pin, J. Bockaert, and L. Fagni.** 1999. A simple method to transfer plasmid DNA into neuronal primary cultures: functional expression of the mGlu5 receptor in cerebellar granule cells Neuropharmacology 38:793-803.
2. **Awaji, T., A. Hirasawa, M. Kataoka, H. Shinoura, Y. Nakayama, T. Sugawara, S. Izumi, and G. Tsujimoto.** 1998. Real-time optical monitoring of ligand-mediated internalization of α1b-adrenoceptor with green fluorescent protein. Mol. Endocrinol. 12:1099-1111.
3. **Barak, L.S., S.S. Ferguson, J. Zhang, C. Martenson, T. Meyer, and M.G. Caron.** 1997. Internal trafficking and surface mobility of a functionally intact β2-adrenergic receptor-green fluorescent protein conjugate. Mol. Pharmacol. 51:177-184.
4. **Chen, C. and H. Okayama.** 1987. High efficiency transformation of mammalian cells by plasmid DNA. Mol. Cell Biol. 7:2745-2752.
5. **Connor, J.X., A.J. Boileau, and C. Czajkowski.** 1998. A $GABA_A$ receptor $\alpha 1$ subunit tagged with green fluorescent protein requires a beta subunit for functional surface expression. J. Biol. Chem. 273:28906-28911.
6. **Cubitt, A.B., R. Heim, S.R. Adams, A.E. Boyd, L.A. Gross, and R.Y. Tsien.** 1995. Understanding, improving and using green fluorescent proteins. Trends Biochem. Sci. 20:448-455.
7. **Cubitt, A.B., L.A. Woollenweber, and R. Heim.** 1999. Understanding structure-function relationships in the Aequorea victoria green fluorescent protein. Methods Cell Biol. 58:19-30.
8. **David-Watine, B., S.L. Shorte, S. Fucile, D. de Saint Jan, H. Korn, and P. Bregestovski.** 1999. Functional integrity of green fluorescent protein conjugated glycine receptor channels. Neuropharmacology 38:785-792.
9. **Grabner, M., R.T. Dirksen, and K.G. Beam.** 1998. Tagging with green fluorescent protein reveals a dis-

tinct subcellular distribution of L-type and non-L-type Ca^{2+} channels expressed in dysgenic myotubes. Proc. Natl. Acad. Sci. USA *17*:1903-1908.
10. **Hirasawa, A., T. Sugawara, T. Awaji, K. Tsumaya, H. Ito, and G. Tsujimoto.** 1997. Subtype-specific differences in subcellular localization of α1-adrenoceptors: chlorethylclonidine preferentially alkylates the accessible cell surface α1-adrenoceptors irrespective of the subtype. Mol. Pharmacol. *52*:764-770.
11. **MacDonald, R.L. and R.W. Olsen.** 1994. $GABA_A$ receptor channels. Annu. Rev. Neurosci. *17*:569-602.
12. **Marshall, J., R., Molloy, G.W. Moss, J.R. Howe, and T.E. Hughes.** 1995. The jellyfish green fluorescent protein: a new tool for studying ion channel expression and function. Neuron *14*:211-215.
13. **Rao, A. and A.M. Craig.** 1997. Activity regulates the synaptic localization of the NMDA receptor in hippocampal Neurons. Neuron *19*:801-812.
14. **Rongo, C., C.W. Whitfield, A. Rodal, S.K. Kim, and J.M. Kaplan.** 1998. LIN-10 is a shared component of the polarized protein localization pathways in neurons and epithelia. Cell *94*:751-759.
15. **Shi, S.H., Y. Hayashi, R.S. Petralia, S.H. Zaman, R.J. Wenthold, K. Svoboda, and R. Malinow.** 1999. Rapid spine delivery and redistribution of AMPA receptors after synaptic NMDA receptor activation. Science *284*:1811-1816.
16. **Van den Pol, A.N. and P.K. Ghosh.** 1998. Selective neuronal expression of green fluorescent protein with cytomegalovirus promoter reveals entire neuronal arbor in transgenic mice. J. Neurosci. *18*:10640-10651.
17. **Wan, Q., Z.G. Xiong, H.Y. Man, C.A. Ackerley, J. Braunton, W.Y. Lu, L.E. Becker, J.F. MacDonald, and Y.T. Wang.** 1997. Recruitment of functional GABA(A) receptors to postsynaptic domains by insulin. Nature *388*:686-690.
18. **Zhu, W.J., J.F. Wang, K.E. Krueger, and S. Vicini.** 1996. δ Subunit inhibits neurosteroid modulation of $GABA_A$ receptors. J. Neurosci. *16*:6648-6656.

6 Neuronal Transfection Using Particle-Mediated Gene Transfer

Harold Gainer, Raymond L. Fields, and Shirley B. House
Laboratory of Neurochemistry, National Institutes of Health, NINDS, Bethesda, MD, USA

OVERVIEW

The ability to transfect differentiated neurons with DNA constructs of various sizes has been limited by the relative refractoriness of these cells, to classical calcium phosphate-DNA coprecipitation, and liposome–carrier-mediated transfection methods. Particle-mediated gene transfer (also called biolistics) represents a relatively simple solution to this problem, requiring little by the way of special technical expertise. Plasmid DNAs of unlimited sizes are used to coat approximately one micron tungsten or gold particles, and these particles are then accelerated into living cells and tissues by a blast of helium gas. A commercial device, referred to as the "Gene Gun" obtained from Bio-Rad, is available for this purpose. Biolistics has many uses, e.g., the production of antibodies after in vivo transfection of foreign DNAs into skin, ex vivo transfection of genes into tissues to be used for transplantation, and for the study of specific gene expression in distinct differentiated neurons in vitro. Cotransfection of multiple and distinct DNA constructs into single cells is easily performed, and this makes this technology of special value for physiological studies of differentiated neurons.

BACKGROUND

The technique of particle-mediated gene transfer (also known as biological ballistics, or in its shortened form, biolistics) was first developed in order to deliver nucleic acids into plant cells by the use of particles coated with foreign DNAs which are accelerated at high velocity (12,30,31,38). In the case of plant cells, the presence of cell walls made conventional techniques, such as electroporation, direct DNA uptake, liposome–carriers, calcium phosphate, and microinjection techniques ineffective and inefficient. The development of the biolistics technique (30,31) permitted the accelerated 1 to 4 µm tungsten microprojectiles to penetrate the cell walls and plasma membranes and to carry the surface-adsorbed foreign DNA into the plant cells' nuclear genome, ultimately resulting in the generation of stable plant transformants (37).

Over the years, a number of devices have been developed to accelerate the microprojectiles. In the first devices, gunpowder was used in a so-called particle gun to generate the pressure to accelerate 4 µm tungsten microprojectiles into the plant cells at 10 to 15 cm distances from the end of the device. Today, the gunpowder has been replaced by a blast of helium gas, and the tungsten particles by 0.6 to 1.6 µm gold particles, which are now routinely used for animal cells. Two devices sold by Bio-Rad Laboratories (Hercules, CA, USA), the PDS 1000/He Biolistic particle delivery system, an in-chamber model that requires a vacuum for transfection, and the newer Helios™ Gene Gun System, which is hand-held and requires no vacuum, are most commonly used. In the PDS 1000 system, the target cells (covered only by a thin layer of medium) are placed in a chamber that is flushed with helium gas and then evacuated to 25 inches of mercury before the gold particles are accelerated into the tissue by a He pressure of about 1000 to 1200 psi. The point of these maneuvers is to reduce the resistance of the ambient air and surface fluid over the tissue to the momentum of the gold particle, and thereby enhance penetration of the coated gold particle into the tissue. The above manipulations are obviously undesirable for the health of the tissue, and in fact, the PDS 1000 system is very inefficient and restricted in its use. In contrast, the Helios Gene Gun System is hand-held in open space (usually in a tissue culture hood for sterility) and in addition to being more flexible [e.g., it can be easily used in in vitro and in vivo experiments (46)], it is faster, less variable, and more efficient in the transfection of neurons (unpublished observations and Reference 45).

Biolistic techniques have been highly effective in transfecting a wide variety of organisms and cell types (6,15,22,27,46). Among these are cells in invertebrates and lower vertebrates (10,11,13,47), bacteria and yeast (26,39), and even subcellular organelles (e.g., for mitochondria genomic transformation, see References 7, 8, and 12). Biolistics has also been prominent in the development of DNA vaccines, genetic immunization strategies (3,27,29,35,43), and in experimental approaches to ex vivo gene therapy of pancreatic islets (19,20, 36). The use of biolistics with pancreatic islets is a particularly interesting case, since cells in this organ do not replicate, thereby precluding the use of retroviruses for ex vivo gene therapy. Application of adenoviruses successfully transduces 50% of islet cells, whereas biolistics only transfects 3% of the cells (19). While relatively inefficient when compared to adenovirus transduction, biolistics was 35 times more effective than liposomal delivery techniques and, most important, did not develop the immune response rejection evoked by the introduction of adenoviral proteins to the system. In fact, despite the relatively low transfection efficiency of the biolistics approach, transplantation of the transfected islets was able to reverse the diabetic state of alloxan-induced diabetes in Balb/c recipient mice (19).

Differentiated neurons are not often able to be transfected with DNA constructs by using conventional methods (see also Chapters 3 through 5 of this book). Viral vectors can be very effective in this regard (see Results and Discussion), however, as in the case of the islets, lower efficiencies may also be adequate in studies on neurons. One of the first uses of biolistics in the nervous system was done as a test of feasibility of using particle-mediated gene transfer for ex vivo gene transfer into brain tissues (25). In this study the authors transfected fetal brain with various promoter–luciferase constructs in vitro and found that the biolistic method produced 100-fold more gene expression than either calcium phosphate coprecipitation, electroporation, or lipofection methods (25). The biolistically transfected tissues were immediately transplanted into caudate or intracortical regions of adult host rats, and luciferase activity could be detected in the site of the transplant up to 2 months after gene transfer. One year after this study was published, two other reports, using the PDS 1000 system, described the use of the biolistic technology on central nervous system tissue in novel experimental paradigms. Arnold et. al (1) described the use of particle-bombardment transfection in combination with organotypic cultures of mice cerebellum to study purkinje cell-specific expression of the calbindin D_{28k} gene as well as glial and granule cell-specific expression of the brain lipid-binding protein gene. Using this strategy, they could identify 5′ flanking regions in these genes that were responsible for this cell type-specific expression in postnatal cerebellum. Lo et. al. (32) exploited the low efficiency of biolistic transfection using the β-galactosidase (Lac-Z) gene in slices of postnatal rat and ferret visual cortex and hippocampus to visualize the Golgi-like 5-bromo-4-chloro-3-indolyl-β-D-galactopyranoside (X-gal) staining of neuronal dendrites, axons, and glial processes. The authors suggested that the use of green fluorescent protein (GFP) reporters in biolistics in the future would be very valuable for developmental studies of living neurons (23).

Following these pioneering papers which illustrated the uses of biolistics to transfect neurons, a wide variety of neurobiological studies using particle-mediated gene transfer were reported. These have included identification of a calcium responsive element regulating expression of calcium binding proteins in Purkinje cells (2), studies of neurotrophin regulation in cortical dendrites and spines (23,33,34), development in embryonic retina (48), and various physiological studies (14,49). Given the versatility of the gene gun, its accessibility, and the wide variety of experimental purposes to which it can be put, it is likely that biolistic transfection will soon become a frequently used tool in many neurobiological studies.

Several technical papers, describing detailed protocols for successful biolistics have been published. Particularly valuable to the reader are the early papers from the Sanford Laboratory (31,37) discussing the various factors that must be considered to get optimal transfection by biolistics. Biewenga et al. (6) discuss the critical parameters for use of the PDS 1000 Instrument, and Wellmann et al. (45) present an optimized protocol for transfection of brain slices and dissociated neurons using the hand-held Helios gene gun system. The two papers provide valuable practical tips and clues for the successful employment of biolistics in neuronal systems. In the sections below, we present detailed protocols for the tissue culture and biolistic transfection of rat hypothalamic neurons in organotypic slice cultures which we have used with success (16,24,40). It should be noted however, that optimization of the biolistic parameters (particularly the helium air pressure and the distance of the gun from the target) must be determined empirically for each new tissue and cell type.

PROTOCOLS

Protocol for Stationary Organotypic Slice Explant Cultures

Materials and Reagents

- McIlwain Tissue Chopper (Brinkmann Instruments, Westbury, NY, USA).
- Tissue culture dishes (150, 100, 60, and 35-mm Falcon®; Becton Dickenson, Bedford, MA, USA).
- Gey's Balanced Salt Solution (Life Technologies, Gaithersburg, MD, USA and Sigma, St. Louis, MO,

USA).
- 50% glucose (Sigma).
- 70% ethanol.
- Breakable blades (Fine Science Tools, Vancouver, BC, Canada and featherblades; Ted Pella, Redding, CA, USA).
- Blade holder (Fine Science Tools and ROBOZ Surgical, Rockville, MD, USA).
- Double edge blades.
- Surgical supplies (No. 5 forceps, fine spatulas, large and small scissors; ROBOZ and Fine Science Tools).
- Millipore Millicell-CM filter inserts (Millipore, Bedford, MA, USA).

Culture Medium

To prepare 200 mL, mix together these reagents from Life Technologies, except BME which is from Sigma:
- 100 mL Eagle Basal Medium (BME).
- 50 mL horse serum.
- 50 mL Hanks Balanced Salt Solution.
- 2 mL 50% glucose (vol/vol).
- 1 mL L-glutamine (200 mM, 100×).

Cover with aluminum foil, medium is light sensitive.

Procedure to Dissect and Culture Rodent Hypothalamus

1. Postnatal 5 to 7-day-old rats or 5 to 10-day-old mice are washed with 70% ethanol and decapitated. Their brains are quickly and aseptically removed under a hood.

   Lateral cuts are made in the skull starting at the foramen magnum and ending at the olfactory lobes. The skull is gently lifted up from the rear exposing the brain. Cuts are then made between olfactory lobes and the frontal cortex and rostral to the cerebellum. The rostral part of the brain is gently lifted to expose the optic nerves, which are then cut prior to removing the brain from the skull. Finally the brain is removed and placed into cold Gey's solution in a 60-mm petri dish containing 0.5% glucose.

2. Carefully remove blood vessels and meninges around hypothalamus and gently straighten out residual optic nerves. These serve as helpful landmarks since the suprachiasmatic nucleus (SCN) is found at the base of optic chiasm.

3. Block out the hypothalamus using a razor blade in a blade holder. Cut away all cortex on both sides lateral to the hypothalamus (from a ventral view) and make a clean horizontal cut 2 to 3 mm above the third ventricle (coronal view).

4. Place the blocked hypothalamus on the tissue chopper so that the hypothalamus is ventral side down on the chopper disk and the optic nerves are abutting the chopper blade. Set the thickness of the slices at 350 to 400 µm for rats and 300 µm for mice.

   Note: Try to avoid excess fluid on chopper disk as this will cause the tissue to be picked up by the blade.

5. Cut the sections as a group, trying to keep them in order, and lift the group of sections onto a spatula and place them into a cold drop of Gey's solution plus glucose in a 100-mm petri dish. Carefully separate the slices using a small spatula and forceps.

   Note: It is critical for the survival of the tissue to be extremely gentle in the transferring and manipulating of the slices at this stage and afterwards.

6. Select the sections of interest and trim

them free of extraneous tissue using a razor blade scalpel, place them in fresh drops of the Gey's/glucose solution, and allow them to sit at 4°C in a refrigerator (1–2 h).

7. Put 1.1 mL of culture media into each 35-mm petri dish. The 35-mm petri dishes are placed inside of a 150-mm petri dish acting as a container to minimalize their handling.

8. Place tissue slices to be cultured onto Millicell-CM filters by using two small spatulas touching each other and by using the capillary forces between them to transfer the slices. Make sure the tissue rests flat on the filters.

Note: Too much fluid on the filter will prevent tissue from adhering to the filter. If this happens, remove the excess fluid or replate (even if it is the next day).

9. Place the Millicell-CM filters containing the slices onto the 1.1 mL of culture media in the 35-mm petri dishes, and incubate at 36°C in 5% CO_2 for 14 to 20 days.

Note: The slices thin to optimal thicknesses for immunohistochemistry (IHC) and in situ hybridization histochemistry after incubation for 10 to 20 days. The slices can survive for 1 to 2 months, but may thin too much over such long periods for subsequent experimental manipulations.

10. Media contains penicillin–streptomycin either for the first 3 days in vitro (DIV) or throughout.

11. Media is changed 3 times a week, with fresh media always in a new 35-mm petri dish in order to maximize the vitality and sterility of the culture.

12. In biolistics protocols, we usually shoot the cultured slices after 4 to 5 DIV, replace the medium, and assay by IHC after a total of 10 DIV.

Protocols for Immunohistochemistry of Organotypic Cultures

Materials and Reagents

- Netwells (Corning Costar, Cambridge, MA, USA).
- Netwell Carriers (Corning Costar).
- Phosphate-buffered saline (PBS) (1×).
- Normal goat serum (NGS) (Sigma).
- ABC Elite kit (Vector Laboratories, Burlingame, CA, USA).
- Rabbit IgG (biotinylated) (Vector Laboratories).
- Mouse IgG (biotinylated) (Vector Laboratories).
- 3,3′ Diaminobenzidine tetrahydrochloride (DAB; Sigma).
- Nickel sulfate (Sigma).
- Imidazole (Sigma).
- Glucose (Sigma).
- NH_4Cl.
- Glucose oxidase (GOD; Calbiochem-Novabiochem, San Diego, CA, USA).
- Permount (Fisher Scientific, Pittsburgh, PA, USA).
- Formaldehyde.
- Tris-buffered saline (TBS) 1×, pH 7.4.
- Triton X-100.
- Nos. 0 or 1 fine sable paint brush.
- Gelatin-coated slides or
- Superfrost-plus slides (Fisher Scientific).
- Xylene.
- 100 % EtOH.
- Americlear Histology clearing solvent (Baxter Healthcare, McGraw Park, IL, USA).
- Coverslips.
- Filter paper.

Cryoprotectant Medium

From Watson et. al. (44a).

1.59 g NaH$_2$PO$_4$·H$_2$O (mono)
5.47 g Na$_2$HPO$_4$ (dibasic)
9.0 g NaCl
300 g Sucrose
10 g Polyvinyl pyrrolidone (PVP-40) (Bio-Rad)
300 mL Ethylene glycol

Bring to 1 L with distilled water and store in refrigerator.

Procedure

1. To fix the sections on the filters, 4% formaldehyde/PBS is placed beneath and on top of the slice on the Millicell filter for 1.5 to 2 hours.
2. The slice cultures are thoroughly but gently rinsed with PBS. If rinsed too hard, it might dislodge and damage the tissue.
3. Use a scalpel blade or scissors to cut out the filters carrying the slices.
4. Place the filters holding the slices in Netwells into the 6-well plate carriers containing PBS (or into cryoprotectant if you are not going to stain tissue right away).

Note: It is very important that the slices on the filters are always fully immersed throughout this and all following procedures.

5. Rinse the slices extremely well in PBS before going into blocking solution.
6. Place the slices into blocking solution (10% NGS, 0.3% Triton X-100, 0.01% Na azide) for 1 to 2 hours.
7. Place into primary antibody (diluted in 1% egg albumin, 0.01% Na azide) and incubate overnight at 4°C.
8. Rinse well in PBS (3 × 10 min) using Netwells and carriers.
9. Apply secondary antibody—rabbit or mouse IgG biotinylated (at 1:500 dilution) in PBS at room temperature for 1.5 to 2 hours. Do not add sodium azide to this mixture, as it will interfere with the peroxidase reaction.
10. Rinse well in PBS (3 × 10 min).
11. Place slices (in Netwells) in ABC solution in the ABC kit (1:600 dilution) in the 6-well plates for 1 to 1.5 hours at room temperature.
12. Rinse tissues well with 2× PBS and rinse with 2× TBS (if you are using NiSO$_4$ in DAB reaction mixture).
13. Place sections (in Netwells) into the DAB reaction mixture.

To make DAB reaction mixture:
 Imidazole 136 mg
 NH$_4$Cl 40 mg
 Glucose 400 mg
 PBS 85 mL or
 TBS 85 mL (when using with NiSO$_4$)

To the above, add 10 mL of filtered DAB (100 mg DAB/20 mL in 1× TBS, pH 7.4).

Then add 500 µL of glucose oxidase (100 U/mL) to above and mix well.

Important: Allow mixture (with or without 70–80 mg NiSO$_4$/100 mL) to sit at least 10 to 15 minutes before use.

14. Gently rock sections (in Netwells) in DAB reaction mixture (time of reaction should be determined empirically).
15. Stop reaction by rinsing thoroughly in PBS (2 × 10 min) or TBS and finally in fresh PBS.
16. Samples can be stored in PBS no longer than 2 days.
17. If you are doing double label IHC, do the NiSO$_4$ step first (this results in a black reaction product). Then rinse well in PBS (3 x 10 min) and block again for 30 minutes in NGS.
18. Place sections into the second primary antibody and incubate overnight at 4°C.

6. Neuronal Transfection Using Particle-Mediated Gene Transfer

19. Repeat the above steps 8 through 15. The final staining reaction for the second antibody is in the same DAB reaction mixture, but without NiSO$_4$ (reaction product is brown in color).

Procedure for Removing Sections from Filters in order to Mount on Slides

1. Set Netwell in water before removing tissue from filter.
2. Place a drop of water on the coated slide.
3. Place the filter on top of a petri dish in a drop of water.
4. Gently ease the tissue off with the camel hair brush.
5. Let the tissue stick to the brush and place it in the drop of water on the slide.
6. Check the slide under the scope after you have removed all of the tissue and gently unfold and straighten out the tissue.
7. Remove excess fluid with filter paper.
8. Allow to air-dry undisturbed overnight.
9. Place slides in water for 15 to 20 minutes to remove salt.
10. Wash in 100% EtOH (2 × for 10–15 min).
11. For slides only, clear in xylene (3 × for 10–15 min).
12. Add permount and place coverslip on slides.

Note: In some cases, the tissue may be too fragile and might have to stay on the filter in order to be processed. In these cases, place sections still on filters and in Netwells into Americlear Clearing Solvent to clear tissue (xylene will cause filters to curl up). Remove section still on filter and place filter on slide, add permount, and place coverslip over filter containing the section. Visualization of the immunostained cells will be useful but less than optimal as compared to the mounted section that can be completely removed from the filter.

Protocol for Biolistics Using the Helios Gene Gun

The Helios Gene Gun is the apparatus used in our laboratory for biolistics. It uses pressurized helium gas to accelerate micron sized gold particles which are coated with plasmid DNA. Several factors are important to consider for effective biolistics. These are:

1. The amount of gold used per cartridge (individual bullet) ranges from 0.125 to 1.0 mg. The standard amount used in our laboratory is 0.5 mg/cartridge. The size of gold particles available from Bio-Rad are 0.6, 1.0, and 1.6 µm in diameter. We have found no difference in transfection efficiency between any of these sizes either in cell cultures or in organotypic slice explants. Our laboratory routinely uses 1.0 µm gold for all transfections.

2. The quantity of nucleic acid used for each transfection can vary over a range of 1.0 ng to 5.0 µg/cartridge. In any new experimental situation, we begin with 1.0 µg/cartridge. This 1.0 µg can be made up of single or multiple nucleic acid constructs. DNA transfection efficiencies in our laboratory are similar when using from 1.0µg and 0.1 µg/cartridge. Further dilution to 0.01 µg results in a noticeable decrease in efficiency.

3. The amount of DNA per milligram gold can range from 0.002 to 10.0 µg. A value of 2 is standard in our laboratory. Optimum transfection efficiencies as well as target specificities are greatly influenced by this value and must be determined empirically for each construct–target used. A table in the Helios Gene Gun Manual gives details for varying this parameter.

4. The pressure and distance to shoot at are key parameters, and these values will have to be determined experimentally for each target. For organotypic slice explants, we use between 120 to 180 psi and between 10 to 20 mm distance between the gun barrel and target. For cell lines and primary dissociated cell cultures, we use 120 psi and 17 to 29 mm.

Materials and Reagents

- Helios Gene Gun.
- Helium hose (Bio-Rad).
- Helium regulator (Bio-Rad).
- Helium tank (grade 4.5, 99.995%).
- Gold Microcarriers (Bio-Rad).
- PVP.
- Fresh 100% EtOH.
- 15-mL Disposable polypropylene centrifuge tubes.
- 1.5-mL Microcentrifuge tubes.
- 0.05 M Spermidine.
- 1 M $CaCl_2$.
- 200-µL and 1-mL pipetman and tips.
- 5-mL, 10-mL pipets and pipet-aid.
- Purified plasmid DNA in distilled water or TE (10 mM Tris, 1 mM EDTA, pH 8.0).
- Ultrasonic cleaner (e.g., Fisher F83, Branson 1210).
- Analytical balance.
- Microfuge.
- Tubing Prep Station (Bio-Rad).
- Nitrogen tank (grade 4.8, 99.998%).
- Gold-Coat tubing (Bio-Rad).
- Nitrogen regulator (Bio-Rad).
- Ultrasonic cleaner.
- Tubing cutter (Bio-Rad).
- Vortex shaker (Fisher Scientific).
- Peristaltic pump (Amersham Pharmacia Biotech, Piscataway, NJ, USA).
- 20-mL Scintillation vial.
- Dessicant.
- Parafilm.
- Timer.

Procedure

Note: An unopened bottle of 100% EtOH must be used each day this procedure is used in order to prevent the absorption of water in EtOH from inhibiting the process.

1. Connect the N_2 supply line to the Tubing Prep Station. Cut a 30-inch length of tubing. Connect a 10-mL syringe with adapter tubing to one end of the Gold-Coat tubing and flush the tubing with EtOH. Insert the tubing into the tubing support cylinder. Be sure that the tubing extends past the O-ring seal. Turn on the nitrogen and set the flow to 0.3 LPM. Purge the tubing for at least 15 minutes to dry the inside before loading the gold.
2. Place 25 mg of 1.0 µm gold into a 1.5-mL microfuge tube.
3. Add 100 µL of 0.05 M spermidine to the gold and vortex mix for 5 seconds.
4. Sonicate for 5 to 10 seconds.
5. Add 50 µg plasmid DNA in 100 µL or less total volume and vortex mix immediately.
6. While vortexing, add 100 µL 1 M $CaCl_2$ dropwise.
7. After all $CaCl_2$ is added, turn vortex mixer off and let tube stand for 10 minutes at room temperature.
8. Spin tube for 10 seconds in a microfuge to pellet the gold.
9. Aspirate off most of the liquid but leave approximately 20 µL.
10. Resuspend the pellet in the remaining supernatant by flicking the tube with your finger. Wash the pellet 3 times

with 1 mL of fresh 100% ethanol each time, spin approximately 10 seconds between each wash and discard the supernatants, leaving 20 µL of the EtOH supernatant.

11. Add 8.75 µL of 20 mg/mL PVP (in EtOH) to a 15-mL screw cap tube.
12. Add 3.491 mL of 100% EtOH to the 15-mL screw cap tube and mix.
13. Transfer the pellet from step 11 in 200 µL of the PVP/EtOH solution to the 15-mL tube.
14. Wash the microfuge tube with 200 µL PVP/EtOH to suspend any remaining gold and transfer to the 15-mL tube.
15. Vortex mix the gold suspension to break up any clumps and then sonicate briefly to break up any remaining clumps.
16. Invert the tube continuously to keep the gold from settling in the 15-mL tube.
17. Remove the cap of the 15-mL tube. Using a 10-mL syringe with adapter tubing on the end, draw the gold suspension all the way into the Gold-Coat tubing. Remove the tubing from the suspension and continue drawing the suspension into the tubing in order to leave an inch of airspace at the end.
18. Bring the tubing to a horizontal position and slide it into the tubing support cylinder of the Tubing Prep Station. Make sure it goes through O-ring.
19. Let stand for 3 minutes to settle the gold particles. Detach the adapter tubing from the syringe and attach it to the tubing on the peristaltic pump. Remove ethanol at the rate of 0.7 inches/seconds.
20. Detach the adapter tubing and rotate the Gold-Coat tubing in the support cylinder 180° and let sit for 5 seconds.
21. Turn Tubing Prep Station on to start rotating the tubing.
22. Rotate the tubing for 20 seconds; then purge with N_2 at 0.3 LPM.
23. Continue drying the tubing while rotating for 5 minutes.
24. Turn off the motor on the Tubing Prep Station. Close the valve on the flowmeter. Remove the tubing from the tubing support cylinder.
25. Inspect the tubing for blank spots and cut them out with a razor blade.
26. Place a desiccant pellet into a 20-mL scintillation vial and place the vial into the bottom of the tubing cutter.
27. Insert one end of the tubing into the tubing cutter and cut tubing into cartridges.
28. If not using the cartridges immediately, seal the lid of the scintillation vial with parafilm and store at 4°C in a closed chamber with desiccant.

Biolistic Transfection of Organotypic Slice Explant Cultures

1. Sterilize the barrel liner and cartridge holder with 70% EtOH.
2. Attach the barrel liner to the Gene Gun.
3. Load cartridges into the cartridge holder (leaving at least one chamber empty) and place in the Gene Gun.
4. Connect the helium hose line to the regulator and Gene Gun.
5. Set the pressure on the regulator. For organotypic cultures we use 120 to 180 psi (exact value is determined empirically).
6. To adjust the pressure in the regulator, keep one cartridge holder position empty. Using this empty position, test the shooting pressure. Adjust if necessary.
7. Cock the cartridge holder to load a cartridge.

8. Position the barrel liner over the target at 10 to 20 mm and shoot the target.
9. After the biolistic run is completed, turn helium supply off and bleed the line. Disconnect the hose line and remove the cartridge holder from the Gene Gun.
10. Place slice cultures into fresh medium.
11. Culture as described above for 4 to 5 days (optimal duration for immunohistochemical assays).

RESULTS AND DISCUSSION

The following example from our laboratory illustrates the use of biolistics in combination with organotypic neuronal tissue culture, in order to elucidate the mechanisms that are responsible for the cell-specific gene expression of oxytocin (OT) and vasopressin (VP) genes in the magnocellular neurons of the hypothalamo-neurohypophysial system (HNS). The absence of homologous cell lines that express these genes required that investigators conduct such experiments using transgenic mice (17,18,44). The number of DNA constructs that would be needed to be evaluated in order to determine the cis elements in these genes that control their cell-specific expression were too large to routinely use the relative expensive and protracted transgenic approach. Hence, an alternative in vitro strategy was sought.

Organotypic Hypothalamic Tissue Cultures

One of the essential criteria for a tissue culture model to study cell-specific gene expression is that the distinct neuronal phenotypes found in vivo can be identified in vitro. In the case of the HNS, this means that the neuronal phenotypes expressing the OT and VP peptides should be found in identifiable nuclei in the in vitro model. This is found in organotypic cultures of hypothalamus derived from neonatal mice and rats (24), and this is illustrated in Figure 1. Using specific antibodies which are markers of the OT and VP phenotypes (i.e., PS-38, against OT-associated neurophysin, OT-Np; and THR, against VP-associated neurophysin, VP-Np) in immunohistochemical assays, it is possible to identify the various nuclei. The paraventricular (PVN), accessory (ACC), and supraoptic nuclei (SON) in Figure 1, A through D, and the suprachiasmatic nucleus in Figure 1E are clearly identifiable in slices cultured for 15 DIV. Higher magnification views of these immunoreactive neurons in Figure 1, C, D, and F show that they resemble differentiated OT and VP neurons with their large cell bodies and robust nonspiny dendritic processes.

Efficacy of Biolistics in Organotypic Cultures

As noted earlier, conventional methods such as calcium–phosphate coprecipitation, electroporation, and lipid-mediated cell transfection are not very effective when used with organotypic cultures. This could be due, in part, to the presence of a reactive astrocyte layer which grows over the cultured slice, and thereby prevents access of the DNA-containing vehicles to the neurons below. One alternative is to use viral vectors, and indeed, we have successfully used adenoassociated viral vectors to transfect large numbers of OT and other neurons in such organotypic cultures (28). However, the production of multiple viral vectors accommodating the large numbers of DNA constructs necessary for promoter–enhancer analysis would be nearly as labor-intensive and time-consuming as the use of transgenic mice for this purpose. In biolistics, however, any DNA construct of any size and in any configuration can be

6. Neuronal Transfection Using Particle-Mediated Gene Transfer

used virtually immediately for transfection. Therefore, biolistics provides a very rapid assay, comparable to the use of cell lines for deletion construct analysis, but with the advantage of being able to be used in the more biologically relevant primary neurons.

In preliminary experiments (42), we used biolistics to transfect cells in station-

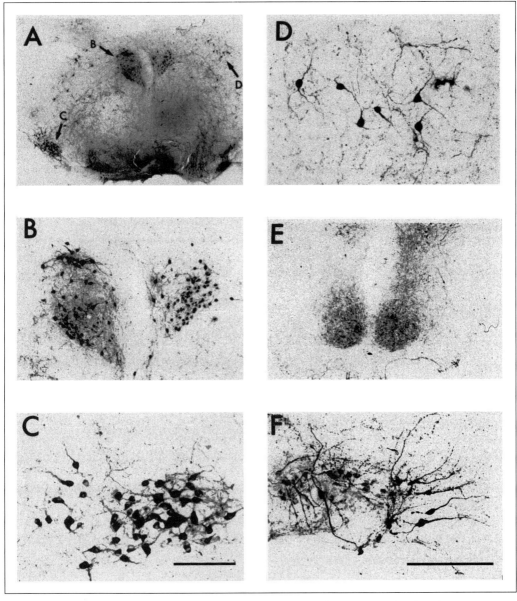

Figure 1. Organotypic cultures. Slice explants of mouse and rat hypothalamus after 15 DIV. Panels A through D are mouse neurons immunostained with PS 36 monoclonal antibody for OT-Np. Panels E and F are immunostained for VP-Np with THR polyclonal antibody (obtained from Dr. Alan Robinson, University of Pittsburgh). (A) A lower power micrograph showing an entire slice explant stained for OT-Np (PS 38) and containing three magnocellular nuclei (see arrows labeled B–D). These nuclei are shown at high power in B to D for PVN (B), SON (C), and ACC (D) nuclei, respectively. (E and F) High magnification views of VP cells in rat hypothalamic slice explants (15 DIV) immunostained using THR antibody. VP cells in rat suprachiasmatic nuclei are shown in panel E, and VP cells in SON are shown in panel F. Scale line in panel F represents 800 µm in panel E and 200 µm in panel F. Scale line in panel C represents 1 mm in panel A, 300 µm in panel B, and 150 µm in panels C and D. Adapted from Reference 24.

77

ary organotypic cultures with a variety of promoters [e.g., cytomegalovirus (CMV), rous sarcoma virus (RSV), glial fibrillary acidic protein (GFAP), α-tubulin, and various α-1$_B$ calcium (N) channel promoter constructs] and reporters (e.g., Lac-Z, luciferase, and GFP). We found that hippocampal and hypothalamic slice explants were easily and effectively transfected by this method. This is illustrated in Figure 2, where a CMV-GFP construct was "shot" using a PDS Bio-Rad System into hippocampal (Figure 2A) and hypothalamic (Figure 2B) slices. Note that the transfection incidence is low and random. Both glia (Figures 2, C and D) and neurons (Figure 2F) are transfected by the CMV-GFP, since the CMV promoter does not distinguish between these cell types. The layer of reactive astrocytes covering the surfaces of the cultured slices are preferentially hit by the gold particles and therefore are the predominant cell types visualized (Figure 2, C and D) when using the CMV promoter.

In contrast, when a nerve-specific enolase (NSE) promoter was fused to a GFP reporter to produce a NSE-GFP construct, which was shot using a Helios Gene Gun apparatus into hypothalamic slices, predominantly neuronal phenotypes were visualized expressing the GFP. Figure 3 shows the multipolar neurons with long processes in both organotypic (Figure 3, A and B) as well as in dissociated (Figure 3, C and D) hypothalamic cultures that were transfected with the NSE-GFP construct and which robustly expressed the GFP reporter.

Use of Biolistics to Assay Vasopressin–Gene Constructs in Hypothalamic Organotypic Cultures

Figure 4 illustrates the results of preliminary experiments using the biolistic technique to transfect hypothalamic neurons in organotypic cultures with VP gene constructs containing either choramphenicol transferase (CAT) (Figures 4, A through C) or GFP (Figures 4D though F) reporters. Previous studies in transgenic mice have suggested that key cis elements in the VP gene for cell-specific expression are located in the downstream 3′ flanking (noncoding) region of the gene (17,18). Consistent with this view is the observation that when VP gene constructs containing the entire gene (including the 3 exons and 2 introns) plus the 2 to 3 kb upstream and 2 to 3 kb downstream flanking regions and with a CAT reporter inserted in exon III are introduced into transgenic mice, the expression of the CAT reporter is highly cell-specific in magnocellular VP neurons (17,44). We used one of these constructs, termed VP III-CAT-2.1 (the CAT was inserted at the end of exon III followed by a 2.1 kb 3′ flanking sequence), as a positive control in biolistic experiments. The results of these experiments are shown in Figure 4, A through C, where CAT expressing cells, visualized by immunocytochemistry (ICC), occur only in areas of the slice where the endogenous VP cells are found (see Figure 1). The specificity of the expression was indicated by the following observations: *(i)* comparable experiments using hippocampal slices, which have no VP-expressing cells, showed no cells that expressed the CAT reporter after transfection with the VP-III-CAT-2.1 construct; *(ii)* the CMV-GFP experiments shown in Figure 2 indicates that mostly glial cells were being penetrated by the gold particles, but yet no glial expression of CAT was found using the VP III-CAT-2.1 construct; *(iii)* the morphology of the neuronal cell types expressing the CAT in the hypothalamic (in Figure 4, A through C slices) resembles the endogenously VP-expressing cells in these organotypic cultures (see Figure 1) and not the predominately multipolar neurons that were visualized when NSE was used as a promoter (Figure 3); and *(iv)*, double label ICC using CAT antibodies and VP marker antibodies (e.g., against VP-N$_p$) showed colocalization between endogenous VP and CAT immunoreactivity (unpublished observations). Given this success using the positive control construct, VP III-CAT-2.1, we then tested a new deletion construct.

The VP gene deletion construct used

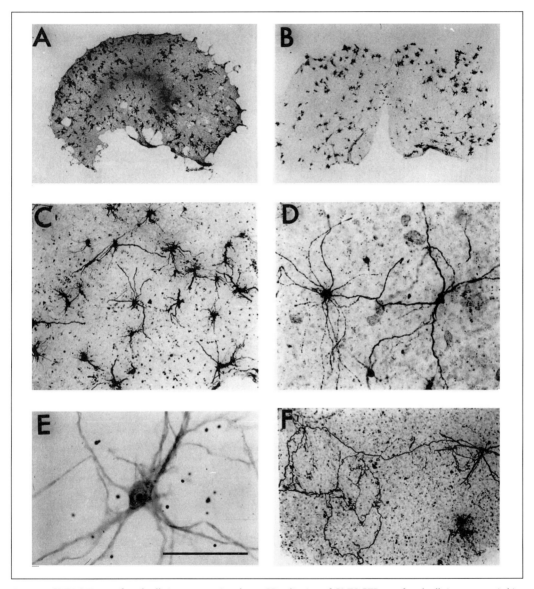

Figure 2. CMV-GFP transfected cells in organotypic cultures. Visualization of CMV-GFP transfected cells in organotypic hippocampal (A,C, and E) and hypothalamic (B,D, and F) cultures. Panel A shows visualization by ICC of CMV-GFP biolistically transfected cells in a hippocampal slice explant culture (scale line in panel E = 2.5 mm). Panel B shows visualization of CMV-GFP transfected cells in a hypothalamic slice explant culture (scale line = 2.0 mm). Panel C shows a higher power photomicrograph of a hippocampal slice explant culture. Notice the variety of cell types transfected (scale line = 400 μm). Panel D shows a higher power photomicrograph of a hypothalamic slice explant culture. Notice the variety of cell types transfected (scale line = 200 μm). Panel E shows a high power view of CMV-GFP transfected cells in a hippocampal slice explant culture (scale line = 40 μm). Notice that the 2 transfected cells have a gold particle in the nucleus. The other gold particles seen in the photomicrograph have not successfully transfected any cells. Panel F shows a CMV-GFP transfected neuron with a long axonal process in a hypothalamic slice explant culture (scale line = 400 μm). Adapted from Reference 42.

contained the same upstream and downstream sequences and exon I in the VP III CAT-2.1 gene, but had exons II and III and introns I and II removed, and the CAT reporter was replaced by GFP, which was fused to exon I. This gene construct was named VP I-GFP-2.1, and the results of its biolistic transfection into hypothalamic slices are shown in Figure 4, D through F. A similar expression pattern in hypothalamic neurons was observed using this deletion construct, with GFP expression occurring exclusively in the relevant nuclear areas (e.g., PVN, Figure 4D and SCN, Figure 4F) that express the endogenous gene in the cultured slice (see Figure 1). Immunocytochemical studies showed colocalization of the VP-Np marker and the expressed GFP (unpublished observations), further indicating correct cell-specific expression by the deletion construct. Thus, it appears that the combined biolistic-organotypic culture approach can be a very practical and effective strategy to study cell-specific gene expression mechanisms in a variety of cell types in the central nervous system.

CONCLUSIONS

Particle-mediated gene transfer has a significant advantage over the conventional

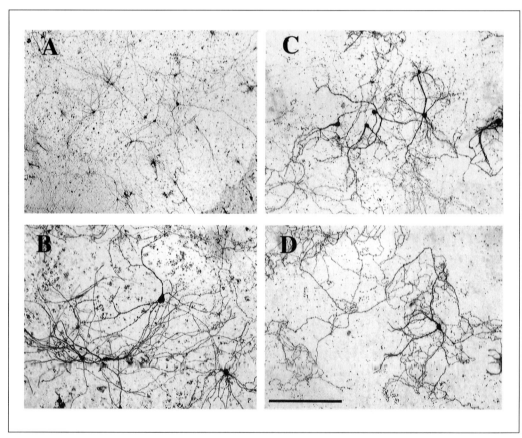

Figure 3. NSE-GFP transfection of neuronal cultures. GFP expression in neurons in organotypic (A and B) and dissociated (C and D) cultures prepared from PN 2 rodent hypothalami 4 days after biolistic transfection with pNSE-GFP constructs (NSE promoter obtained from Dr. Freda Miller, McGill University). Scale line in panel D represents 400 μm in panel A, and 200 μm in panels B through D.

procedures that are used for transfection, in that it is applicable to all cell types and can be used in most experimental circumstances. Only viral vectors have a comparable applicability (but not all viral vectors are equally effective in all animal species), however, the viral vector technology is much more labor-intensive, more time-consuming, requires special biosafety facilities, may exhibit biotoxicity, and often elicits a deleterious immune response from the host when used in vivo. A second major

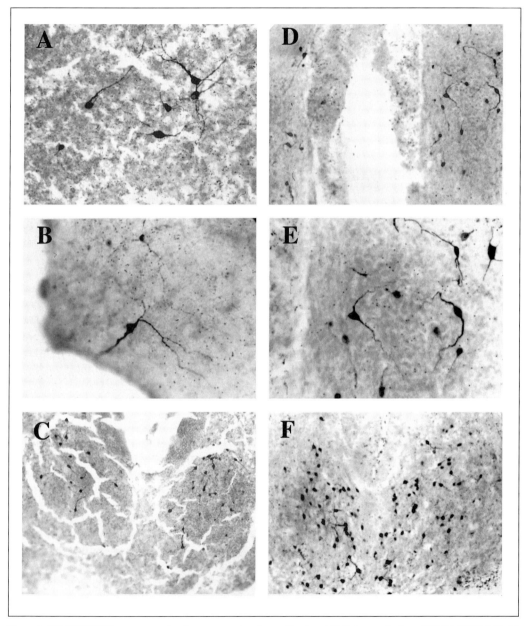

Figure 4. VP-CAT/EGFP gene expression in organotypic cultures. Results of biolistic transfection of VP III-CAT-2.1 (A–C) and VP-I-EGFP-2.1 (D–F) constructs into organotypic hypothalamic cultures. Analysis of CAT or GFP expression was made by IHC in PVN or SON (panels A, B, D, and E) and SCN (panels C and F) regions of the cultures.

advantage of biolistics is that only supercoiled plasmid DNA is needed to coat the gold particles, and hence, nonmolecular biologists can easily and rapidly incorporate this technology into their experimental repertoires. The only disadvantage of the biolistic methodology, is its relatively low efficiency (the usual being about 1%–3% of cells transfected), however, a recent report claimed transfection efficiencies of 10% with dissociated cerebellar granule cells and hippocampal neurons cultured on coverslips (45). It is clear that in two circumstances, viral vectors would be the preferred method for transfection, i.e., where large (>50%) transfection efficiencies are needed, and in vivo in deep regions (e.g., subcortical nuclei) of the central nervous system where the gold particles can neither penetrate nor be selectively targeted.

There is, however, a unique feature of biolistics that provides an important opportunity for experimental biologists. This derives from the fact that multiple plasmid DNA copies are adsorbed to the gold particle's surface. It has been estimated that approximately 200 plasmid DNA copies can coat a single one micron gold particle (4,5,27,41). Hence, penetration of a cell nucleus would deliver multiple copies to a single cell and would be likely to produce hyperexpression of the transfected gene relative to the endogenous gene. This could possibly explain why the same construct (e.g., VP-III-CAT-2.1) when used in transgenic mice did not produce detectable CAT expression in the SCN in vivo (17,44), but could produce robust expression in neurons of the SCN (in Figure 4C) when transfected by biolistics. Presumably, the integrated DNA constructs in the transgenic mice (even when present as multiple concatenated copies) had lower transcriptional efficiencies than the same DNA constructs when used in biolistics. Even more important, is the fact that a variety of plasmid DNAs can be placed on the same gold particle, thereby providing an easy technique for colocalization of more than one type of DNA construct in any transfected cell. Arnold et. al. (1) transfected coated gold particles with a mixture of two distinct constructs with different reporters and found, following biolistic transfection, that 97% of the 130 cells studied expressed both reporters. In contrast, when each construct coated a different population of gold particles, and the gold particles were subsequently mixed and shot into cells, there was no colocalization of reporters. This property of coating gold particles with two distinct genes allowed us to do quantitative biolistic assays of gene expression by using one of the colocalized constructs as an internal control (42). Although similar experiments have not been done in animal cells, the cotransfection of plant cells with as many as 12 (21) or 13 (9) different plasmid DNAs has been reported. Eighty five percent of the transgenic plants contained more than 2 and 17% more than 9 of the transgenes. If such cotransfection is also possible in animal cells (e.g., neurons), then it should be feasible to perturb a neuron by transfecting several specific wild-type or mutant genes, and to simultaneously mark the transfected neuron with a GFP reporter in order to identify it in living cultures for imaging and/or physiological analysis.

ACKNOWLEDGMENTS

The authors wish to thank Drs. S.-W. Jeong and Abraham Thomas for their contributions in the early stages of this work, Ms. Sharon Key for her help in making the figures, and Ms. Sophia D. Jackson for typing this manuscript.

REFERENCES

1. Arnold, D., L. Feng, J. Kim, and N. Heintz. 1994. A

6. Neuronal Transfection Using Particle-Mediated Gene Transfer

1. strategy for the analysis of gene expression during neural development. Proc. Natl. Acad. Sci. USA *91*:9970-9974.
2. **Arnold, D.B. and N. Heintz.** 1997. A calcium responsive element that regulates expression of two calcium binding proteins in Purkinje cells. Proc. Natl. Acad. Sci. USA *94*:8842-8847.
3. **Barry, M.A. and S.A. Johnston.** 1997. Biological features of genetic immunization. Vaccine *15*:788-791.
4. **Baker, S.M., P.G. Okkema, and J.A. Jaehning.** 1984. Expression of the *Saccharomyces cerevisiae* GAL7 gene on autonomously replicating plasmids. Mol. Cell Biol. *4*:2062-2071.
5. **Baker, S.M., S.A. Johnston, J.E. Hopper, and J.A. Jaehning.** 1987. Transcription of multiple copies of the yeast GAL7 gene is limited by specific factors in addition to GAL4. Mol. Gen. Genet. *208*:127-134.
6. **Biewenga, J.E., O.H. Destree, and L.H. Schrama.** 1997. Plasmid-mediated gene transfer in neurons using the biolistics technique. J. Neurosci. Methods *71*:67-75.
7. **Butow, R.A. and T.D. Fox.** 1990. Organelle transformation: shoot first, ask questions later. TIBS *15*:465-468.
8. **Butow, R.A., R.M. Henke, J.V. Morgan, S.M. Belchar, and P.S. Perlman.** 1996. Transformation of *Saccharomyces cerevisiae* mitochondria using the biolistic gun. Methods Enzymol. *264*:265-279.
9. **Chen, L., P. Marmey, N.J. Taylor, J.-P. Brizard, C. Espinoza, P. D'Cruz, H. Huet, S. Zhang, A. de Kochko, R.N. Beachy, and C.M. Faquet.** 1998. Expression and inheritance of multiple transgenes in rice plants. Nat. Biotechnol. *16*:1060-1064.
10. **Cheng, F.-M. and K.E. Joho.** 1994. Effect of biolistic particle size on the efficiency of transfection of oocytes in Xenopus ovary tissue. Nucleic Acids Res. *22*:3265-3269.
11. **Cheng, L., P.R. Ziegelhoffer, and N.-S. Yang.** 1993. *In vivo* promoter activity and transgene expression in mammalian somatic tissues evaluated by using particle bombardment. Proc. Natl. Acad. Sci. USA *90*:4455-4459.
12. **Daniell, H.** 1997. Transformation and foreign gene expression in plants mediated by microprojectile bombardment. Methods Mol. Biol. *62*:463-489.
13. **Davis, R.E., A. Parra, P.T. LoVerde, E. Ribeiro, G. Glorioso, and S. Hodgson.** 1999. Transient expression of DNA and RNA in parasitic helminths by using particle bombardment. Proc. Natl. Acad. Sci. USA *96*:8687-8692.
14. **Fernandez-Fernandez, J.M., N. Wanaverbecq, P. Halley, M.P. Caulfield, and D.A. Brown.** 1999. Selective activation of heterologously expressed G-protein-gated K+ channels by M_2 muscarinic receptors in rat sympathetic neurons. J. Physiol. *515*:631-637.
15. **Furth, P.A.** 1997. Gene transfer by biolistic process. Mol. Biotechnol. *7*:139-143.
16. **Gahwiler, B.H., M. Capogna, D. Debanne, R.A. McKinney, and S.M. Thompson.** 1997. Organotypic slice cultures: a technique has come of age. Trends Neurosci. *20*:471-477.
17. **Gainer, H.** 1998. Cell-specific gene expression in magnocellular oxytocin and vasopressin neurons. Adv. Exp. Med. Biol. *449*:15-27.
18. **Gainer, H. and S. Wray.** 1994. The cellular and molecular biology of oxytocin and vasopressin, p. 1099-1129. In E. Knobil and J.D. Ncill (Eds.), Physiology of Reproduction, 2nd ed. Raven Press, New York.
19. **Gainer, A.L., G.S. Korbutt, R. Rajotte, G.I. Warnock, and J.F. Elliott.** 1996. Successful biolistic transformation of mouse pancreatic islets while preserving cellular function. Transplantation *61*:1567-1571.
20. **Guo, Z., A.S.F. Chong, S. Jandeska, W.H. Sun, Y. Tian, W. Podlasek, J. Shen, D. Mital, S. Jensik, and J.W. Williams.** 1997. Gene gun-mediated gene transfer and expression in rat islets. Transplant. Proc. *29*:2209-2210.
21. **Hadi, M.Z., M.D. McMullen, and J.J. Finer.** 1996. Transformation of 12 different plasmids into soybean via particle bombardment. Plant Cell Reports *15*:500-505.
22. **Hapala, I.** 1997. Breaking the barrier: methods for reversible permeabilization of cellular membranes. Crit. Rev. Biotechnol. *17*:105-122.
23. **Horch, H.W., A. Krüttgen, S.D. Portbury, and L.C. Katz.** 1999. Destabilization of cortical dendrites and spines by BDNF. Neuron *23*:353-364.
24. **House, S.B., A. Thomas, K. Kusano, and H. Gainer.** 1998. Stationary organotypic cultures of oxytocin and vasopressin magnocellular neurons from rat and mouse hypothalamus. J. Neuroendocrinol. *10*:849-861.
25. **Jiao, S., L. Cheng, J.A. Wolff, and N.-S. Yang.** 1993. Particle bombardment-mediated gene transfer and expression in rat brain tissues. Biotechnology *11*:497-502.
26. **Johnston, S.A. and M.J. DeVit.** 1996. Biolistic transformation of yeasts. Methods Mol. Biol. *53*:147-153.
27. **Johnston, S.A. and D.-C. Tang.** 1994. Gene gun transfection of animal cells and genetic immunization. Methods Cell Biol. *43*:353-365.
28. **Keir, S.D., S.B House, J. Li, X. Xiao, and H. Gainer.** 1999. Gene transfer into hypothalamic organotypic cultures using an adeno-associated virus vector. Exp. Neurol. *160*:313-316.
29. **Kilpatrick, K.E., T. Cutler, E. Whitehorn, R.J. Drape, M.D. Macklin, S.M. Witherspoon, S. Singer, and J.T. Hutchins.** 1998. Gene gun delivered DNA-based immunizations mediate rapid production of murine monoclonal antibodies to the Flt-3 receptor. Hybridoma *17*:569-576.
30. **Klein, T.M., E.D. Wolf, R. Wu, and J.C. Sanford.** 1987. High-velocity microprojectiles for delivering nucleic acids into living cells. Nature *327*:70-73.
31. **Klein, T.M., T. Gradziel, M.E. Fromm, and J.C. Sanford.** 1988. Factors influencing gene delivery into *zea mays* cells by high velocity microprojecters. Biotechnology *6*:559-563.
32. **Lo, D.C., A.K. McAllister, and L.C. Katz.** 1994. Neuronal transfection in brain slices using particle-mediated gene transfer. Neuron *13*:1263-1268.
33. **McAllister, A.K., L.C. Katz, and D.C. Lo.** 1997. Opposing roles for endogenous BDNF and NT-3 in regulating cortical dendritic growth. Neuron *18*:767-768.
34. **McAllister, A.K., D.C. Lo, and L.C. Katz.** 1995. Neurotrophins regulate dendritic growth in developing visual cortex. Neuron *15*:791-803.

35. Robinson, H.L. and C.A. Torres. 1997. DNA vaccines. Semin. Immunol. *9*:271-283.
36. Rodriquez-Rilo, H.L., W.R. Paljug, J.R.T. Lakey, M.J. Taylor, and D. Grayson. 1998. Biolistic bioengineering of pancreatic beta-cells with fluorescent green protein. Transplant Proc. *30*:465-468.
37. Sanford, J.C. 1990. Biolistic plant transformation. Physiol. Plant *79*:206-209.
38. Sanford, J.C., F.D. Smith, and J.A. Russell. 1993. Optimizing the biolistic process for different biological applications. Methods Enzymol. *217*:483-509.
39. Smith, F.D., P.R. Harpending, and J.C. Sanford. 1992. Biolistic transformation of prokaryote-factors that effect biolistic transformation of very small cells. J. Gen. Microbiol. *36*:239-248.
40. Stoppini, L., P.A. Buchs, and D. Muller. 1991. A simple method for organotypic cultures of nervous tissue. J. Neurosci. Methods *37*:173-182.
41. Tanelian, D.L., M.A. Barry, S.A. Johnston, T. Le, and G. Smith. 1997. Controlled gene gun delivery and expression of DNA within the cornea. BioTechniques *23*:484-488.
42. Thomas, A., D.S. Kim, R.L. Fields, H. Chin, and H. Gainer. 1998. Quantitative analysis of gene expression in organotypic slice-explant cultures by particle-mediated gene transfer. J. Neurosci. Methods *84*:181-191.
43. Turner, J.G., J. Tan, B.E. Crucian, D.M. Sullivan, O.F. Ballester, W.S. Dalton, N.-S. Yang, J.K. Burkholder, and H. Yu. 1998. Broadened clinical utility of gene gun-mediated, granulocyte-macrophage colony-stimulating factor cDNA-based tumor cell vaccines as demonstrated with a mouse myeloma model. Hum. Gene Ther. *9*:1121-1130.
44. Waller, S.J., A. Ratty, J.P. Burbach, and D. Murphy. 1998. Transgenic and transcriptional studies on neurosecretory cell gene expression. Cell Mol. Neurobiol. *18*:149-171.
44a. Watson, R.E., Jr., S.J. Wiegand, R.W. Clough, and G.E. Hoffman. 1986. Use of cryoprotectant to maintain long-term peptide immunoreactivity and tissue morphology. Peptides *7*:155-159.
45. Wellmann, H., B. Kaltschmidt, and C. Kaltschmidt. 1999. Optimized protocol for biolistic transfection of brain slices and dissociated cultured neurons with a hand-held gene gun. J. Neurosci. Methods *92*:55-64.
46. Williams, R.S., S.A. Johnston, M. Riedy, M.J. DeVit, S.G. McElligott, and J.C. Sanford. 1991. Introduction of foreign genes into tissues of living mice by DNA-coated microprojectiles. Proc. Natl. Acad. Sci. USA *88*:2727-2730.
47. Wilm, T., P. Demel, H.-U. Koop, H. Schnabel, and R. Schnabel. 1999. Ballistic transformation of caenorhabditis elegans. Gene *229*:31-35.
48. Wong, W.T., J.R. Sanes, and R.O.L. Wong. 1998. Developmentally regulated spontaneous activity in the embryonic chick retina. J. Neurosci. *18*:8839-8852.
49. Xu, T., S. Finkbeiner, D.B. Arnold, A.J. Shaywitz, and M.E. Greenberg. 1998. Ca^{2+} influx regulates BDNF transcription by a CREB family transcription factor-dependent mechanism. Neuron *20*:709-726.

7 Analysis of Gene Expression in Genetically Labeled Single Cells

Stefano Gustincich, Andreas Feigenspan, and Elio Raviola
*Department of Neurobiology, Harvard Medical School,
Boston, MA, USA*

OVERVIEW

A combination of transgenic technology and single-cell reverse transcription polymerase chain reaction (RT-PCR) has been used to study gene expression in dopaminergic amacrine (DA) cells of the mouse retina. Because there are only 900 DA cells, and they cannot be distinguished from neighboring neurons on the basis of their morphology, we labeled them with human placental alkaline phosphatase (PLAP) by introducing into the mouse genome PLAP cDNA under the control of the promoter of the gene for tyrosine hydroxylase (TH), the rate-limiting enzyme for dopamine biosynthesis. Because PLAP is an enzyme that resides on the outer surface of the cell membrane, we can identify DA cells after dissociation of the retina by immunocytochemistry in the living state. Cells are then patch clamped and harvested for single-cell RT-PCR analysis of gene expression. Here, we describe the preparation of the fluorescent antibody E6-Cy3 to specifically detect PLAP-expressing cells, methods to obtain short-term cultures of solitary neurons from mouse retinas, and techniques to detect gene expression in individual neurons. Properties and pitfalls of single-cell RT-PCR are described and discussed.

BACKGROUND

Analysis of Neural Networks in the Adult Brain

To date, an impressive body of knowledge exists on the physiological events that underlie visual perception. This is due, in part, to the fact that the visual input can be controlled with great precision and, in part, to our adequate understanding of the first stages of visual processing. In this context, the retina has been a particularly attractive object of study because of its physical location, the distinctive morphology of its neurons, and the regularity of its architecture. Interesting and intellectually stimulating hypotheses on the physiological organization of the retina have been deduced from output measurements, which are the responses of ganglion cells, after manipulation of the neural networks and subsequent modeling. This top-down paradigm is today the favored approach to the understanding of neural networks in general. In the case of the retina, it relies on the assumption that this neural center contains a limited number of neuronal cell types: photoreceptor, bipolar, horizontal, amacrine, and ganglion cells.

The reality, however, is far more complex. Nine different types of cone bipolars were identified in the rat retina (11). A combination of photochemical and Golgi staining methods described at least 28 different types of amacrine cells in the rabbit (24). Finally, DeVries and Baylor were able to assign rabbit ganglion cells to 11 distinct physiological classes (8). This extraordinary complexity is not restricted to the retina. Electrophysiological characterization of inhibitory interneurons identified at least 16 different functional types in the hippocampus (31). On the other hand, the Golgi method revealed 50 anatomical types of local circuit neurons in the striate cortex of the monkey, which led to the suggestion that there could be as many as 100 different cell types in each layer of the neocortex (34).

At the heart of this fundamental issue is the definition of cell type, intended here as a class of neurons that receives a unique synaptic or sensory input, carries out a specific set of computations, and transmits the products of its activity to a unique set of postsynaptic neurons or peripheral effector cells. It is possible, therefore, that each neuronal type expresses a unique constellation of neuroactive molecules, ion channels, and receptors. Often, but not necessarily everywhere, each neuronal type has a unique morphology dictated by the need for ordered sampling of the visual world, efficient utilization of space, and uniformity in the neuronal geometry. In this case, shape of a neuron generally reflects the specificity of its synaptic connections.

As a result, the various neuronal types express unique combinations of genes, which can be explored with the techniques of molecular biology. In a bottom-up approach, neurons are identified on the basis of cell type-specific markers, and their complement of molecular constituents is subsequently analyzed. This knowledge is then used to design appropriate physiological experiments to clarify their function in the computations carried out by the networks of which they are components (25).

Gene Expression in the Nervous System

Our underestimation of the complexity of neural networks derives mainly from the lack of markers specific for the various cell types and of methods to assign gene expression either at the mRNA or protein level to identified neuronal populations.

When a new gene is identified and its cDNA cloned, a first step in its characterization is the analysis of its pattern of expression in various tissues. In the literature, vast numbers of northern blots were performed on brain RNA. However, the detection of the mRNA of interest in the whole brain is not very informative; this gene can be moderately expressed in most if not all the neurons, or it can be highly expressed in a single large cell population while absent in all others. Even less informative is the failure to detect any given species of mRNA. This occurs for many genes expressed in the central nervous system. A gene can in fact be transcribed at a very high level in a rare cell type only. Northern analysis of RNAs purified from specific regions of the brain reduces, but does not eliminate, this averaging effect. With in situ hybridization experiments on brain sections, it is often impossible to recognize the types of neurons that contain a species of mRNA; most of the neurons are too small, too closely packed, and their shape is not visible in its entirety. Furthermore, this technique is not very sensitive, and the analysis is limited to a very small number of genes per experiment.

Double immunostaining with a cell-specific marker is nowadays the standard technique to identify the neurons that express an antigen of interest. Markers are usually neurotransmitters and their synthetic enzymes, peptides, cytoskeletal, and calcium-binding proteins. The number of such markers, however, is limited, and they are rarely distributed throughout the cytosol, thus allowing recognition of the cell shape. Furthermore, they can be expressed in more than one cell type in the same area of the brain, thus adding new difficulties to the interpretation of the pattern of expression. In the retina, for instance, antibodies directed against glutamic acid decarboxylase (GAD), the GABA synthesizing enzyme, recognize most of the amacrine cells without discriminating among the 28 different types.

Single-Cell Analysis of Gene Expression

Besides immunocytochemistry, other routine methods to study the presence of receptors and channels in neurons are intracellular recordings followed by injection of dyes or enzymes and patch clamping of identified cells in slices, after dissociation or in primary cultures (25).

In recent years, however, the patch pipet has also been exploited to collect cytoplasm from morphologically identified neurons. This material is then the substrate of a series of biochemical reactions that lead to the description of the pattern of expression of selected genes at a single-cell level. Subsequently, biophysical and pharmacological properties of ligand and voltage-gated ion channels in single identified neurons can be correlated with the expression of specific mRNAs (28,30,35).

During a patch clamp experiment in the whole-cell configuration, the rupture of the membrane patch establishes a low resistance pathway between the interior of the cell and the lumen of the recording pipet, thus allowing the dialysis of the cell interior with the electrode solution. At the end of the recording, negative pressure is applied, and this causes the entry of the cell or its contents into the pipet tip. The tip is finally broken into a microfuge tube to harvest the cell. In the original version of the single-cell RT-PCR technique, the solution filling the microelectrode contains all the reagents necessary for first-strand cDNA synthesis (22,37).

Two methods have been used to detect the presence of specific mRNAs in the harvested cell. After random primed first-strand cDNA synthesis, PCR amplification with specific primers is carried out in the same tube. The amplified DNA is then analyzed by agarose gel electrophoresis. A second round of PCR amplification may be required to detect low abundance genes or to synthesize larger amounts of material for further analysis. The choice of the PCR primers determines which gene is analyzed. This method was successfully applied to study the expression of single genes or families of genes. A very powerful extension of this method is called multiplex PCR; after harvesting the cell content, a first round of PCR is performed with multiple primer pairs. Then, several secondary PCRs are carried out, each containing only one of the previously used pairs of primers (22).

This technique has demonstrated all its potential in a series of elegant papers on the molecular constituents of native glutamate receptors (1). In a first-round PCR, degenerated primers specific for regions highly conserved in all AMPA subunits subtypes were used to amplify mRNAs expressed in the neuron of interest. Several, second-round PCRs were then assembled using subunit-specific primers to detect the subunits expressed and their spliced forms. The amplification products were then directly sequenced to detail the state of editing of the messages or cloned to quantify the relative abundance of the various subunits. These works, on cerebellar Purkinje and granule cells in culture (22), in type I and II hippocampal cells in culture (4), and in a variety of identified cells in brain slices (14,19), have clearly demonstrated that differential regulation of gene expression leads to functional diversity in the receptors of native neurons. For instance, the presence of the GluR2 subunit is related to the calcium permeability of the receptor–ionophore complex (14,19).

A second method is based on the activity of the T7 RNA polymerase (6,37). First-strand cDNA is synthesized with a primer that contains the promoter sequence for this phage enzyme. After the harvest of the cell contents and the incubation with reverse transcriptase, a second-strand synthesis reaction is performed. The double-stranded cDNA is in turn the template for an in vitro transcription reaction in the presence of a radiolabeled ribonucleotide. The amplified RNA can then be used as a probe to hybridize cDNA clones immobilized on a nylon membrane. This method has two advantages: *(i)* the T7 amplification step is linear, thus maintaining the correct proportion between different mRNAs; and *(ii)* a theoretically unlimited number of cDNAs can be detected simultaneously. This technique was successfully applied on a number of different cell types. It requires, however, more manipulations than single-cell RT-PCR, and it does not provide information on common cell-specific processing, such as splicing and editing. Furthermore, reverse northern blots can be subject to a high frequency of false positives derived from cross-hybridization of noncomplementary genes that contain common sequences.

Appendix I shows a list of genes whose expressions were assessed with single-cell RT-PCR and the neuronal cell types that were harvested. In addition to the description of multisubunit ligand-gated channels, single-cell analysis was applied to the study of the pattern of splicing of voltage-gated channels to identify the subtypes in a family of metabotropic receptors and to detect cell-specific markers in order to confirm the identity of the investigated neuron.

Single-cell RT-PCR is significantly more sensitive than in situ hybridization and permits, therefore, the detection of transcripts present in low abundance. Its absolute sensitivity, however, is still uncertain (see Results and Discussion).

Identification of Specific Cell Types for Single-Cell RT-PCR

In many instances, single-cell RT-PCR has followed conventional patch clamp recordings. Neuronal cytoplasm was harvested from cells in slices, primary cultures, and after acute dissociation. Unfortunately, the same limitations that restrict the electrophysiological analysis to a relatively small number of identified cell types also hold true for the analysis of gene expression. Only well recognizable neurons can be studied, and the diversity among types is clearly underestimated. For instance, typical classifications were "spiking and non-spiking" or "agonist-sensitive and agonist-insensitive" neurons. Furthermore, electrophysiological recordings are often obtained from neurons whose identity in the intact tissue is not known.

Rare populations are probably never encountered by chance during recording in slices, or they cannot be identified unequivocally without the injection of a fluorescent dye such as Lucifer yellow at the end of the experiment. Primary cultures are prepared from many areas of the brain, but most of the cells lose their distinctive morphology after a few days in vitro. Furthermore, it is unclear whether neurons after several days in culture are still representative of their phenotypes in vivo.

Enzymatic digestion and mechanical trituration of regions of the nervous system is a powerful method to obtain short-term cultures of solitary neurons from both young and adult animals. The majority of the cells is damaged during dissociation, but some of them retain highly characteristic morphological traits that facilitate their identification, such as cerebellar Purkinje cells, hippocampal pyramidal cells, retinal rods, and rod bipolars (25).

Nonspecific fluorescent markers, like 4′6-diamidine-2-phenylindole (DAPI) and acridine orange, are sometimes applied to the tissue maintained in vitro to label the soma of specific neuronal populations. Other dyes can be injected in vivo and accumulated by retrograde transport.

Transgenic technology is a powerful tool for stable ectopic expression of a gene in a tissue or cell type of interest. Therefore, a neuronal type can be labeled by linking the promoter of a cell type-specific gene to a reporter gene whose product can be detected with common microscopic techniques. This approach is dependent upon the identification of the correct genomic sequences that drive gene expression in specific cell types and the use of an appropriate reporter.

There is a large amount of literature on the structure of the regulatory elements of genes expressed in the nervous system and the effects of the introduction into the mouse genome of cloned genomic fragments acting as promoters for a visible reporter (29). In some cases, such as with the promoter for the tyrosine hydroxylase gene, the expression of the reporter is spatially and temporally regulated as the wild-type gene (2,16). In these cases, entire populations of specific neuronal types are labeled in the genetically modified mouse line. More frequently, no expression, low level expression, and expression limited to a fraction of the relevant cell population are the outcomes. These results are interpreted as a consequence of the absence of important regulatory sequences in the construct or as position effects caused by endogenous sequences near the insertion site of the transgene. Interestingly, in some transgenic lines the reporter is expressed in cell populations that do not contain the endogenous product of the gene whose promoter was used in the construct. When these ectopias are stable throughout subsequent generations, cell types can be labeled for which no specific markers are currently available (15).

Because the fundamental goal of this approach is to carry out electrophysiological recordings in neurons that belong to specific populations or to study gene expression at the single-cell level, neurons must be identified in the living state. Therefore, an ideal reporter must be either spontaneously fluorescent, detected by a fluorescent probe, or capable of generating a fluorescent reaction product. Specific neuronal types become thus visible in the intact tissue, in slices, or after enzymatic digestion and mechanical trituration of the neural center in which they reside. Furthermore, because the data obtained from gene expression will be used to deduce the function of a neuronal type as a component of a network, an ideal marker should be visible with both light and electron microscopes and stain neurons in their entirety. Information about the morphology of the cells (shape and size), their location (position of cell bodies and processes), and distribution (coverage) can thus be obtained with the light microscope and integrated with data on their synaptic connections after a study with the electron microscope.

Unfortunately, at the state of the art, there is no single reporter that fulfills all the above requirements. A reporter gene frequently used is *LacZ*, that codes for the enzyme β-galactosidase (β-gal) (23). Cells expressing β-gal become fluorescent in the living state after incubation with fluorescein-di-β-D-galactopyranoside, but the reaction product, fluorescein, leaks out of the cell very rapidly. In fixed tissue, the enzyme generates a blue precipitate from a halogenated indolyl derivative of galactose; perikarya are visible as well as the initial portion of large processes, but neither dendritic arborizations nor the axon terminals are stained in their entirety. Furthermore, the product of the histochemical reaction is not sufficiently electron dense to permit identification of fine processes with the electron microscope.

A new fluororogenic substrate for β-lactamase has been recently synthesized (40). However, no transgenic mouse line has been produced to date that expresses this reporter. Furthermore, no substrate is currently available for stable light and electron microscope preparations.

A green fluorescent protein (GFP) from the jellyfish *Aequorea victoria* is another commonly used reporter (5,36). Its primary sequence was extensively modified for use in mammals (41). The mutation S65T gives enhanced brightness and an excitation–emission spectrum similar to that of fluorescein. Furthermore, translation efficiency increased dramatically when the sequence of the cDNA coding region was adapted to the codon usage of mammals, and a canonical Kozak sequence was included for initiation of translation. Visualization of GFP-expressing cells with the electron microscope must rely on postembedding immunocytochemistry, which is a less than ideal way to study the synaptic connections of the stained cells. There are reports that GFP-induced fluorescence can be efficiently photo-converted, but large amounts of protein are required. The main shortcoming of this reporter is that a relatively large cytoplasmic concentration is required for its detection (10^5–10^6 molecules/cell), a quantity that is often difficult to obtain when a cell type-specific mammalian promoter is used. Recently, however, GFP was successfully expressed in transgenic mice under the control of a mammalian promoter in gonadotropin-releasing hormone neurons (33) and in retinal dopaminergic cells (Gustincich and Raviola, unpublished results).

PLAP was originally introduced as a marker to follow the fate of progenitors cells in the mouse retina and in other regions of the central nervous system (13). PLAP is normally present in high concentration in the microvilli of the syncytiotrophoblast, and it is reexpressed ectopically in a variety of tumors in the adult. Because it is linked to the outer cell surface, it was exploited for immunodetection of cancer cells in patients. Therefore, numerous monoclonal antibodies exist which recognize various antigenic regions of the protein (7). The monoclonal antibody E6, that belongs to the IgG2b, *k* subtype, was produced against butanol-extracted acetone-precipitated human placentas and is highly specific for both native and cDNA-synthesized PLAP protein (3,9,20). As the enzyme is located on the outer surface of the cell membrane, PLAP-expressing neurons can be identified by immunocytochemistry in the living state. Thus, cells can be harvested for molecular biology and their voltage-gated and ligand-gated channels studied with the patch clamp technique. To prevent a secondary antibody to induce patching and subsequent endocytosis of the primary antibody–antigen complex, purified E6 was directly coupled to the monofunctional reactive dye Cy3™. The Cy3-conjugated antibody (E6-Cy3) did not significantly cross-link its antigen, because the fluorescent marker neither formed patches at the cell surface nor was internalized by endocytosis in cultured ψ-2-DAP cells, which synthesize human PLAP as a result of retroviral infection. In contrast, dramatic endocytosis was observed when the cultures were stained with polyclonal anti-PLAP antibodies, or when incubation with the monoclonal antibody was followed by treatment with polyclonal antimouse IgGs. Furthermore, GABA-gated currents were studied in PLAP-carrying rod bipolars and found to be indistinguishable from those recorded from morphologically identified wild-type mouse rod bipolars (16).

Because of its high pH optimum (pH 10.0–10.5), PLAP biological activity is probably minimal in the normal environment of the developing and adult nervous systems (27,32). We and others have shown that PLAP ectopic expression interferes neither with the development nor with the electrophysiological properties of various neuronal populations (13,16). PLAP is a sturdy enzyme; it is not inactivated by fixation with both formaldehyde and low concentrations of glutaraldehyde. Furthermore, in contrast with other endogenous phosphatases, its activity survives exposure to a temperature of 65°C for 30 minutes. Thus, after heat inactivation, only ectopically expressed PLAP is stained both at the light and electron microscopes by common histochemical techniques for phosphatase activity. Because of its connection to the bilayer by a glican–inositol phosphate tail, PLAP diffuses throughout the neuronal surface. As a result, the product of the histological reaction outlines the entire neuronal surface; dendritic arbors are stained in their entirety, and axons can be followed to their termination in distant regions of the brain or body. For the light microscope, a purple product is generated upon hydrolysis of 5-bromo-4-chloro-3-indolyl phosphate in presence of nitro blue tetrazolium chloride as an electron acceptor. The histochemical method for electron microscopy derives from the old Gomori technique, based on the hydrolysis of organic phosphates in the presence of calcium ions. With a lead citrate modification, PLAP enzymatic activity generates a dense reaction product on the outer aspect of the cell membrane without obscuring the underlying cytoplasm. Thus, synaptic connections can be studied (16,26).

Because of the unique possibility of using both immunolabeling and enzymatic reactions, PLAP-carrying neurons can therefore be identified in the living state for molecular and electrophysiological analysis and, after chemical fixation, for light and electron microscopic studies.

A Case Study: Dopaminergic Amacrine Cells of the Retina

Dopamine (DA) subserves fundamental functions in the nervous system; it facilitates and controls initiation of movement, regulates reward behaviors and affective states, and is involved in the response to stress. Dysfunction of the DA system has been implicated in the neurobiology of addictions and in a vast array of clinical conditions, such as Parkinson's disease, tardive dyskinesia, and schizophrenia. Thus, an understanding of the mechanisms that control the release of this modulator may have important consequences for interventions in motor and affective disorders.

In the retina, DA is synthesized by either amacrine or interplexiform cells. When the vertebrate retina is illuminated, DA extracellular levels increase and modulate many of the events that lead to neural adaptation to light (39). It has been shown that there is a significant decrease of DA cells in aged human and rat retinas compared to younger individuals. Furthermore, retinal DA cells degenerate in patients affected by Parkinson disease, suggesting a common mechanism of degeneration in mesencephalic and retinal DA neurons (10).

DA is the first neuroactive molecule in the synthetic pathway for catecholamines, and tyrosine hydroxylase (TH) is the rate-limiting enzyme for its synthesis. Immunocytochemical detection of TH is the most sensitive tool to identify DA cells. However, progress in the study of these neurons has been slow because these cells are very rare (approximately 450 neurons in the mouse retina) and for the lack of reliable methods to label them in the living state (16,38).

We have exploited the properties of PLAP to study the dopaminergic neurons of the mouse retina. To this purpose, we linked its cDNA to a promoter sequence of the gene for TH (2). This construct was introduced into the mouse genome, and a line of transgenic animals was obtained in which PLAP was expressed on the outer surface of the cell membrane of DA cells (16). PLAP-labeled DA cells were studied with the light and electron microscopes and, after enzymatic digestion and mechanical trituration of the retina, patch clamped for single-cell RT-PCR studies of gene expression or for electrophysiological experiments (12,16,17). Thus, their repertory of mRNAs could be correlated with their physiological properties and with their synaptic connections. In this way, a large amount of information was obtained on a very rare type of neuron.

In the following sections, we illustrate the techniques to study gene expression at a single-cell level in PLAP-containing neurons of transgenic mice. First, we describe the culture conditions of E6-Hy hybridoma cells, the purification of E6 antibody, and the coupling reaction to the fluorochrome Cy3. Then, we illustrate the preparation of short-term cultures of solitary neurons from adult mouse retinas and the harvesting of PLAP-expressing cells for single-cell RT-PCR. Because we know from immunocytochemistry that the promoter of the TH gene directs the expression of PLAP to all neurons of the intact retina which contain TH (16), we analyzed the presence of TH mRNA by single-cell RT-PCR in solitary dopaminergic neurons harvested after dissociation and labeling with E6-Cy3. In the protocol section, we report the conditions for detecting this transcript in single cells. In the next section, the experimental results are presented and discussed. They represent a reliable measure of the sensitivity and reproducibility of our method for single-cell RT-PCR. We also comment upon results obtained with other mRNAs, such as those for the $GABA_A$ receptor subunits (17). Furthermore, we discuss properties and limits of the single-cell RT-PCR technique when it is applied to solitary neurons and to experiments of simultaneous detection of several transcripts in the same cell.

The characterization of the transgenic mouse used in this study was presented elsewhere as well as the details on the protocols used for light and electron microscopy (16).

PROTOCOLS

Protocol for the Preparation of the Antibody E6-Cy3

We describe here all the steps that lead to the production of E6-Cy3 as they are routinely carried out in our laboratory, from growing E6-Hy hybridoma cells to the coupling reaction. Because most of them are common knowledge, they can be modified according to the needs and habits of each laboratory (18). We recommend the production of at least 1 L of supernatant to obtain approximately 3 to 5 mg of protein. The step in Ultra Low-IgG medium is required to avoid copurification of the antibodies present in the serum. They can in fact cause an inefficient labeling of the E6 antibody and undesirable nonspecific staining during in vivo experiments. The most delicate step is the coupling reaction between the amino groups of the lysines present in the antibody and the activated groups of the dye. Many factors can affect its efficiency: temperature, buffer, pH, length of time, and the ratio between the concentrations of antibody and dye. After the purification, a dialysis step is needed to change the buffer solution from Tris-HCl to phosphate to avoid the reaction of the dye with the amino groups contained in the Tris molecule. The alkaline pH is used to keep all the amino groups of the protein in the reactive form. We find convenient to lyophilize the antibody after the dialysis so that its concentration can be optimally adjusted for the coupling reaction. The uncoupled E6 can be stored for months at 4°C. It must not be frozen. When 1 mg of E6 is coupled to half tube of Cy3, we obtain approximately 1.5 mL of E6-Cy3 antibody that can be stored at 4°C for several months. This sample is tested for specificity and brightness on retinal sections at 1:1000/1:5000 dilutions, and it is used 1:100 in in vivo experiments.

Materials and Reagents

- HGM (hybridoma growth medium): 10% fetal bovine serum, MEM (minimum essential medium) sodium pyruvate 1 mM, L-glutamine 2 mM, penicillin 100 U/mL, streptomycin sulphate 100 µg/mL in RPMI 1640 (all from Life Technologies, Gaithersburg, MD, USA).
- AHM (antibody harvesting medium): 10% Ultra Low IgG fetal bovine serum, MEM sodium pyruvate 1 mM, L-glutamine 2 mM, penicillin 100 U/mL, streptomycin sulphate 100 µg/mL in RPMI 1640 (all from Life Technologies).
- Protein A Sepharose® 4 Fast Flow (Amersham Pharmacia Biotech, Piscataway, NJ, USA).
- 1 M Tris-HCl, pH 8.0 (autoclaved).
- 10 Kd cutoff Slide-A-Lyzer dialysis cassettes (Pierce Chemical, Rockford, IL, USA).
- Glycine 0.1 M, pH 2.5 (filtered).
- Carbonate buffer 1 M, pH 9.3 (filtered).
- Monofunctional Cy3 dye (Amersham Pharmacia Biotech).
- G25 Sepharose (Amersham Pharmacia Biotech).

Procedure

Culturing Hybridoma Cells and Harvesting the Antibody

1. Resuspend cells from the cryotube in 10 mL HGM. Spin for 5 minutes at 400× g. Discard supernatant. Resuspend pellet in 10 mL HGM. Spin again for 5 minutes at 400× g. Discard supernatant. Resuspend pellet in 5 mL HGM. Take 100 µL and measure cell concentration with the Trypan blue exclusion method: refringent cells alive, blue cells dead. Add HGM to an optimal concentration of 2×10^5 cells/mL. Incubate the flasks at 37°C in 5% CO_2, 95% O_2 atmosphere.

2. Count the number of cells regularly. The culture must be kept at a concentration between 10^5 and 10^6 cells/mL (3–4 days) and should never exceed 10^6 cells/mL. E6 hybridoma cells tend to stick to the bottom of the flask; shake it vigorously when cells are collected for counting or splitting.

3. Expand the culture to a volume of 100 mL with a concentration of 10^6 cells/mL. Cells are then collected and spun for 5 minutes at 400× g. Discard the supernatant. Resuspend 10^8 cells in 1 L of AHM (10^5 cells/mL). Count cell concentration every day. When the culture reaches 10^6 cells/mL, incubate the culture for 2 more days.

4. Collect the medium. Spin for 15 minutes at 1000× g. Pool the supernatant (1 L), add 0.1% sodium azide, and

store at 4°C.

Purification of the Antibody

1. Wash 1 mL of Protein A Sepharose 4 Fast Flow with 10 mL 0.1 M Tris-HCl, pH 8.0 (1 mL drained resin binds 35 mg human IgG). Repeat twice.

2. Insert a piece of packing foam at the bottom of a 20-mL sterile syringe (check the flow with water) and mount the syringe as a chromatography column. Wash with water. Resuspend the resin in 10 mL 0.1 M Tris-HCl, pH 8.0 and gently add it to the syringe. Wash the column with 40 mL of 0.1 M Tris-HCl, pH 8.0.

3. Add 110 mL 1 M Tris-HCl, pH 8.0 to the 1 L of supernatant (note the change in color). Before the resin dries, add 10 mL of the supernatant/Tris solution to the column.

4. Close the mouth of the syringe with a rubber stopper. This must have a hole through which a plastic tube is inserted. The other end of the tube is submerged in a bottle containing the supernatant/Tris solution. Raise the bottle to fill the syringe and keep adjusting the height so that the fluid in the syringe does not overflow.

5. Pass all the supernatant/Tris solution through the column twice (24–36 h). **Note:** Never let the column dry up.

6. When all the supernatant/Tris solution has passed through the column, add 2 x 20 mL of 0.1 M Tris-HCl, pH 8.0. Then, add 2 x 20 mL of 10 mM Tris-HCl, pH 8.0.

7. Let the fluid sink to the top level of the resin and add 0.1 M glycine, pH 2.5, in small aliquots. Recover approximately 900-µL aliquots of eluted solution in microfuge tubes containing 100 µL of 1 M Tris-HCl, pH 8.0. Collect 10 aliquots and store them at 4°C.

8. Regenerate the column with 2 additional washes of 20 mL each of 0.1 M glycine, pH 2.5, and resuspend the resin in 10 mL water. Wash it twice in a Falcon® tube (Becton Dickinson, Bedford, MA, USA) and store it at 4°C.

Electrophoresis, Spectrophotometric Reading, and Dialysis

We usually carry out polyacrylamide gel electrophoresis on 2 µL of each aliquot (18). After the gel is stained with Coomassie® blue, the aliquots that contain the antibody are pooled. The antibody solution is then dialyzed for at least 48 hours against phosphate-buffered saline (PBS) in a 10 Kd cutoff Slide-A-Lyzer dialysis cassette. The buffer is changed at least twice a day. The solution is then collected, and the protein concentration is measured with a spectrophotometer, reading the absorbance at 280 nm. The antibody is then lyophilized to a concentration of at least 1.5 mg/mL.

Coupling

1. Prepare 1 mg of dialyzed antibody in 700 µL PBS.

2. Add 100 µL of 1 M carbonate buffer, pH 9.3.

3. Dissolve the content of a single tube of Cy3 dye in 400 µL of PBS. Take 200 µL and add them to the antibody solution. Wrap the tube in aluminum foil and stir by rotation for 30 minutes at 30°C.

4. Put the tube on ice.

5. During the reaction, prepare the sepharose column. Suspend the G25 resin in an equal volume of PBS to a final volume of 50 mL. Pour the resuspended resin into a 50-mL Falcon tube and wash it with PBS. Spin down for 1

minute at 400× *g*. Repeat twice. Prepare a 10-mL sterile syringe with foam at the nipple (check the flow by adding water), and mount it as a chromatography column. Wash with water. Resuspend the resin in 25 mL of PBS and gently add aliquots to the syringe. When an approximate 8-mL resin column has formed, continue to add PBS until the coupling reaction is complete.

6. Before the resin dries, add the antibody solution. After the entire sample has entered the column, add 100-μL aliquots of PBS. When a dye-free upper portion of the column becomes visible, fill the syringe with PBS. During chromatography, two thick red bands will become visible. Collect, in an Eppendorf® tube, the volume containing the bottom band.

7. Store the E6-Cy3 antibody at 4°C in a tube wrapped in aluminum foil.

Protocol for Dissociation of the Retina and Harvesting of Neurons

This protocol has been optimized to detect a very small population of neurons that are located in the middle of the retina; DA cell bodies occupy the most vitreal tier of the inner nuclear layer (INL) and the dendrites spread in the stratum S1 of the inner plexiform layer (16). The procedure must be drastic enough to free these cells from the surrounding tissue, but it should not compromise their viability, and it should preserve some of the dendrites for the study of postsynaptic receptors. It is especially important that PLAP at the cell surface is not completely digested; although the dissociated cells are never as intensely stained as in the intact retina, enough enzyme must survive papain digestion to permit easy identification of the labeled neurons.

Papain probably acts only at the inner and outer limiting membranes of the retina; therefore, the enzymatic digestion must be followed by trituration (21).

Because DA cells are very rare, the entire cell suspension must be plated. We usually distribute our preparation among 6 small petri dishes. We can detect an average of 5 DA cells per petri dish in a carpet of thousands of unlabeled neurons. Thirty minutes are the minimum amount of time to allow the solitary neurons to sediment on the glass coverslip at the bottom of the dish. The dishes must be pretreated for at least 1 hour with a solution of concavalin A to promote cell adhesion to the glass.

For single-cell RT-PCR experiments on other retinal cell types, the dissociation procedure has to be modified. Rods, rod bipolars, and ganglion cells are more numerous than DA cells and probably easier to obtain by dissociation of the retina. Best results are obtained by decreasing the digestion step to 20 min and limiting the number of trituration events. For neurons of the central nervous system, the dissociation is carried out on vibratome sections. After trituration, we centrifuge only the supernatant, discarding the fragments that fall at the bottom of the tube. The solitary neurons are then plated at the lowest concentration possible taking into account the frequency of the cell type of interest. We were not successful with protocols that include an incubation step in a very low amount of calcium ions. Furthermore, such treatment may have undesirable effects on gene expression.

When harvesting neurons for single-cell RT-PCR experiments, our main concern is the preservation of the integrity of the mRNA and the elimination of RNA originating from other cells or cDNA contaminants present in the laboratory. Recording chambers are washed with hydrogen peroxide prior to concavalin A treatment; aerosol-resistant pipet tips, tubes, and glass for electrodes are autoclaved; reagents are stored at -20°C in single-use aliquots. Cells are harvested, and solutions are prepared in rooms that are different from the laboratory in which the PCRs and agarose gel electrophoresis are carried out. We pretreat the electrodes with dimethyl-chloro-sylane to decrease the affinity of the glass for the RNA present in the medium. Furthermore, we apply positive pressure to the solution within the electrode from the moment this is immersed into the bath and until it touches the cell membrane. After the seal is formed, we monitor its resistance throughout the harvesting procedure to control its integrity. The cell is then lifted and positioned in front of a flux of fresh extracellular solution for 1 min. All these operations occur under constant visual control (17).

In the literature, neurons floating in the medium have been collected for single-cell RT-PCR using a large-bore suction pipet. This technique can only be applied when the tissue of interest contains a nearly homogeneous cell population or the cell type under study can be easily recognized on account of its shape. Furthermore, the collection must be fast, because of the high mortality of the cells when they do not adhere to a solid substrate. More importantly, the probability that undesirable cells or cellular debris are sucked into the pipet is very high.

We find it convenient to carry out cDNA synthesis and PCR amplification in a day different from that of harvesting. Therefore, we store single cells at -80°C in the presence of RNase inhibitor. This procedure also decreases the chances of contamination. Lysis of the neurons takes place when they are thawed.

Alternatively, the harvested cells can be momentarily kept on ice in a solution containing RNase inhibitor and the detergent Nonidet® P-40 (NP40), which lyses only the plasma membrane without affecting the nuclear envelope. Furthermore, 0.5%
NP40 does not inhibit reverse transcriptase.

Materials and Reagents

- Coating solution: concavalin A (Sigma, St. Louis, MO, USA) 1 mg/mL in PBS, prepared fresh.
- EBSS medium: (Earle's Balanced Salt Solution; Sigma) enriched with glucose (10 mg/mL).
- 10× Papain-activating stock solution: 1 mM L-cysteine, 0.5 mM EDTA in EBSS medium, stored in 500-µL aliquots at -20°C.
- 20× DNase I stock solution: 200 U/mL DNase I (Worthington Biochemical, Lakewood, NJ, USA) in EBSS medium, stored in 500-µL aliquots at -80°C.
- Digestion buffer: 20 U/mL papain (Worthington Biochemical) in 5 mL of EBSS medium containing 1× papain-activating solution and 1× DNase I solution.
- 10× OVOBSA stock solution: 1 mg/mL ovomucoid inhibitor (Worthington Biochemical) and 1 mg/mL bovine serum albumin (Sigma) in EBSS medium, stored in 500-µL aliquots at 4°C.
- Trituration buffer: 5 mL of EBSS medium containing 1× DNase I solution and 1× OVOBSA solution.
- MEM (Sigma).
- Extracellular solution: NaCl 137 mM, KCl 5.4 mM, $CaCl_2$ 1.8 mM, $MgCl_2$ 1 mM. 4-(2-hydroxethyl)-1-piperazineethanesulfonic acid (HEPES) 5 mM, glucose 20 mM. Autoclaved and stored at 4°C.
- Intracellular (electrode) solution: KCl 130 mM, EGTA 0.5 mM in HEPES 10 mM, pH 7.4. Autoclaved and stored at 4°C.

- Electrodes: borosilicate glass electrodes (1.65 mm OD, 1.2 mm ID; A-M Systems, Carlsborg, WA, USA) are immersed in a 5% solution of dimethyl-chloro-sylane (Sigma) in chloroform for at least 1 hour. They are then carefully washed, baked at 80°C overnight, and autoclaved. Patch pipets are constructed using a horizontal 2-stage electrode puller (BB-CH, Mecanex, Geneva, Switzerland); the electrode resistance ranges from 5 to 7 MΩ. Electrodes are connected to the amplifier via an Ag/AgCl wire. The electrode holder and the headstage are mounted on a piezoelectric remote-controlled device attached to a tridimensional micromanipulator (Burleigh Instruments, Fishers, NY, USA). The intracellular solution is filtered and loaded into the electrode.
- Harvesting solution: 2 µL contains 0.4 µL of 5× first-strand buffer (KCl 375 mM, $MgCl_2$ 15 mM, Tris-HCl 250 mM, pH 8.3), 10 mM dithiothreitol (DTT) and 4 U RNAGUARD (Amersham Pharmacia Biotech).

Procedure

Dissociation of the Retina

1. Eyes are enucleated from anesthetized, 1 to 6-month-old mice and opened at the equator. After removal of cornea, lens, and vitreous body, the eyecups are transferred to a petri dish containing EBSS and cut in half. The 4 specimen are then moved into a 10-mL Falcon tube containing 5 mL of digestion buffer and incubated for 40 minutes at 37°C with gentle agitation. Prior to use, the enzyme is activated for 30 minutes at 37°C by adding 20 U/mL papain to 1× papain-activating solution. At the moment of immersing the specimens, the digestion buffer is completed by adding an equal amount of 1× DNase I stock solution.
2. At the end of the digestion, eyecups are transferred into a petri dish containing 2 mL of trituration buffer. The retinas are carefully detached from the pigment epithelium and moved to a Falcon tube containing 1 mL of fresh trituration buffer. For trituration, they are slowly sucked into and immediately expelled from a fire-polished Pasteur pipet and this process is repeated about 10 times. The supernatant containing a suspension of dissociated cells is transferred into a new Falcon tube. The remaining retinal pieces are resuspended in 1 mL of trituration buffer and triturated 5 more times using a Pasteur pipet whose bore had been reduced by fire polishing.
3. Cell suspensions are pooled and centrifuged at 1000 rpm for 5 minutes. The pellet is then resuspended in 2 mL MEM. The antibody E6-Cy3 is diluted 1:200 in 2 mL MEM, and this solution is added to the cell suspension.
4. Aliquots of the cell suspension are deposited in the recording chambers. These are prepared by machining a hole 20 mm in diameter in the bottom of a 35-mm petri dish and gluing a circular glass coverslip 25 mm in diameter to the margin of the hole with Sylgard 184 (Dow Corning, Midland, MI, USA). Finally, the surface of the coverslip is treated with concanavalin A solution for at least 1 hour prior to transfer of the cell suspension. The retinal neurons are allowed to sediment on the glass for 30 minutes at 37°C in 5% CO_2, 95% O_2 atmosphere.

Harvesting of Neurons

1. Thirty minutes after plating, cells are washed 5 times and kept in the extracellular solution for the rest of the

experiment. After positioning the recording chamber on the stage of an inverted microscope, E6-Cy3-stained neurons are identified by scanning the coverslip in epifluorescence (535 nm excitation filter; 610 nm barrier filter). Immediately after cell identification, the UV beam is turned off, and the rest of the procedure is carried out in visible light with Nomarski optics.

2. When the patch pipet is immersed into the bath, positive pressure is continuously applied to the fluid within to avoid entry of cellular debris. After gigaseal formation on the surface of the neuron and disruption of the patch membrane, the cellular contents are aspirated into the pipet by gently applying negative pressure until the residual cell ghost became stuck to the pipet tip. The electrode is then positioned in front of a perfusion tube with an internal diameter of 0.5 mm. The cell is washed for 1 min with fresh extracellular solution applied by pressure ejection. The electrode is then lifted from the bath and removed from the holder.

3. The tip of the pipet is finally broken into an Eppendorf tube containing 2 µL of harvesting solution. After brief centrifugation, the tube is frozen on dry ice and stored at -80°C. Harvesting lasts from 30 to 90 minutes from the time of plating.

Protocol for Single-Cell RT-PCR

Our protocol consists of a random-primed cDNA synthesis step followed by 2 successive rounds of PCR (Figure 1) (17). The first amplification reaction occurs in presence of forward (F) and reverse (R) primers, whereas in the second round, two primers internal to the first amplification product (FN and RN, N for nested) are used to increase specificity. Primer sequences are located in different exons to avoid amplification of genomic DNA and are all designed to anneal at 55°C (first round) or 60°C (second round), so that different cDNAs can be amplified together in a single multiplex experiment. Primers are usually 22 to 24 nucleotides long with a guanosine and cytosine content of 50% to 60%. 5′ and 3′ end of the primers are not complementary to prevent the formation of hairpins or primer dimers. The amplified fragment must be kept short, preferably in the range of 150 to 400 bp. Each couple of primers is first tested in a RT-PCR experiment from 100 ng of retina total RNA. These amplifications are carried out for 40 cycles at the appropriate annealing temperature. The RT-PCR products are digested with suitable restriction enzymes to confirm specificity. RNA-dependency of the amplification is established by omitting RT during first-strand cDNA synthesis.

Neurons of interest, bystander cells and controls (supernatant and first-strand mixture in absence of a cell) are examined simultaneously in each experiment. Furthermore, six is the highest number of cells that are analyzed at any given time to minimize contamination.

The repertory of PCR products should be the same when the number of cycles is increased from 28 + 30 cycles to 40 + 35 cycles. However, the number of cycles must be kept as low as possible to avoid contamination and increase in number of nonspecific signals.

The number of cycles required, however, depends on a variety of factors, but especially the abundance of the amplified transcript. In most cases, a single round of 40 cycles is sufficient.

Many different RT-PCR strategies are reported in the literature. The first-strand synthesis can be directed by oligo(dT) primers, random examers or antisense primers that recognize specific target sequences. The cDNA thus obtained can then be amplified with specific primers for a single-round PCR or with two-round PCR, where the first amplification occurs in presence of degenerated oligonucleotides and the second with primers specific for the various transcripts.

Several methods of detection have been employed, including ethidium bromide staining, Southern blotting, and direct autoradiography of the amplified bands through incorporation of a radioactive nucleotide or primer in the PCR.

Materials and Reagents

- First-strand mixture: 18 µL contain 3.6 µL of 5× first-strand buffer (KCl 375 mM, MgCl$_2$ 15 mM, Tris-HCl 250 mM, pH 8.3); 36 U RNA-GUARD; 4 ng Pd(N)$_6$, 1 mM each dNTPs (from Amersham Pharmacia Biotech), 4 mM DTT, 200 U Reverse Transcriptase SUPERSCRIPT™ II H- (both from Life Technologies).
- Primers: for the amplification of TH mRNA, first-round F primer was 5′-CTGGCCTTCCGTGTGTTTCAG-TG-3′, which hybridizes with TH cDNA at nt 915 (GenBank® Accession No. M69200), and the corresponding R primer was 5′-CCGGCTGGTA-GGTTTGATCTTGG-3′, which hybridizes with TH cDNA at nt 1296. The first-round RT-PCR product was 382 bp long. The second-round FN primer was 5′-AGTGCACACAGTA-CATCCGTCAT-3′, which hybridizes with TH cDNA at nt 934, and the corresponding RN primer was 5′-GC-TGGTAGGTTTGATCTTGGTA-3′, which hybridizes with TH cDNA at nt 1293. The final nested RT-PCR product was 360 bp long and contained a unique *Sac*I site.
- First-round PCR mixture: 80 µL contains 10 µL of 10× RT-PCR buffer (KCl 350 mM, MgCl$_2$ 9 mM, Tris-HCl 100 mM, pH 8.3), 40 pM each F and R primers, 1 U *Taq* DNA polymerase (Roche Molecular Biochemicals).
- Second-round PCR mixture: 49 µL contains 5 µL of 10× PCR buffer (KCl 500 mM, MgCl$_2$ 15 mM, Tris-HCl 100 mM, pH 8.3), 200 µM each dNTPs, 20 pM each FN and RN primers, 1 U *Taq* DNA polymerase.
- QIAquick™ gel extraction kit (Qiagen, Valencia, CA, USA).
- Original TA Cloning® Kit (Invitrogen, Carlsbad, CA, USA).

Procedure

First-Strand cDNA Synthesis

1. Tubes containing single cells are incubated for 1 minute 30 seconds at 65°C.
2. After cooling in ice for 2 minutes, 18 µL of First Strand Reaction Mixture is added to each tube, and first-strand cDNA is synthesized at 42°C for 1 hour. At the end of the reaction, tubes are cooled on ice.

Two Rounds PCR

Here, we describe the procedure that we used for amplification of TH mRNA.

1. Eighty microliters of the first-round PCR mixture are added to the tubes. First-round amplification comprises: 2 minutes 30 seconds at 94°C; 10 cycles each consisting of 30 seconds at 94°C, 30 seconds at 55°C, and 1 minute at 72°C; 18 cycles each consisting of 30 seconds at 94°C, 30 seconds at 55°C, and 1 minute at 72°C, with a 5 second extension time added to the final step of each cycle beginning from the second cycle; 10 minutes at 72°C.
2. One microliter of the first-round PCR products is used as a template for

second-round PCRs and mixed with 49 µL of the second-round PCR mixture solution. Amplification is carried out as following: 2 minutes 30 seconds at 94°C; 10 cycles each consisting of 30 seconds at 94°C, 30 seconds at 60°C, and 1 minute at 72°C; 20 cycles each consisting of 30 seconds at 94°C, 30 seconds at 60°C, and 1 minute at 72°C, with a 5 second extension time added to the final step of each cycle beginning from the second cycle.

Analysis of the Results

1. Ten microliters of the second-round PCR products are analyzed by electrophoresis in 1% Seakem® GTG® and 1% NuSieve® GTG agarose (FMC BioProducts, Rockland, ME, USA), containing 0.5 µg/mL ethidium bromide.
2. The remaining PCR products are purified and desalted using the Qiagen kit. DNA is then digested with *Sac*I to check for specificity.
3. When the experiments at the single-cell level are completed, a single-cell RT-PCR product is cloned into the pCR-II vector by using the Original TA Cloning Kit. Both strands are sequenced using the dideoxy-chain termination method. DNA similarities are examined by matching the query sequence to database entries using the basic local alignment search tool (BLAST) algorithm. We rigorously avoided cloning the PCR fragments before the end of the single-cell RT-PCR experiments, because the presence of a recombinant sequence in the laboratory increases the possibility of contamination.

RESULTS AND DISCUSSION

E6-Hy hybridoma cells were grown until they reached a concentration of 10^6 cells/mL in 1 L of AHM. The supernatant was collected, and the E6 antibody was purified with a protein A-sepharose column. After a 2-day dialysis, the antibody was concentrated to 2 mg/mL. One milligram of E6 was coupled to half the tube of monofunctional Cy3, and the E6-Cy3 antibody was collected from a G25 chromatography column.

Staining (1:1000) of cryostat sections of the retina of the transgenic animals showed that all cells positive to the histochemical method for PLAP activity were also stained by E6-Cy3 and immunoreactive for TH (Figures 2 and 3).

After enzymatic digestion and mechanical trituration of the transgenic retina, DA neurons were identified in the living state after staining with E6-Cy3 (Figure 3, inset). After establishing giga-seal with a patch electrode, the cell was harvested into the pipet by applying negative pressure and washed with fresh extracellular solution. The tip of the pipet was then broken into a PCR tube containing RNase inhibitor and stored. Large unlabeled neurons, possibly ganglion cells, were also collected for further analysis. During cell harvest, a sample of the supernatant in the recording chamber was collected and used as a control to rule out amplification of mRNA from floating or dead cells.

Single neurons were then denatured, and first-strand cDNA was synthesized from each of them in the presence of random primers. Another control reaction was carried out in the absence of a cell. The cDNAs were in turn the template for two rounds of PCR, and the PCR products were finally analyzed by restriction enzyme digestion and agarose gel electrophoresis.

A 360-bp product specific for the TH transcript was amplified in 23 out of 28 labeled cells (82%), whereas unlabeled neurons were consistently negative (Figure 4). An aliquot of the reaction products from

7. Gene Expression in Genetically Labeled Single Cells

labeled cells was purified and cut with *Sac*I to prove the specificity of the amplification. As expected, the digestion with this restriction enzyme produced fragments of 271 and 89 bp (Figure 4). When the experiments at the single-cell level were completed, a single-cell RT-PCR product was cloned and sequenced to confirm its identity.

The combination of transgenic technology with single-cell RT-PCR proved to be a

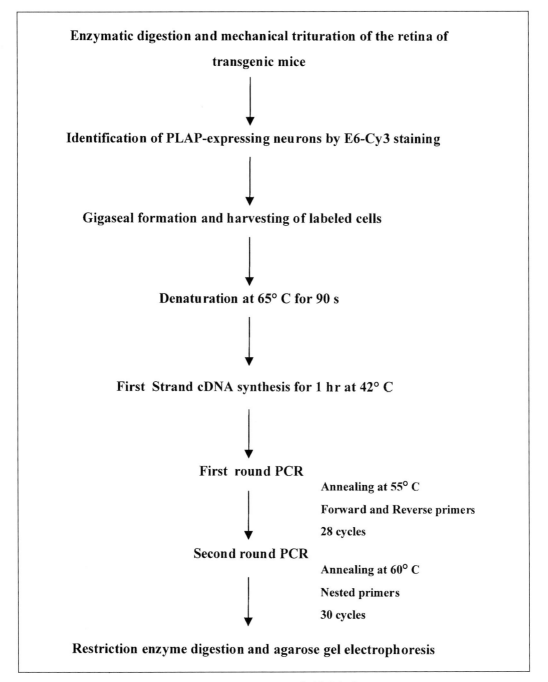

Figure 1. Flow chart of a single-cell RT-PCR experiment on genetically labeled solitary neurons.

powerful tool to detect gene expression in a very small cell population; DA cells represent in fact less than 0.005% of the neurons of the mouse retina.

We know from immunocytochemistry that in the retina of this transgenic mouse line, all PLAP-expressing neurons contain TH (16). Therefore, the proportion of single cells that yielded a specific amplification product represents an estimate of the efficiency of our single-cell RT-PCR technique (82%). Either loss of the cell, deterioration of the message, or inefficient amplification easily explains failures. These data must be taken into account when no amplification product is obtained with a pair of primers specific for other messages.

Negative results should be considered with caution, because the sensitivity of the technique remains to be determined. Harvesting efficiency and integrity of the mRNA cannot be measured for every cell, and the PCR itself is susceptible to a certain amount of internal variability.

With a technique based on processing individual cells and in absence of false positive results, we have adopted a rate of success higher than 70% as an indication that a given gene was expressed throughout an entire neuronal population.

Theoretically, the simultaneous amplification of the gene of interest and another, high-abundance gene would confirm the specificity of the reaction. However, we consistently failed in our effort to demonstrate in the very same cell the presence of messages for both TH and less abundant transcripts such as those for the $GABA_A$ receptor subunits. Amplification of the subunit mRNAs was, in fact, competitively suppressed by the reaction for TH.

Because we have previously shown that DA cells express $GABA_A$ receptors, the single-cell RT-PCR technique was applied to a study of the subunit composition of this ligand-gated channel (16,17). At least 10 DA cells were individually examined for each of the 13 $GABA_A$ subunit transcripts for a total of 161 cells. DA cells express mRNA for the $\alpha 1$, $\alpha 3$, $\alpha 4$, $\beta 1$, $\beta 3$, $\gamma 1$, $\gamma 2S$, and $\gamma 2L$ subunits. All these subunit mRNAs were successfully detected in at least 70% of the cells tested. Messages for the $\alpha 2$, $\alpha 5$, $\beta 2$, $\gamma 3$, δ, and ε subunits were not detected in DA cells. To confirm that the absence of these subunit transcripts was not due to low sensitivity of our technique, we showed that they were present in unlabeled bystander neurons.

Then, we examined whether all DA cells contained the 7 $GABA_A$ receptor subunit transcripts, or they could be assigned to subpopulations each expressing a different repertory of $GABA_A$ receptor mRNAs. To this purpose, we resorted to multiplex semi-nested RT-PCR; in this method, first-strand cDNA was synthesized from single cells and amplified in the presence of primers for all subunit transcripts detected previously. Then, the product of the first-round amplification was distributed among 7 different second-round reactions, each specific for a single subunit cDNA. Seven DA cells were examined, and 4 of them contained the transcripts of all 7 subunits. In 3 cells, the reaction failed, and we did not obtain any PCR product. Large unlabeled neurons, possibly ganglion cells, expressed different repertories of subunit mRNAs.

Because little is known of the turnover of mRNAs in intact organs or after dissociation, we were concerned that our findings in the isolated cells did not reflect the true situation in the intact retina. Especially worrisome, in this respect, was the suppression of synaptic inputs that could alter the expression of postsynaptic receptor genes. Our approach to this problem was twofold. First, we investigated whether the repertory of transcripts varied over time; we began to observe loss of some of the messengers at 4 hours 30 minutes. As a result, cell harvesting was limited to the first 90 minutes after dissociation. To test whether the experimental conditions induced a new pattern of gene expression, we shut off de novo transcription immediately after trituration by blocking the activity of RNA polymerase II with α-amanitin. Multiplex seminested RT-PCR analysis proved that the pattern of gene expression remained the same after inhibition of mRNA synthesis. Immunocytochemistry with subunit-specific antibodies showed that all subunits detected by RT-PCR after dissociation were expressed in the intact retina (17).

These findings prove that single-cell RT-PCR on solitary neurons is a reliable method to study gene expression in a specific cell population of the adult brain. When carried out in combination with a physiological analysis by patch clamping, it will speed up and focus the biophysical and pharmacological research and elucidate the molecular species responsible for the cells' behavior. On the other hand, genetic labeling permits the study of types of neurons that cannot be recognized on the basis of their shape or size. These new data will address the role of these cell types in the computations carried out by the neural networks of which they are components.

TROUBLESHOOTING

Staining with the E6-Cy3 Antibody

After the coupling reaction, a new E6-Cy3 preparation must be tested at a dilution of 1:1000 on brain sections of transgenic mice containing the PLAP-expressing cells of interest. Two problems may arise: *(i)* nonspecific staining of cells that do not express PLAP; and *(ii)* PLAP-expressing cells are not labeled, or they are weakly stained. E6 is very sensitive to overcoupling, and, for this reason, we use half of the quantity of the dye which is recommended by the manufacturer. When the antibody is over-coupled, it usually loses specificity, and in mouse retinal sections it reacts intensely with nuclei in the INL. For this reason, we recommended the purification of adequate amounts of antibody. The coupling reaction can then be repeated with decreasing quantity of dye until specific staining is obtained. On the other hand, when the antibody is insufficiently coupled, it does not stain adequately PLAP-expressing cells in vivo. This may be caused by copurification of large quantities of protein from the hybridoma supernatant (a remedy would be to use the Ultra Low IgG fetal bovine serum), presence of Tris-HCl in the buffer (a remedy would be dialysis for 2 additional days), denaturation of the antibody during the coupling (test the E6 batch at a dilution of 1:1000 on brain sections followed by an antimouse secondary).

When the batch of E6-Cy3 is satisfactory, many other factors may influence the staining of cells after dissociation. If the staining is weak or absent, the antibody may be old, and it should be tested on sections. Papain must be washed very carefully after the digestion step during the dissociation because contaminant traces of the enzyme may degrade the antibody. If nonspecific staining is observed, the quality of the dissociation may be poor, for dead cells bind the antibody. Alternatively, papain has completely digested the PLAP present at the neuronal surface; then, the concentration of the enzyme must be decreased or the incubation time shortened.

Neuronal Viability in Short-Term Cultures

The presence of a high number of cells with well-preserved processes is an indicator of the good quality of the dissociation. The number of trituration events and the length of the incubation in papain are crucial steps to obtain optimal results. Furthermore, results can vary with the different batches of the enzyme provided by the manufacturer. When the quality of the dissociation deteriorates, papain is the prime suspect. If the change of papain has no effects, we make new petri dishes; repeated use of the same dishes or lack of appropriate care in handling them can cause the accumulation of undesired contaminants. When this occurs, differences in cells' viability between dishes are observed. If the problem persists, we prepare new media carefully, avoiding heavy metal contamination.

Absence of Single-Cell RT-PCR Products

The failure of obtaining single-cell RT-PCR products is a common outcome when a new cell type or a new transcript is studied. Furthermore, it may occasionally happen even when the protocol is normally successful. Most common causes are: *(i)* the experiment was not carried out in an RNase-free environment; *(ii)* reverse transcriptase did not work properly; *(iii)* primers were degraded; and *(iv)* cells were dying. The standard requirements when working with RNA are familiar to molecular biologists. Recording chambers must be washed with hydrogen peroxide, tubes and glass for electrodes must be baked or autoclaved, and gloves must always be worn. The reverse transcriptase is stored in a StrataCooler® (Stratagene, La Jolla, CA, USA) and is handled very gently. Primers are stored at -20°C in single-use aliquots to avoid degradation by repeated freezing and thawing. When a new experiment is designed, primers and PCR conditions must be tested first on 100 ng of total RNA from the tissue of interest. Finally, cells must be harvested during the first hour after plating only from optimal preparations.

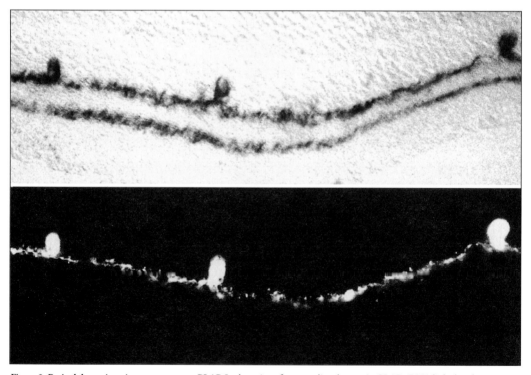

Figure 2. Retinal dopaminergic neurons express PLAP. In the retina of a mouse line that carries PLAP cDNA linked to the promoter for TH, type I catecholaminergic cells express PLAP. In a vertical section of the retina, a neuron is positive to the histochemical reaction for PLAP activity (top panel) that exhibits TH-like immunoreactivity (bottom panel) (adapted from Reference 16).

7. Gene Expression in Genetically Labeled Single Cells

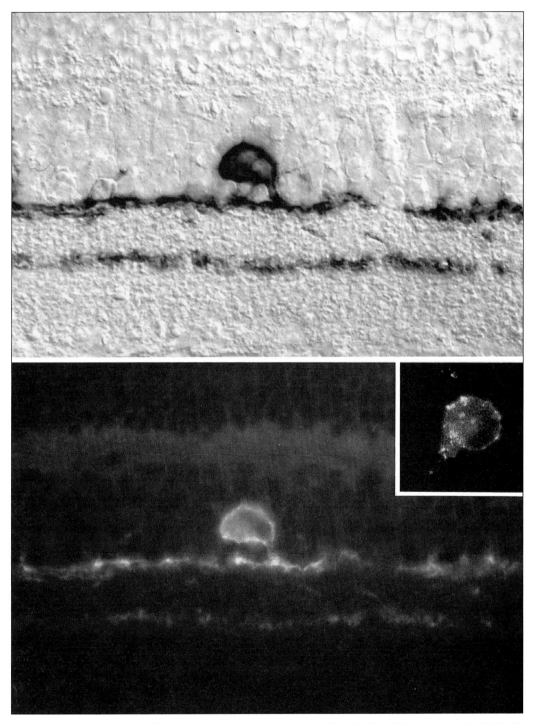

Figure 3. E6-Cy3 specifically stains PLAP-expressing neurons. A monoclonal antibody for PLAP (E6), conjugated to Cy3 (E6-Cy3), is specific for DA cells, because the very cells that are stained by the histochemical method for PLAP (top panel) also bind E6-Cy3 (bottom panel). Inset: solitary DA cells are identified in the living state by staining with the E6-Cy3 antibody (adapted from Reference 16).

Nonspecific RT-PCR Products

Nonspecific bands commonly appear when no target template is present in the reaction mixture. Nonspecific bands may also be associated with the specific amplification product. In this case, the number of cycles should be reduced, or the sequence of the primers should be changed.

RT-PCR Products in the Control Reaction

There are three major sources of contamination: *(i)* mRNA from dead cells present in the medium of the short-term culture is accidentally collected during cell harvesting; *(ii)* template DNA is present in the tube as a result of cross-contamination during handling of the RT-PCR reactions; and *(iii)* PCR products of previous reactions are reamplified.

To avoid collection of contaminating mRNA from the medium, cells must be washed several times with extracellular solution before harvesting. Electrodes must be pretreated with sylane. Positive pressure is applied to the solution within the electrode from the moment this is immersed into the bath and until it touches the cell membrane. After the seal is formed, the resistance must be continuously monitored, and all operations must occur under constant visual control. Cells must be washed for 1 min with fresh extracellular solution. If the seal is poor, cells are discarded. Cells are harvested only when the dissociation is satisfactory. When the dissociation is poor, the percentage of dead cells is very high, and it is difficult to achieve a good seal, thus increasing the possibility of contamination. An aliquot of the medium is always collected and processed in parallel.

During the single-cell RT-PCR, gloves are the most common source of contamination. This occurs when tubes are opened.

Because single-cell RT-PCR is a very sensitive technique, amplification of PCR products from previous experiments can be a problem. With no exception, cells are harvested and solutions prepared in rooms

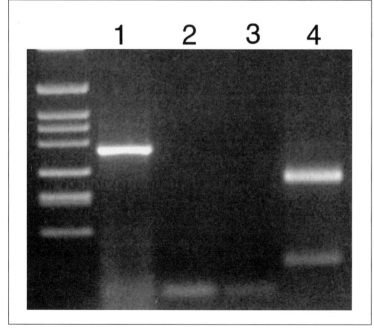

Figure 4. **Single-cell RT-PCR.** Ethidium-stained agarose gel electrophoresis of single-cell RT-PCRs for TH transcript. Lane 1, a single DA cell; lane 2, an unlabeled cell; lane 3, a control reaction where the cell was omitted; lane 4, *Sac*I digestion of the products of lane 1. Markers are molecular weights VI (from Roche Molecular Biochemicals). A single 360-bp product is visible in lane 1. A restriction enzyme digestion produces fragments of 271 and 89 bp, confirming the identity of the amplified band.

that are different from the laboratory in which PCRs are carried out. The analysis of the results is carried out in a third area. We strongly suggest the cloning of the PCR products only at the end of the series of single-cell RT-PCR experiments.

REFERENCES

1. **Audinat, E., B. Lambolez, and J. Rossier.** 1996. Functional and molecular analysis of glutamate-gated channels by patch-clamp and RT-PCR at the single cell level. Neurochem. Int. *28*:119-136.
2. **Banerjee, S.A., P. Hoppe, M. Brilliant, and D.M. Chikaraishi.** 1992. 5′ Flanking sequences of the rat tyrosine hydroxylase gene target accurate tissue-specific, developmental, and transsynaptic expression in transgenic mice. J. Neurosci. *12*:4460-4467.
3. **Berger, J., A.D. Howard, L. Gerber, B.R. Cullen, and S. Udenfriend.** 1987. Expression of active, membrane-bound human placental alkaline phosphatase by transfected simian cells. Proc. Natl. Acad. Sci. USA *84*:4885-4889.
4. **Bochet, P., E. Audinat, B. Lambolez, F. Crepel, J. Rossier, M. Iino, K. Tsuzuki, and S. Ozawa.** 1994. Subunit composition at the single-cell level explains functional properties of a glutamate-gated channel. Neuron *12*:383-388.
5. **Chalfie, M., Y. Tu, G. Euskirchen, W.W. Ward, and D.C. Prasher.** 1994. Green fluorescent protein as a marker for gene expression. Science *263*:802-805.
6. **Estee, J., P. Crino, and J. Eberwine.** 1999. Preparation of cDNA from single cells and subcellular regions, p. 3-18. In S.M. Weissman (Ed.), Methods of Enzymology (303): cDNA Preparation and Characterization. Academic Press, San Diego.
7. **De Groote, G., P. De Waele, A. Van De Voorde, M. De Broe, and W. Fiers.** 1983. Use of monoclonal antibodies to detect human placental alkaline phosphatase. Clin. Chem. *29*:115-119.
8. **DeVries, S.H. and D.A. Baylor.** 1997. Mosaic arrangement of ganglion cell receptive fields in rabbit retina. J. Neurophysiol. *78*:2048-2060.
9. **De Waele, P., G. De Groote, A. Van De Voorde, W. Fiers, J.-D. Franssen, P. Herion, and J. Urbain.** 1982. Isolation and identification of monoclonal antibodies directed against human placental alkaline phosphatase. Arch. Int. Physiol. Biochim. *90*:B21.
10. **Djamgoz, M.B., M.W. Hankins, J. Hirano, and S.N. Archer.** 1997. Neurobiology of retinal dopamine in relation to degenerative states of the tissue. Vision Res. *37*:3509-3529.
11. **Euler, T., H. Schneider, and H. Wassle.** 1996. Glutamate responses of bipolar cells in a slice preparation of the rat retina. J. Neurosci. *16*:2934-2944.
12. **Feigenspan, A., S. Gustincich, B.P. Bean, and E. Raviola.** 1998. Spontaneous activity of solitary dopaminergic cells of the retina. J. Neurosci. *18*:6776-6789.
13. **Fields-Berry, S.C., A.L. Halliday, and C.L. Cepko.** 1992. A recombinant retrovirus encoding alkaline phosphatase confirms clonal boundary assignment in lineage analysis of murine retina. Proc. Natl. Acad. Sci. USA *89*:693-697.
14. **Geiger, J.R., T. Melcher, D.S. Koh, B. Sakmann, P.H. Seeburg, P. Jonas, and H. Monyer.** 1995. Relative abundance of subunit mRNAs determines gating and Ca^{2+} permeability of AMPA receptors in principal neurons and interneurons in rat CNS. Neuron *15*:193-204.
15. **Gustincich, S., D.K. Wu, L.J. Koopman, and E. Raviola.** 1996. A transgenic approach to the study of neural networks in the retina. Invest. Ophthalmol. Vis. Sci. *37*:S1060.
16. **Gustincich, S., A. Feigenspan, D.K. Wu, L.J. Koopman, and E. Raviola.** 1997. Control of dopamine release in the retina: a transgenic approach to neural networks. Neuron *18*:723-736.
17. **Gustincich, S., A. Feigenspan, W. Sieghart, and E. Raviola.** 1999. Composition of the GABA(A) receptors of retinal dopaminergic neurons. J. Neurosci. *19*:7812-7822.
18. **Harlow, E. and D. Lane.** 1988. Antibodies: A Laboratory Manual. CSH Laboratory Press, Cold Spring Harbor, NY.
19. **Jonas, P., C. Racca, B. Sakmann, P.H. Seeburg, and H. Monyer.** 1994. Differences in Ca^{2+} permeability of AMPA-type glutamate receptor channels in neocortical neurons caused by differential GluR-B subunit expression. Neuron *12*:1281-1289.
20. **Kam, W., E. Clauser, Y.S. Kim, Y.W. Kan, and W.J. Rutter.** 1985. Cloning, sequencing, and chromosomal localization of human term placental alkaline phosphatase cDNA. Proc. Natl. Acad. Sci. USA *82*:8715-8719.
21. **Lam, DM.** 1972. Biosynthesis of acetylcholine in turtle photoreceptors. Proc. Natl. Acad. Sci. USA *69*:1987-1991.
22. **Lambolez, B., E. Audinat, P. Bochet, F. Crepel, and J. Rossier.** 1992. AMPA receptor subunits expressed by single Purkinje cells. Neuron *9*:247-258.
23. **MacGregor, G.R., A.E. Mogg, J.F. Burke, and C.T. Caskey.** 1987. Histochemical staining of clonal mammalian cell lines expressing *E. coli* beta-galactosidase indicates heterogeneous expression of the bacterial gene. Somat. Cell Mol. Genet. *13*:253-65.
24. **MacNeil, M.A., J.K. Heussy, R.F. Dacheux, E. Raviola, and R.H. Masland.** 1999. The shapes and numbers of amacrine cells: matching of photofilled with golgi-stained cells in the rabbit retina and comparison with other mammalian species. J. Comp. Neurol. *413*:305-326.
25. **Masland, R.H. and E. Raviola.** 2000. Confronting complexity: strategies for understanding the microcircuitry of the retina. Ann. Rev. Neurosci. *23*:249-284.
26. **Mayahara, H., H. Hirano, T. Saito, and K. Ogawa.** 1967. The new lead citrate method for the ultracytochemical demonstration of activity of non-specific alkaline phosphatase (orthophosphoric monoester phosphohydrolase). Histochemie *11*:88-96.
27. **McComb, R.B. and G.N. Bowers, Jr.** 1972. Study of optimum buffer conditions for measuring alkaline phosphatase activity in human serum. Clin. Chem.

28. Monyer, H. and B. Lambolez. 1995. Molecular biology and physiology at the single-cell level. Curr. Opin. Neurobiol. *5*:382-387.
29. Oberdick, J., R.J. Smeyne, J.R. Mann, S. Zackson, and J.I. Morgan. 1990. A promoter that drives transgene expression in cerebellar Purkinje and retinal bipolar neurons. Science *248*:223-226.
30. O'Dowd, D.K. and M.A. Smith. 1996. Single-cell analysis of gene expression in the nervous system. Measurements at the edge of chaos. Mol. Neurobiol. *13*:199-211.
31. Parra, P., A.I. Gulyas, and R. Miles. 1998. How many subtypes of inhibitory cells in the hippocampus? Neuron *20*:983-993.
32. Posen, S., C.J. Cornish, M. Horne, and P.K. Saini. 1969. Placental alkaline phosphatase and pregnancy. Ann. NY Acad. Sci. *166*:733-774.
33. Spergel, D.J., U. Kruth, D.F. Hanley, R. Sprengel, and P.H. Seeburg. 1999. GABA- and glutamate-activated channels in green fluorescent protein-tagged gonadotropin-releasing hormone neurons in transgenic mice. J. Neurosci. *19*:2037-2050.
34. Stevens, C.F. 1998. Neuronal diversity: too many cell types for comfort? Curr. Biol. *8*:R708-R710.
35. Sucher, N.J. and D.L. Deitcher. 1995. PCR and patch-clamp analysis of single neurons. Neuron *14*:1095-1100.
36. Tsien, R.Y. 1998. The green fluorescent protein. Ann. Rev. Biochem. *67*:509-544.
37. Van Gelder, R.N., M.E. von Zastrow, A. Yool, W.C. Dement, J.D. Barchas, and J.H. Eberwine. 1990. Amplified RNA synthesized from limited quantities of heterogeneous cDNA. Proc. Natl. Acad. Sci. USA *87*:1663-1667.
38. Versaux-Botteri, C., J. Nguyen-Legros, A. Vigny, and N. Raoux. 1984. Morphology, density and distribution of tyrosine hydroxylase-like immunoreactive cells in the retina of mice. Brain Res. *301*:192-197.
39. Witkovsky, P. and A. Dearry. 1991. Functional roles of dopamine in the vertebrate retina. Progr. Retinal Res. *11*:247-292.
40. Zlokarnik, G., P.A. Negulescu, T.E. Knapp, L. Mere, N. Burres, L. Feng, M. Whitney, K. Roemer, and R.Y. Tsien. 1998. Quantitation of transcription and clonal selection of single living cells with beta-lactamase as reporter. Science *279*:84-88.
41. Zolotukhin, S., M. Potter, W.W. Hauswirth, J. Guy, and N. Muzyczka. 1996. A "humanized" green fluorescent protein cDNA adapted for high-level expression in mammalian cells. J. Virol. *70*:4646-4654.

Appendix I

Analysis of gene expression at a single-cell level on cells from the mammalian nervous system. Single-cell RT-PCR experiments are classified according to the function of the gene product whose mRNA expression was analyzed. For each experiment the following information is given: first author of the paper and year of publication (the complete reference is listed alphabetically below); species (R = Rat; M = Mouse); mRNAs identified (see below for abbreviations); method of preparation (C = primary Culture; S = Slice; D = Dissociation); neuronal types that were analyzed (n. = neuron).

RECEPTORS

Glutamate

Lambolez et al. (1992)	R	GluR1-4	C	cerebellum Purkinje n.
Bochet et al. (1994)	R	GluR1-4	C	hippocampus n. type II
Jonas et al.(1994)	R	GluRA-D	S	neocortex pyramidal and non-p. n.
Geiger et al. (1995)	R	GluRA-D	S	CA3 pyramidal n.
				Hilar mossy cells
				dentate gyrus granule cells
				layer V pyramidal cells
				dentate gyrus basket cells
				Hilar interneurons
				neocortex layer V non-pyramidal cells
				medial nucleus of the trapezoid body
				Bergmann glial cells
Ruano et al. (1995)	R	GluR5-7, KA1-2	C	hippocampus n.
Zhang et al. (1995)	R	GluR1-4	D	retina ganglion cell
Audinat et al. (1996)	R	GluR1-4	S	neocortex pyramidal and non-p. n.
Garaschuk et al. (1996)	R	GluRA-D NR2A-C	S	CA1 pyramidal n.
Lambolez et al. (1996)	R	GluR1-4	S	neocortex pyramidal and non-p. n.
Tempia et al. (1996)	R	GluR1-4	S	cerebellum Purkinje n.
Wang & Grabowski (1996)	R	NMDAR1	S	cerebellum granule n.
				cerebellum Purkinje n.
Angulo et al. (1997)	R	GluR1-4	S	cortex nonpyramidal n.
Flint et al. (1997)	R	NR2A-D	S	somatosensoy cortex n.
Plant et al. (1997)	R	NR2A-D	S	forebrain medial septal inhibitory n.
Ying et al. (1997)	R	GluR1-4	C	hippocampus n.

Chen et al. (1998)	R	GluR1-4	D	striatum n.	
Das et al. (1998)	M	NR1, NR3A	D	cerebrocortical n.	
Porter et al. (1998)	R	GluR1-4 GluR5-7, KA1-2 NR2A-D	S	neocortex interneurons	
Stefani et al. (1998)	R	GluR1-4 GluR5-7, KA1-2	D	striatum medium spiny n.	
Zawar et al. (1999)	R	NR2A-C	S	CA1 pyramidal n. stratum radiatum interneurons glial cells dentate gyrus granule cells	
Tehrani et al. (2000)	R	mGluR1-8	S	retina ganglion cell	

GABA

Grigorenko & Yeh (1994)	R	β1-3	D	retina rods retina ganglion cells retina Muller glia	
Wang & Grabowski (1996)	R	γ2	S	cerebellum granule n.	
Yeh et al. (1996)	R	α1-6, β1-3, γ1-3, δ, ρ1-2	D	retina bipolar cells retina ganglion cells	
Ruano et al. (1997)	R	α1-6, β1-3, γ1-3, δ	S	cerebellum Purkinje n. cortex pyramidal n	
Yan et al. (1997)b	R	α1-4, β1-3, γ1-3, δ	D	neostriatum cholinergic interneurons	
Berger et al. (1998)	R	α1-5, β1-3, γ1-3	S	hippocampus basket n.	
Gustincich et al. (1999)	M	α1-5, β1-3, γ1-3, δ, ε	D	retina dopaminergic n.	
Guyon et al. (1999)	R	α1-6, β1-3, γ1-3, δ	D	substantia nigra dopaminergic n.	

ACh

Yan et al. (1996)	R	mAChR: m1-5	D	neostriatum cholinergic interneurons	
Poth et al. (1997)	R	nAChR: α2-5,7, β2-4	C	intracardiac parasympathetic n.	
Vysokanov et al. (1998)	R	mAChR: m1	D	medialprefrontal cortex pyramidal n. medialprefrontal cortex interneurons	
Porter et al. (1999)	R	nAChR: α2-7, β2-4	S	cortex pyramidal n. cortex interneurons	
Stewart et al. (1999)	R	mAChR: m1-5	D/C	sensorimotor cortex pyramidal n.	

Other Receptors

Surmeier et al. (1996)	R	dopamine	D	neostriatum medium spiny n.	

		(D1$_{a\text{-}b}$, D2-4)		
Neumann et al. (1997)a	R	IFNγ-R; TNFαR	C	hippocampus
Neumann et al. (1997)b	R	IFNγ-R	C	dorsal root ganglion
Yan et al. (1997)a	R	dopamine (D1$_{a\text{-}b}$, D2-4)	D	neostriatum cholinergic interneurons
Yan et al. (1997)b	R	dopamine (D1-5)	D	neostriatum cholinergic interneurons
Futami et al. (1998)	R	SP	S	substantia nigra dopaminergic n. cerebellum Purkinje n.
Hurbin et al. (1998)	R	VP (V1$_{a\text{-}b}$-2)	D	supraoptic nucleus magnocellular n.
Schmidt-Ott et al. (1998)	M	ET (ET$_{A\text{-}B}$)	S	cerebellum Purkinje n. Bergmann glial cells
Vysokanov et al. (1998)	R	dopamine (D4) serotonin (5-HT2a, 5-HT2c, 5-HT7)	D	medialprefrontal cortex pyramidal n. medialprefrontal cortex interneurons
Cardenas et al. (1999)	R	serotonin (5-HT$_{1D}$, 5-HT$_{2C}$, 5-HT$_7$)	D	dorsal root ganglion n.
Skynner et al. (1999)	M	ERα-β	S	gonadotropin-releasing hormone n.
Tsuchiya et al. (1999)	R	serotonin (5HT3)	S	spinal cord dorsal horn n.
Zhu et al. (1999)	R	AT$_{1A}$	C	hypothalamus

CHANNELS

Sodium

Vega-Saenz de Miera et al. (1997)		RBI, Scn8a	D	cerebellum Purkinje n.

Calcium

Song et al. (1996)	R	α1$_{C\text{-}D}$	D	neostriatum medium spiny n.
Yan et al. (1996)	R	α1$_{A\text{-}E}$	D	neostriatum cholinergic interneurons
Mermelstein et al. (1997)	R	α1$_{A\text{-}D}$	D	neostriatum medium spiny n.
Plant et al. (1998)	R	α1$_{A\text{-}S}$	S	facial nucleus motorneurons
Glasgow et al. (1999)	R	α1$_{A\text{-}E}$; α2; β$_{1\text{-}4}$	D	hypothalamus magnocellular n.
Mermelstein et al. (1999)	R	α1, β$_{1\text{-}4}$	D	neostriatum medium spiny n. cortex pyramidal n.

Potassium

Massengill et al. (1997)	M	Kv3.1	C	cortex n.
Martina et al. (1998)	R	Kv1.1, Kv1.2, Kv1.4, Kv1.6, Kv2.1, Kv2.2 Kv3.1; Kv3.2, Kv 4.1, Kv 4.2, Kv 4.3	S	dentate gyrus basket cell CA1 pyramidal n.
Mermelstein et al. (1998)	R	IRK1-3	D	nucleus accumbens projection n.
Song et al. (1998)	R	Kv1.1, Kv1.2,	D	neostriatum cholinergic interneurons

		Kv1.4, Kv1.5, Kv3.4, Kv 4.1, Kv 4.2, Kv 4.3, Kv β1, Kv β2, Kv β3		
Baranauskas et al. (1999)	R	Kv2.1, Kv2.2, Kv 3.1, Kv 3.2	D	globus pallidus n.
Zawar et al. (1999)	R	Kir6.1, Kir6.2, SUR1, SUR2	S	CA1 pyramidal n. stratum radiatum interneurons glial cells dentate gyrus granule cells
Tkatch et al. (2000)	R	Kv 4.1; Kv 4.2; Kv 4.3	D	neostriatum medium spiny n. neostriatum cholinergic interneurons globus pallidus n. basal forebrain cholinergic n.

CELL-TYPE MARKERS
Enzymes

Bochet et al. (1994)	R	GAD65	C	hippocampus n. type II
Jonas et al. (1994)	R	GAD65	S	neocortex pyramidal and non-p. n.
Yan et al. (1996)	R	ChAT	D	neostriatum cholinergic interneurons
Cauli et al. (1997)	R	GAD65; GAD67;ChAT	S	sensorymotor cortex nonpyramidal n. cerebellum Purkinje n.
Ceranik et al. (1997)	R	GAD67	S	dentate gyrus interneurons
Plant et al. (1997)	R	GAD65	S	forebrain medial septal inhibitory n.
Yan et al. (1997)a	R	ChAT	D	neostriatum cholinergic interneurons
Yan et al. (1997)b	R	ChAT	D	neostriatum cholinergic interneurons
Comer et al. (1998)	R	TH; PNMT	D	rostral ventrolateral medulla
Lipski et al. (1998)	R	TH; PNMT	D	rostral ventrolateral medulla
Porter et al. (1998)	R	GAD65; GAD67 ChAT	S	neocortex interneurons
Song et al. (1998)	R	ChAT	D	neostriatum cholinergic interneurons
Tkatch et al. (1998)	R	GAD67; ChAT	D	globus pallidus n. basal forebrain n.
Vysokanov et al. (1998)	R	GAD67	D	medialprefrontal cortex pyramidal n. medialprefrontal cortex interneurons
Baranauskas et al. (1999)	R	GAD67; ChAT	D	globus pallidus n.
Fusco et al. (1999)	R	ChAT	D	striatum interneurons striatum projection n. corticostriatal projection n. globus pallidus projection n. nucleus basalis n.
Gustincich et al. (1999)	M	TH	D	retina dopaminergic n.
Guyon et al. (1999)	R	TH; GAD65; GAD67	D	substantia nigra dopaminergic n.

Kawai et al. (1999)	R	TH; PNMT	D	rostral ventrolateral medulla
Porter et al. (1999)	R	GAD65; GAD67	S	cortex pyramidal n. cortex interneurons

Neuropeptides

Song et al. (1996)	R	ENK; SP	D	neostriatum medium spiny n.
Surmeier et al. (1996)	R	ENK; SP	D	neostriatum medium spiny n.
Cauli et al. (1997)	R	VIP; CCK; SS; NPY	S	sensorymotor cortex nonpyramidal n. cerebellum Purkinje n.
Chen et al. (1998)	R	ENK; SP	D	striatum n.
Hurbin et al. (1998)	R	VP; OT	D	supraoptic nucleus magnocellular n.
Mermelstein et al. (1998)	R	ENK; SP	D	nucleus accumbens projection n.
Porter et al. (1998)	R	VIP	S	neocortex interneurons
Song et al. (1998)	R	ENK; SP	D	neostriatum cholinergic interneurons
Stefani et al. (1998)	R	ENK; SP	D	striatum medium spiny n.
Fusco et al. (1999)	R	ENK; SP	D	striatum interneuron striatum projection n. corticostriatal projection n. globus pallidus projection n. nucleus basalis n.
Glasgow et al. (1999)	R	OT; VP; Dyn; Gal; CCK; CRH	D	hypothalamus magnocellular n.
Kelz et al. (1999)	M	ENK; SP	D	striatum n.
Mermelstein et al. (1999)	R	ENK; SP	D	neostriatum medium spiny n. cortex pyramidal n.
Porter et al. (1999)	R	VIP; CCK; SS; NPY	S	cortex pyramidal n. cortex interneurons
Tsuchiya et al. (1999)	R	ENK	S	spinal cord dorsal horn n.

Calcium-Binding Proteins

Wang & Grabowski (1996)	R	CB	S	cerebellum granule n. cerebellum Purkinje n.
Cauli et al. (1997)	R	CR; CB; PV	S	sensorymotor cortex nonpyramidal n. cerebellum Purkinje n.
Porter et al. (1998)	R	CR	S	neocortex interneurons
Vysokanov et al. (1998)	R	CR; CB; PV	D	medialprefrontal cortex pyramidal n. medialprefrontal cortex interneurons
Baranauskas et al. (1999)	R	PV	D	globus pallidus n.
Glasgow et al. (1999)	R	CB	D	hypothalamus magnocellular n.
Porter et al. (1999)	R	CR; CB; PV	S	cortex pyramidal n. cortex interneurons

Other Markers

Lambolez et al. (1992)	R	GFAP	C	cerebellum Purkinje n.
Gahring et al. (1996)	M	NSE	C	cortex n.
Neumann et al. (1997)a	R	MAP2; GFAP	C	hippocampus
Neumann et al. (1997)b	R	GFAP	C	dorsal root ganglion
Comer et al. (1998)	R	NSE	D	rostral ventrolateral medulla

Kawai et al. (1999)	R	NSE	D	rostral ventrolateral medulla
Lipski et al. (1998)	R	NSE	D	rostral ventrolateral medulla
Skynner et al. (1999)	M	GFAP	S	gonadotropin-releasing hormone n.
Tsuchiya et al. (1999)	R	NSE	S	spinal cord dorsal horn n.

MISCELLANEA

Chiang et al. (1994)	R	NOS	C	hippocampus n.
Crepel et al. (1994)	R	NOS	S	cerebellum granule n. cerebellum Purkinje n.
Baba-Aissa et al. (1996)	R	SERCA2-3	S	cerebellum Purkinje n.
Gahring et al. (1996)	M	TNFα	C	cortex n.
Wang & Grabowski (1996)	R	clathrin; NCAM; hnRNP A1-2	S	cerebellum granule n. cerebellum Purkinje n.
Wang & Wu (1996)	R	$G_{\alpha q}$; $G_{\alpha 11}$	D	substantia nigra dopaminergic n.
Wu & Wang (1996)	R	$G_{\alpha q}$; $G_{\alpha 11}$	D	substantia nigra dopaminergic n.
Neumann et al. (1997)a	R	GAPDH; MHC I; β2-microglobulin; TAP1; TAP2 calnexin	C	hippocampus
Neumann et al. (1997)b	R	GAPDH; CD4 IFNγ	C	dorsal root ganglion
Comer et al. (1998)	R	GAPDH; NAT	D	rostral ventrolateral medulla
Schmidt-Ott et al. (1998)	M	ET (ET1-3) ET-converting enzyme (ECE1-2)	S	cerebellum Purkinje n. Bergmann glial cells
Shen et al. (1998)	M	Rps4; Snrpn; Ncam; F3cam	S	hippocampus pyramidal n.
Tkatch et al. (1998)	R	ACh Vtr.; GABA Vtr.	D	globus pallidus n. basal forebrain n.
Vysokanov et al. (1998)	R	CaMKII	D	medialprefrontal cortex pyramidal n. medialprefrontal cortex interneurons
Zhu et al. (1998)	R	PLA_2; AT_2	C	hypothalamus and brainstem
Brocke et al. (1999)	R	CaMKII (α, $α_B$, β, β′, $β_e$, $β'_e$, γ variants, $δ_{A-C}$)	S	CA1 pyramidal n.
Fusco et al. (1999)	R	huntingtin	D	striatum interneurons striatum projection n. corticostriatal projection n. globus pallidus projection n. nucleus basalis n.
Glasgow et al. (1999)	R	GAPDH	D	hypothalamic magnocellular n.
Skynner et al. (1999)	M	GnRH	S	gonadotropin-releasing hormone n.
Zhu et al. (1999)	R	CaMKII; PKCα	C	hypothalamus

7. Appendix I

ABBREVIATIONS

ACh = Acetylcholine; ACh Vtr. = Acetylcholine Vesicular transporter; AT = Angiotensin Type receptor; CaMKII = Ca^{2+}/calmodulin-dependent kinase II; CB = Calbindin; CCK = Cholecystokinin; ChAT = Choline Acetyltransferase; CR = Calretinin; Dyn = Dynorphin; ENK = Enkephalin; ER = Estrogen Receptor; ET = Endothelin; GABA = γ-aminobutyric acid; GABA Vtr. = GABA Vesicular transporter; GAD = Glutamic Acid Decarboxylase; Gal = Galanin; GAPDH = Glyceraldehyde-3-Phosphate Dehydrogenase; GFAP = Glial Fibrillary Acidic Protein; GluR(1-4) = α-amino-3-hydroxy-5-methyl-4-isoxazolepropionate-type ionotropic Glutamate Receptor; GnRH = Gonadotropin-Releasing Hormone; hnRNP = heterogeneous nuclear Ribonucleoprotein; IFNγ = Interferon γ; IFNγ-R = Interferon γ Receptor; IRK = Inwardly Rectifying Potassium; KA = Kainate-type ionotropic Glutamate Receptor; Kir = K^+ channel Inward Rectifier; mAChR = muscarinic Acetylcholine Receptor; MAP = Microtubule Associated Protein; mGluR = metabotropic Glutamate Receptor; MHC = Major Histocompatibility Complex; nAChR = nicotinic Acetylcholine Receptor; NAT = Noradrenaline Transporter; NCAM = Neural Cell Adhesion Molecule; NOS = Nitric Oxide Synthase; NPY = Neuropeptide Y; NR2 = *N*-methyl-D-aspartate-type ionotropic Glutamate Receptor 2; NSE = Neurone-Specific Enolase; OT = Oxytocin; PKC = Protein Kinase C; PLA_2 = Phospholipase A_2; PNMT = Phenylethanolamine-*N*-methyltransferase; PV = Parvalbumin; RBI = Rat Brain I; Scn8a = Sodium Channel α subunit; SERCA = Sarco(endo)plasmic reticulum Ca^{2+} ATPase; SP = Substance P; SS = Somatostatin; SUR = Sulphonylurea Receptor; TH = Tyrosine Hydroxylase; TNFα-R = Tumor Necrosis Factor α Receptor; VIP = Vasoactive Intestinal Peptide; VP = Vasopressin.

REFERENCES

1. Angulo et al. 1997. J. Neurosci. *17*:6685.
2. Audinat et al. 1996. J. Physiol. (Paris) *90*:331.
3. Baba-Aissa et al. 1996. Brain Res. Mol. Brain Res. *41*:169.
4. Baranauskas et al. 1999. J. Neurosci. *19*:6394.
5. Berger et al. 1998. J. Neurosci. *18*:2437.
6. Brocke et al. 1999. J. Biol. Chem. *274*:22713.
7. Cardenas et al. 1999. J. Physiol. *518*:507.
8. Cauli et al. 1997. J. Neurosci. *17*:3894.
9. Ceranik et al. 1997. J. Neurosci. *17*:5380.
10. Chen et al. 1998. Neuroscience *83*:749.
11. Chiang et al. 1994. Mol. Brain Res. *27*:183.
12. Comer et al. 1998. Brain Res. Mol. Brain Res. *62*:65.
13. Crepel et al. 1994. Neuropharmacology *33*:1399.
14. Das et al. 1998. Nature *393*:377.
15. Flint et al. 1997. J. Neurosci. *17*:2469.
16. Fusco et al. 1999. J. Neurosci. *19*:1189.
17. Futami et al. 1998. Brain Res. Mol. Brain Res. *54*:183.
18. Gahring et al. 1996. Neuroimmunomodulation *3*:289.
19. Garaschuk et al. 1996. J. Physiol. *491*:757.
20. Geiger et al. 1995. Neuron *15*:193.

21. Glasgow et al. 1999. Endocrinology *140*:5391.
22. Grigorenko and Yeh. 1994. Vis. Neurosci. *11*:379.
23. Guyon et al. 1999. J. Physiol. *516*:719.
24. Gustincich et al. 1999. J. Neurosci. *19*:7812.
25. Hurbin et al. 1998. Endocrinology *139*:4701.
26. Jonas et al. 1994. Neuron *12*:1281.
27. Kawai et al. 1999. Brain Res. *830*:246.
28. Kelz et al. 1999. Nature *401*:272.
29. Lambolez et al. 1992. Neuron *9*:247.
30. Lambolez et al. 1996. Proc. Natl. Acad. Sci. USA *93*:1797.
31. Lipski et al. 1998. Am. J. Physiol. *274*:R1099.
32. Martina et al. 1998. J. Neurosci. *18*:8111.
33. Massengill et al. 1997. J. Neurosci. *17*:3136.
34. Mermelstein et al. 1997. Neuroreport *8*:485.
35. Mermelstein et al. 1999. J Neurosci *19*:7268.
36. Mermelstein et al. 1998. J. Neurosci. *18*:6650.
37. Neumann et al. 1997. J. Exp. Med. *185*:305.
38. Neumann et al. 1997. J. Exp. Med. *186*:2023.
39. Plant et al. 1998. J. Neurosci. *18*:9573.
40. Plant et al. 1997. J. Physiol. *499*:47.
41. Porter et al. 1999. J. Neurosci. *19*:5228.
42. Porter et al. 1998. Eur. J. Neurosci. *10*:3617.
43. Poth et al. 1997. J. Neurosci. *17*:586.
44. Ruano et al. 1995. Neuron *14*:1009.
45. Ruano et al. 1997. Eur. J. Neurosci. *9*:857.
46. Schmidt-Ott et al. 1998. J. Cardiovasc. Pharmacol. *31*:S364.
47. Seeburg et al. 1995. Recent Prog. Horm. Res. *50*:19.
48. Shen et al. 1998. Mol. Gen. Med. *63*:96.
49. Skynner et al. 1999. Endocrinology *140*:5195.
50. Song et al. 1996. J. Neurophysiol. *76*:2290.
51. Song et al. 1998. J. Neurosci. *18*:3124.
52. Stefani et al. 1998. Dev. Neurosci. *20*:242.
53. Stewart et al. 1999. J. Neurophysiol. *81*:72.
54. Surmeier et al. 1996. J. Neurosci. *16*:6579.
55. Tehrani et al. 2000. Invest. Ophthalmol. Vis. Sci. *41*:314-319.
56. Tempia et al. 1996. J. Neurosci. *16*:456.
57. Tkatch et al. 1998. Neuroreport *9*:1935.
58. Tkatch et al. 2000. J. Neurosci. *20*:579.
59. Tsuchiya et al. 1999. Neuroreport *10*:2749.
60. Vega-Saenz de Miera et al. 1997. Proc. Natl. Acad. Sci. USA *94*:7059.
61. Vysokanov et al. 1998. Neurosci. Lett. *258*:179.
62. Wang and Wu. 1996. Brain Res. Mol. Brain Res. *36*:29.
63. Wang and Grabowski. 1996. RNA *2*:1241.
64. Wu and Wang. 1996. J. Neurochem. *66*:1060.
65. Yan et al. 1996. J. Neurosci. *16*:2592.
66. Yan et al. 1997. J. Neurophysiol. *77*:1003. Y
67. Yan et al. 1997. Neuron *19*:1115.

68. Yeh et al. 1996. Vis. Neurosci. *13*:283.
69. Ying et al. 1997. J. Neurosci. *17*:9536.
70. Zawar et al. 1999. J. Physiol. *514*:327.
71. Zhang et al. 1995. Neuroscience *67*:177.
72. Zhu et al. 1998. J. Neurosci. *18*:679.
73. Zhu et al. 1999. J. Neurophysiol. *82*:1560.

8 Immunocytochemistry and In Situ Hybridization: Their Combinations for Cytofunctional Approaches of Central and Peripheral Neurons

Marc Landry[1,2] and André Calas[3]
[1]*Laboratoire de Biologie de la Différenciation et du Développement, Université Victor Ségalen and* [2]*INSERM EPI 9914, Institut François Magendie, Bordeaux;* [3]*Laboratoire de Cytologie, Institut des Neurosciences, Université Pierre et Marie Curie, Paris, France, EU*

OVERVIEW

The birth of new concepts in chemical neurotransmission, like cotransmission, volume transmission, or neuronal versatility, frequently occurred thanks to the advances in neurocytochemichal technologies. First, the identification and in situ localization of neurotransmitters, related enzymes, and receptors greatly benefited from the development of more and more specific and sensitive cytochemical techniques. For instance, after histochemistry of acetylcholinesterase, the chemical neuroanatomy of monoamine neurons originated from histofluorescence methods. It is the development, however, of immunocytochemistry (ICC) at optic and electron microscope levels which has given an universal and versatile tool for such cartographies, first of neuropeptides, then of enzymes of neurotransmitter metabolism, and finally of the neurotransmitters themselves after the pioneering work of Steinbusch and colleagues (36) for serotonin. However these mappings, completed by that of receptors through ligand binding or ICC, does not allow, for example, the assumption that identified protein molecules are truly synthetized within sites where they are detected.

The in situ hybridization (ISH) of mRNAs, to which our purpose will be restricted (excluding ISH on chromosomes), leads to the detection and cellular localization of another step of the protein and peptide synthesis.

Thus, the comparison of results obtained by both techniques, ICC and ISH, is able to bring invaluable data about the cellular expression of a given peptide or protein, inter- or intracellular relationship between different molecules, and also functional activity of chemically identified neurons.

This is exemplarily the case for the combination of both kinds of labelings on the same material. Moreover, the results of combining these different techniques allow their mutual validation, since their respective artifacts are different.

The aim of the present review is to describe and comment on some protocols combining different ISH procedures, or ISH and ICC, and to develop for each of them their abilities and limitations as well as their possible contributions to cellular neurobiology.

BACKGROUND

Considered individually, each labeling technique, ICC or ISH, constitutes a long chain of histochemical steps whose quality of the final result strongly depends on the weakest link. This is even more true for their combination, which raises not only additive difficulties but also specific problems due to the coupling itself; the readibility of results and the compatibility of histochemical treatments and namely their possible interferences in giving rise to excess or deficiency errors. The use of different consecutive sections (12), to perform ICC and ISH reactions separately, removes their possible interferences and only leaves open the compatibility of histochemical treatments prior to ISH and ICC themselves, like fixation and embedding. In order to compare labelings at the cellular level, mirror sections have been used (41), and the signals observed are generally of the same kind.

However, although being more difficult, the combination of labelings on the same section appears to be both more elegant and rich in information. Compatibility and possible interactions between reagents involved in the labeling techniques will be exposed and discussed further. Here, we wish only to give a rapid overview of the different markers currently used for ICC and ISH detections. A dark precipitate arises from the development of the activity of an enzyme coupled to an antibody. The most commonly used enzymes are alkaline phosphatase and peroxidase, but others can be also considered. Fluorescent dyes give sharp signals and are useful for confocal microscope analysis. Alternatively, antigens as well as reporter molecules may be detected by silver-enhanced ultra small gold particle-conjugated antibodies. Autoradiographic procedures are mostly restricted to ISH studies, although radioactive ICC detections have also been reported. Conversely, postembedding immunogold protocols are mainly carried out for antigen detection by ICC at the electron microscope level (Chapter 10), and only a few ISH studies have been reported, most of them related to viral messengers (see Reference 10).

Here, we shall focus on some protocols that are either widely used or specifically designed for more restricted purposes thanks to the emergence of new and potentially powerful detection systems. Taken together, these protocols cover a wide range of possible applications and allow the consideration of various difficulties of the multiple labeling methods.

PROTOCOLS

General Comments

Some standard protocols for multiple labelings are proposed below. We did not describe protocols for labeling probes used in ISH, since our purpose was restricted to the combination of techniques. Moreover, the need for multiple labeling influences the choice of the reporter molecule but does not modify the way to label probes. The composition of the solutions is not modified by the coupling, and common hybridization solutions or incubation buffers can be used. Of course, the various

8. Immunocytochemistry and In Situ Hybridization

treatments must avoid degradation of mRNAs from tissues or antigens prior to their detection (also discussed in section entitled Radioactive and Enzymatic Double ISH).

Materials and Reagents

The following materials and reagents are used for any of the ISH protocols presented in this review:

- Tissue sections or cell cultures.

Labeling

- Probes:
  Oligoprobes: 25 to 30 pmol are labeled with radioactive dNTP; 100 pmol are labeled with nonradioactive (digoxigenin, biotin, fluorescein) dNTP.
  Riboprobes: 1 µg template is transcribed.
- Terminal transferase or RNA polymerase (Roche Molecular Biochemicals, Mannheim, Germany)
- Labeling buffer (Roche Molecular Biochemicals).
- Probe purification: ethanol, LiCl, EDTA, or DNase I (Sigma, St. Louis, MO, USA).

Hybridization

- Bovine serum albumin (BSA; Sigma).
- Ficoll® (Sigma).
- Polyvinylpyrrolidone (PVP; Sigma).
- N-lauroylsarcosine (Sigma).
- Dextran sulfate (Sigma).
- Denatured salmon testis DNA (Sigma).
- Dithiothreitol (DTT; Sigma).
- Deionized formamide (Sigma).
- Yeast RNA (Sigma).
- Sodium dodecyl sulfate (SDS; Sigma).
- Formamide (Sigma).

Hybridization Buffers

Oligoprobes: 50% deionized formamide, 4× standard saline citrate (SSC) (1× SSC = 0.15 M NaCl, 0.015 M NaCitrate), 1× Denhardt's solution (0.02% BSA, 0.02% Ficoll, 0.02% PVP), 0.02 M NaPO$_4$ (pH 7.0), 1% N-lauroylsarcosine, 10% dextran sulphate, 500 mg/L denatured salmon testis DNA, 20 mM DTT if using radioactively labeled probes.

Riboprobes: 50% deionized formamide, 5× SSC, yeast RNA (50 µg/mL), 1% SDS, 20 mM DTT if using radioactively labeled probes.

Probes

Oligoprobes: radioactively labeled oligoprobes are incubated at 0.5 nM; biotin- or digoxigenin-labeled oligoprobes are used at 0.5 (lowly expressed mRNAs) to 10 (highly expressed mRNAs) nM.

Riboprobes: 10% of the labeled probe is diluted in 500 µL of hybridization solution.

Posthybridization Washings

Oligoprobes: 4 times for 15 minutes at 55°C (SSC 1×), then 30 minutes at room temperature.

Riboprobes: 2 times for 30 minutes at 70°C (50% formamide; 5× SSC, 1% SDS), 2 times for 30 minutes at 65°C (50% formamide, 2× SSC).

Other reagents, specific of a given technique, are listed below, together with the corresponding protocol.

Protocol for Radioactive and Enzymatic Double ISH (Protocol #1)

General Comments

This protocol is well suited for cryostat sections (Figure 1a) or cell cultures (Figure

1b) and can be performed with oligo- or riboprobes on unfixed (5,8,24,45) or fixed (23,25,37,46) tissue. No pretreatments are required on cryostat sections, but cells may need to be permeabilized. Preparations must be finally mounted without dehydration to avoid fading of the nitro blue tetrazolium/5-bromo-4-chloro-3-indolyl phosphate (NBT/BCIP) precipitate.

Materials and Reagents

- ^{35}S- and digoxigenin-labeled nucleotides.
- Tris buffers: buffer A: 0.1 M Tris, pH 7.5, 1 M NaCl, 2 mM $MgCl_2$; buffer B: 0.1 M Tris, pH 9.5, 0.1 M NaCl, 5 mM $MgCl_2$.
- Alkaline phosphatase-conjugated antidigoxigenin antibody (Roche Molecular Biochemicals).
- BSA.
- Alkaline phosphatase substrate (NBT and BCIP) (Roche Molecular Biochemicals).
- Photographic emulsion (Amersham Pharmacia Biotech, Piscataway, NJ, USA).
- Developer and fixer.

Procedure

See flow chart in Protocol 1.

1. Unfixed or fixed (generally with 4% paraformaldehyde) samples are frozen on dry ice or isopentane cooled at -30°C.
2. Sections are cut with a cryostat and thaw-mounted on slides.
3. Hybridization is carried out with all dioxigenin-labeled and radioactive probes together.
4. Perform posthybridization washings according to the type of probes used (see above).
5. Preincubate slides with buffer A containing 0.5% to 1% BSA (buffer A-BSA) for 30 minutes at room temperature.
6. Incubate with alkaline phosphatase-conjugated antidigoxigenin antibody diluted 1:5000 in buffer A-BSA overnight at 4°C.
7. Rinse in buffers A and B.
8. Incubate with the NBT/BCIP alkaline phosphatase substrate diluted in buffer B at room temperature in the dark.
9. Process sections for autoradiography.
10. Mount and observe.

Protocol for Double Fluorescent ISH (Protocol #2)

General Comments

Although scarcely used, the double fluorescent ISH provides quick and easily readable results that can be analyzed with a confocal microscope (Figure 2, b,d, and f). The protocol described below takes advantadge of the high specificity and sensitivity of the 2-hydroxy-3-naphtoic acid-2′-phenylanilide phosphate (HNPP) detection set (Figure 2, a and c, and Figure 3a) (16,18). The precipitation of the dephosphorylated form of HNPP is ensured by Fast Red TR (Roche Molecular Biochemicals), and the combined HNPP/Fast Red TR is a highly fluorescent product. The accumulation of the precipitate, and hence the signal, can be triggered by repeated addition of fresh substrate solution.

Materials and Reagents

- Biotin- and digoxigenin-labeled nucleotides.
- Tris buffers: buffer A: 0.1 M Tris, pH 7.5, 1 M NaCl, 2 mM $MgCl_2$; buffer C: 100 mM Tris-HCl, 150 mM NaCl, 10 mM $MgCl_2$, pH 8.0.

8. Immunocytochemistry and In Situ Hybridization

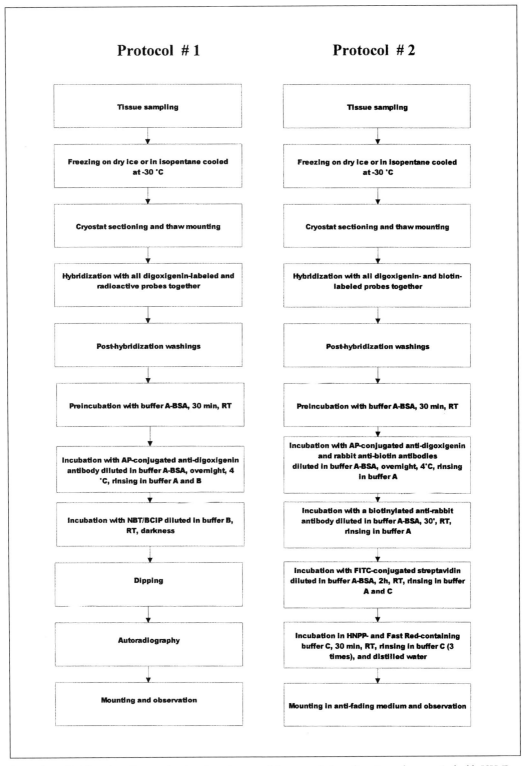

Protocol 1 and 2. Flow chart diagram showing the main steps of the protocols for radioactive and enzymatic double ISH (Protocol #1) and double fluorescent ISH (Protocol #2).

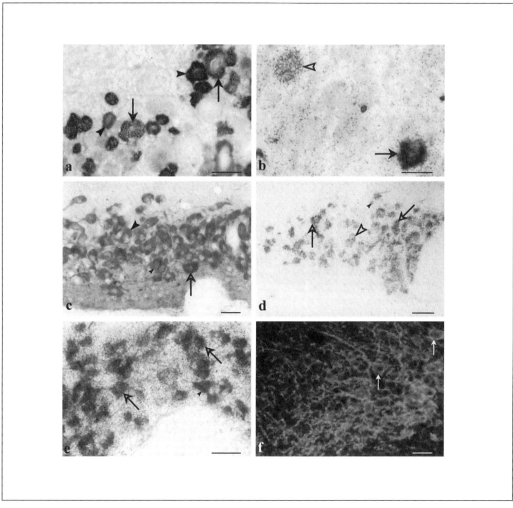

Figure 1. Double labelings on sections of peripheral (a) and central (c–f) rat nervous system and on primary cell cultures (b). (a) Dorsal root ganglia sections have been hybridized with digoxigenin- and radioactively labeled probes complementary to calcitonin gene related peptide (CGRP) (purple precipitate) and galanin receptor 1 (silver grains) mRNA, respectively (Protocol #1). Double labeled (arrows) and neurons containing CGRP only are visualized in normal rats. (b) Primary culture of adult dorsal root ganglia double stained for neuropeptide tyrosine (enzymatic precipitate) and its Y1 receptor (radioactive signal) by using the same protocol. Large neurons are double labeled (arrow), whereas others display only the radioactive signal (open arrowhead). Thus, the Y1 receptor can be considered both a pre- and postsynaptic receptor. (c) Double detection of heterologous gene products: double labeling combining two enzymatic detections (modified from Protocol #5). Galanin peptide is visualized by immunoperoxidase, whereas the development of alkaline phosphatase activity leads to the detection of oxytocin mRNA in the supraoptic nucleus of a lactating female rat hypothalamus. Upon such a stimulation, some neurons appear doubly labeled beside two distinct populations of cells singly labeled either for galanin (small arrowhead) or oxytocin mRNA (large arrowhead). (d) Double detection of heterologous gene products: double labeling combining radioactive ISH detection of vasopressin mRNA to immunoperoxidase for tyrosine hydroxylase (TH) in hypothalamic supraoptic nucleus of a dehydrated rat (modified from Protocol #5). Most neurons are double labeled (open arrows), although some cells remain single labeled either for TH (small arrowhead) or vasopressin mRNA (open arrowhead). Note that double labeled cells are easier to distinguish than when using a double enzymatic detection. (e) Double detection of homologous gene products: double labeling combining radioactive ISH detection of vasopressin mRNA to immunoperoxidase for vasopressin peptide in the rat hypothalamic supraoptic nucleus (modified from Protocol #5). Most neurons are double labeled (open arrows). However, rare cells remain single labeled for vasopressin and can be considered as resting neurons storing peptide (small arrowhead). (f) Double fluorescent immunolabeling using the CARD method to detect neuropeptide tyrosine Y1 receptor containing neurons (green) and neuropeptide tyrosine containing afferent fibers (red) in the rat arcuate nucleus (see Protocol #6). Yellow color corresponds to the superimposition of both fluorescences and indicates double labelings, i.e., contacts between fibers and neurons (arrows). The neuropeptide tyrosine containing fibers establish contact on soma and processes of Y1 receptor expressing neurons. Scale bars: 25 µm. (See color plate A5.)

8. Immunocytochemistry and In Situ Hybridization

- Alkaline phosphatase-conjugated antidigoxigenin antibody.
- BSA.
- Rabbit antibiotin antibody (Roche Molecular Biochemicals.
- Appropriate biotin–streptavidin fluorescent reagents e.g., streptavidin fluorescein isothiocyanate (FITC; Vector, Burlingame, CA, USA).
- Alkaline phosphatase fluorescent substrate (HNPP) (Roche Molecular Biochemicals).

Procedure

See flow chart in Protocol 2.

1. Unfixed or fixed (generally with 4% paraformaldehyde) samples are frozen on dry ice or isopentane cooled at -30°C.
2. Sections are cut with a cryostat and thaw-mounted on slides.
3. Hybridization is carried out with all dioxigenin- and biotin-labeled probes together.
4. Perform posthybridization washings according to the type of probes used (see above).
5. Preincubate slides with buffer A containing 0.5% to 1% BSA (buffer A-BSA) for 30 minutes at room temperature.
6. Incubate with alkaline phosphatase-conjugated antidigoxigenin antibody diluted 1:5000 in buffer A-BSA overnight at 4°C.
7. Incubate with a biotinylated antirabbit antibody diluted in buffer A-BSA for 30 minutes at room temperature, then rinse in buffer A.
8. Incubate with FITC-conjugated streptavidin diluted in buffer A-BSA for 2 hours at room temperature, then rinse in buffers A and C.
9. Incubate in HNPP/Fast Red containing buffer C for 30 minutes at room temperature, then rinse in buffer C (3 times).
10. Rinse in distilled water.
11. Mount in antifading medium and observe.

Protocol for Double ISH at the Electron Microscope Level (Protocol #3)

General Comments

The major problems are, first, to permeabilize the tissue while preserving a good morphology, and second, to maintain, as far as possible, the sensitivity of the ISH procedures despite treatments required for the observation at the electron microscope level (Figure 4b). This method is likely to be applicable not only to vibratome sections but also to slices and cells.

Material and Reagents

- Phosphate buffer (PB), pH 7.4, 0.1 M and phosphate-buffered saline (PBS), pH 7.4, 0.1 M.
- 4% Paraformaldehyde in PBS.
- Biotin- and digoxigenin-labeled nucleotides.
- BSA.
- Rabbit antibiotin antibody.
- Appropriate biotin–streptavidin reagents.
- Gelatin (Sigma).
- Normal goat serum (NGS; Sigma).
- Gold-conjugated antidigoxigenin antibody (British BioCell International, Cardiff, Wales, UK).
- 3,3′-diaminobenzidine (DAB; Sigma).
- Silver enhancement kit (Amersham Pharmacia Biotech).
- Osmium tetroxyde (Electron Microscopy Sciences, Fort Washington, PA, USA).

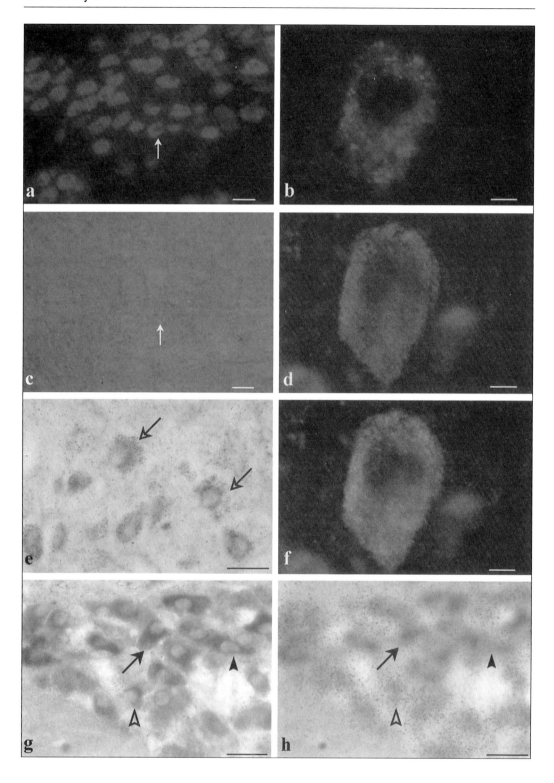

- Epon (Electron Microscopy Sciences).
- Lead citrate (Electron Microscopy Sciences).

Procedure

See flow chart in Protocol #3. Further details on electron microscopy (EM) preparation protocols can be found in Chapter 10.

1. Perfuse the animal with an appropriate fixative (usually 4% paraformaldehyde).
2. Sample areas of interest and postfix in the same fixative.
3. Cut 50 to 70 μm Vibratome® sections, and subject them to permeabilization pretreatments.
4. Hybridize with all digoxigenin- and biotin-labeled probes together.
5. Perform posthybridization washings according to the type of probes used (see above).
6. Preincubate sections with PBS containing 0.5% to 1% BSA (PBS-BSA) for 30 minutes at room temperature.
7. Incubate with an antibiotin antibody diluted in PBS-BSA overnight at 4°C, then rinse in PBS.
8. Preincubate in blocking solution (PBS-BSA, gelatin, and NGS) for 2 hours at room temperature.
9. Incubate with 1 nm gold-conjugated antidigoxigenin antibody diluted 1:50 in the blocking solution for 30 minutes at room temperature, then rinse in PBS.
10. Postfix in 4% paraformadehyde in PBS for 10 minutes at room temperature, then rinse in PBS.
11. Incubate with a biotinylated-conjugated antirabbit antibody diluted in PBS-BSA for 30 minutes at room temperature, then rinse in PBS.
12. Incubate with a peroxidase-conjugated avidin–biotin complex diluted in PBS-BSA for 2 hours at room temperature, then rinse in PBS and Tris buffer.
13. Develop peroxidase reaction, then rinse in Tris buffer and water.
14. Perform silver enhancement, then rinse in water and PB.
15. Postfix in OsO_4 1% in PB for 10 minutes at room temperature.
16. Embed.
17. Cut ultrathin sections and counterstain them with lead citrate before EM observation.

Figure 2. Multiple labelings on sections of the rat peripheral (a and c) and central (b and d–h) nervous system. (a) The HNPP detection set gives a red fluorescence with exciter at 568 nm line leading to identify galanin mRNA containing neurons (arrow) in the axotomized superior cervical ganglia. (c) No green fluorescence is detected on the same section with exciter at 488 nm line. Such a specificity allows the combination of fluorescent ISH to galanin mRNA visualized by the HNPP detection set (b) with a fluorescein isothiocyanate (FITC)-conjugated–streptavidin visualization of biotin-labeled probes complementary to vasopressin mRNA (d) in hypothalamic magnocellular neurons of a salt-loaded rat supraoptic nucleus. The analysis at the confocal microscope and the merge image (f) lead to identify specific subcellular compartments involved in the synthesis of either galanin (perinuclear localization, red fluorescence) or vasopressin (green fluorescence, peripheral localization). The coupling between ICC with an immunoperoxidase protocol and ISH for vasopressin mRNA (e) or oxytocin mRNA (g) can be performed by using a silver-enhanced gold particle method (e) (Protocol #4) or a double enzymatic detection (g) (Protocol #5) identical to that of Figure 1c. Double labeled cells (arrows) are easily visualized with the former protocol. The latter needs to calibrate the development of the enzyme activity to be able to distinguish both colors. Triple labeling is shown in panels g and h (Protocol #5) and combine ICC for galanin (brown precipitate) to double radioactive and enzymatic ISH for vasopressin and oxytocin mRNAs, respectively. Both micrographs correspond to the same field, but different focuses are necessary to identify the three markers that are not in the same plane on the preparation. Some neurons are triple labeled (arrow). Besides, neurons containing oxytocin only (filled arrowhead), cells double labeled for vasopressin mRNA and galanin (open arrowhead) are seen. Scale bars: 50 μm (panels a, c, e, h, and g); 5 μm (panels b, d, and f). (See color plate A6.)

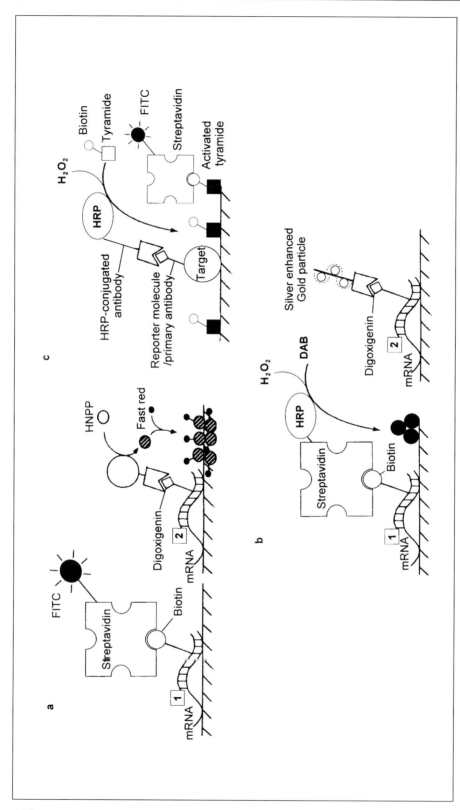

Figure 3. Schematic representation of double nonradioactive ISH protocols (a and b) and tyramide detection (c). (a) Two different mRNAs (1 and 2) are visualized by a double fluorescent ISH method (see Protocol #2) using HNPP combined to Fast red and FITC as reporter molecules. (b) Two different mRNAs (1 and 2) are visualized by a double ISH method compatible with electron microscope analysis (see Protocol #3) using peroxidase and silver-enhanced gold particles as reporter molecules. (c) Principle of the detection of a nonradioactive probe or a tissue antigen by the CARD method using FITC as a reporter molecule. Abbreviations: DAB, 3,3′-diaminobenzidine; FITC, fluorescein isothiocyanate; HNPP, 2-hydroxy-3-naphtoic acid-2′-phenylanilide phosphate; HRP, horseradish peroxidase.

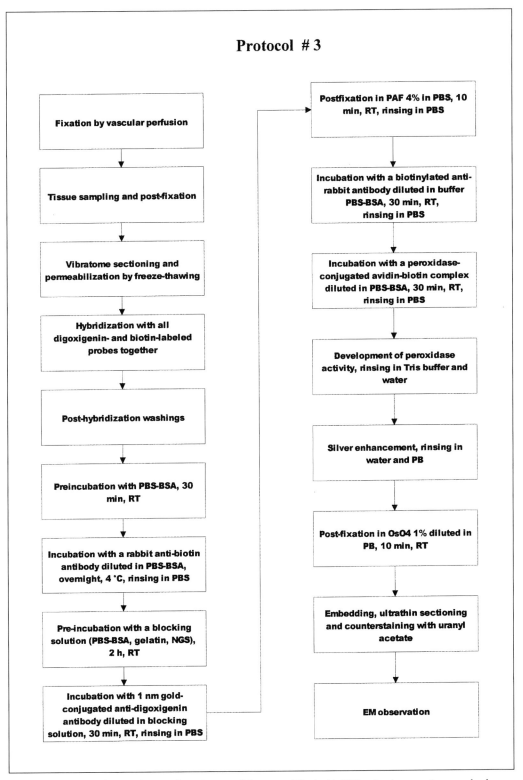

Protocol 3. Flow chart diagram showing the main steps of the protocol for double ISH at the electron microscope level.

Protocol for Double Labeling Combining Gold Particles ISH and Enzymatic ICC at the Electron Microscopic Level (Protocol #4)

General Comments

Beside problems similar to those of Protocol #3, the fixation must be suitable for both ICC and ISH, and antigenicity must remain unaltered. The reporter molecules are identical to those of the previous protocol, and the observation may be carried out either at the light (Figure 2e) or electron microscopic (Figure 4a) level.

Material and Reagents

- PB, pH 7.4, 0.1 M, and PB, pH 7.4, 0.1 M.
- 4% Paraformaldehyde in PB.
- Digoxigenin-labeled nucleotides.
- BSA.
- Appropriate primary and secondary antibodies.
- Appropriate biotin–streptavidin reagents.
- Gelatin.
- NGS.
- Gold-conjugated antidigoxigenin antibody.
- DAB.
- Silver enhancement kit.
- Osmium tetroxyde.
- Epon.
- Lead citrate.

Procedure

See flow chart in Protocol #4. Further details on EM preparation protocols can be found in Chapter 10.

1–6. See Protocol #3.

7. Incubate with a rabbit primary antibody diluted in PBS-BSA overnight at 4°C, then rinse in PBS.

8-17. See Protocol #3.

Protocol for Triple Labeling Combining Enzymatic ICC and Radioactive and Enzymatic Double ISH (Protocol #5)

General Comments

Such a multiple labeling needs to take care of possible interferences between the various procedures. It can be applied to cryostat sections and cells. The sensitivity of the different techniques is likely to be decreased by the multiplication of steps. The use of floating instead of thaw-mounted sections is likely to reduce this loss of sensitivity. A coating may be necessary, as in Protocol #1, but in the given example (Figure 2, g and h), the high amount of mRNA allows to shorten the incubation time in NBT/BCIP, which results in an absence of chemography and maintains the high specificity of each labeling. A double labeling protocol combining radioactive ISH and enzymatic ICC can be derived from this method by omitting the enzymatic ISH (Figure 1, d and e) (21,22,35,41).

Materials and Reagents

- PB, pH 7.4, 0.1 M, and PBS, pH 7.4, 0.1 M.
- 4% paraformaldehyde in PBS.
- ^{35}S- and digoxigenin-labeled nucleotides.
- BSA.
- Appropriate primary and secondary antibodies.
- Appropriate biotin–streptavidin reagents. Tris buffers: buffer A: 0.1 M Tris, pH 7.5, 1 M NaCl, 2 mM $MgCl_2$; buffer B: 0.1 M Tris, pH 9.5, 0.1 M NaCl, 5 mM $MgCl_2$.
- Alkaline phosphatase-conjugated antidigoxigenin antibody.
- DAB.
- Tris buffer, 0.05 M, pH 7.6.
- Alkaline phosphatase substrate (NBT/BCIP).

8. Immunocytochemistry and In Situ Hybridization

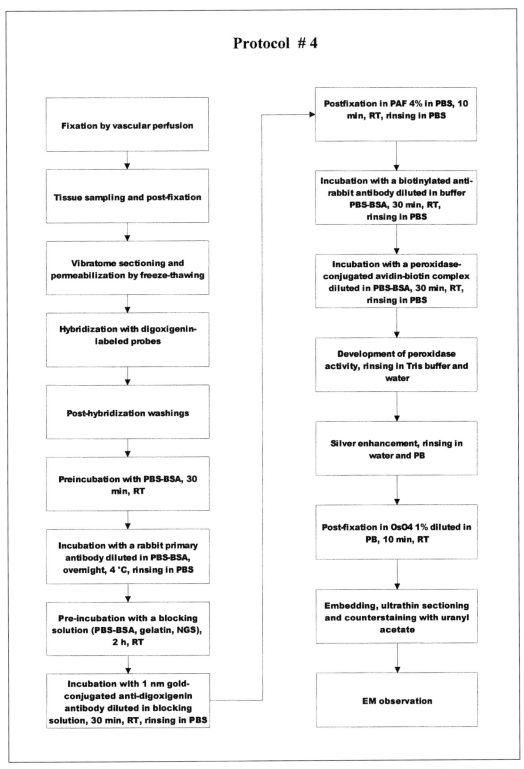

Protocol 4. Flow chart diagram showing the main steps of the protocol for double labeling combining gold particles ISH and enzymatic ICC at the electron microscopic level.

- Photographic emulsion.
- Developer and fixer.

Procedure

1. Perfuse the animal with an appropriate fixative (usually 4% paraformaldehyde), remove the areas of interest, and postfix them in the same fixative.
2. Cut cryostat sections and thaw-mount them on slides.
3. Carry out hybridization with all dioxigenin- and biotin-labeled probes together.
4. Perform posthybridization washings according to the type of probes used (see above).
5. Preincubate sections with buffer A containing 0.5% to 1% BSA (buffer A-BSA) for 30 minutes at room temperature.
6. Incubate with a mixture of primary and an antidigoxigenin antibodies diluted in buffer A-BSA overnight at 4°C, then rinse in buffer A.
7. Incubate with a biotinylated-conjugated antirabbit antibody diluted in buffer A-BSA for 30 minutes at room temperature, then rinse in buffer A.
8. Incubate with a peroxidase-conjugated avidin–biotin complex diluted in PBS-BSA for 1 hour at room temperature, then rinse in buffer A and Tris buffer.
9. Develop the peroxidase reaction, then rinse in Tris buffer and buffer A and B.
10. Incubate with NBT/BCIP diluted in buffer B at room temperature in the dark.
11. Perform autoradiography.
12. Mount and observe.

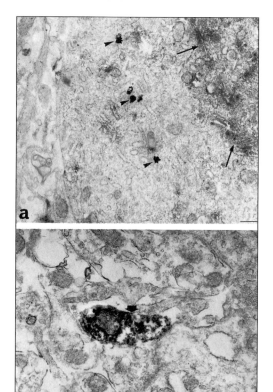

Figure 4. Double (a) (Protocol #4) and single (b) (Protocol #6) detection at the electron microscopic level. (a) Vasopressin mRNA is detected with digoxigenin-labeled probes visualized by the silver-enhanced gold particle method (arrowheads). Galanin peptide is identified by immunoperoxidase (arrows). Both labelings are visualized in the same magnocellular neuron of the rat supraoptic nucleus but they occupy different subcellular domains. Galanin is stored in a perinuclear area, whereas vasopressin mRNA is prominent in the Nissl bodies at the periphery of the cell body. (b) An undilated section of an axon (short arrow) arising from hypothalamic magnocellular neurons is labeled within the internal layer of the median eminence of a salt-loaded rat. Scale bar: 0.5 µm.

Protocol for Tyramide Amplification (Protocol #6)

General Comments

This detection system is based on the reaction of peroxidase with hydrogen peroxide and the phenolic part of tyramide, which produces a quinone-like structure bearing a radical on the C_2 group. This activated tyramide then covalently binds to tyrosine in close vicinity to the peroxidase (3,43,44). Such a reaction can be applied either to ICC (Figure 1f) or ISH (Figure 4b) detection with a similar protocol after

8. Immunocytochemistry and In Situ Hybridization

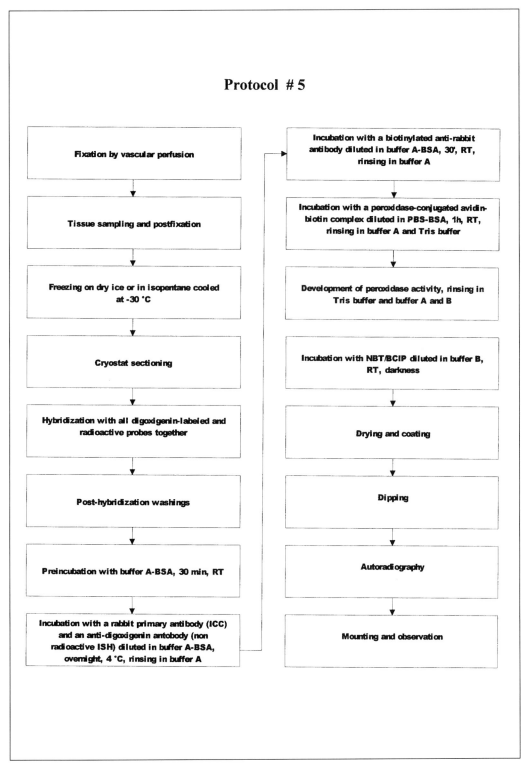

Protocol 5. Flow chart diagram showing the main steps of the protocol for triple labeling combining ICC and a radioactive and enzymatic double ISH.

peroxidase is bound to the target via an antibody. The tyramide can be conjugated to a fluorochrome (Figure 1f) or a hapten (Figure 4b). Hence, direct or indirect (see below and Figure 3c) detection of enzymatically deposited tyramides is possible and can be coupled to another labeling (Figure 1f) (5,14).

Material and Reagents

- Digoxigenin-labeled nucleotides.
- Appropriate primary and secondary antibodies.
- Tyramide amplification kit.
- Appropriate biotin–streptavidin reagents.
- Tris buffer: 0.1 M Tris-HCl, pH 7.5, 0.15 M NaCl.
- DAB.
- Tris buffer, 0.05 M, pH 7.6.
- TNB: Tris-NaCl blocking buffer: 0.1 M Tris-HCl, pH 7.5, 0.15 M NaCl, 0.05% Tween® 20.
- TNT: Tris-NaCl Tween buffer: 0.1 M Tris-HCl, pH 7.5, 0.15 M NaCl, 0.5% DuPont Blocking reagent.

Procedure

See flow chart in Protocol #6.

1. Perform hybridization with a digoxigenin-labeled probe and/or incubation with the primary antibody of the ICC detection.
2. Preincubate in TNB for 30 minutes at room temperature.
3. Incubate with a peroxidase-conjugated secondary or an antidigoxigenin antibody diluted in TNB for 30 minutes at room temperature, then rinse in TNT.
4. Incubate with biotinylated tyramide diluted in its proper amplification diluent for 10 minutes at room temperature, then rinse in TNT.
5. Incubate with peroxidase-conjugated streptavidin diluted in TNB for 2 hours at room temperature, then rinse in TNT and Tris buffer.
6. Develop the peroxidase reaction, then rinse in Tris buffer.
7. Continue the steps corresponding to the coupled reaction.

The biotinylated tyramide can be replaced by a fluorochrome-conjugated tyra-

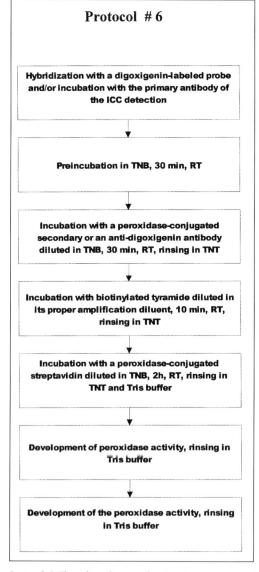

Protocol 6. Flow chart diagram showing the main steps of the protocol for tyramide amplification.

RESULTS AND DISCUSSION

General Comments

Preparation of the material

Fixation

Fixation is the first and a crucial step for ISH and ICC, since it must allow both a good morphological preservation and a good efficiency of the histochemical reactions, two criteria often being contradictory. The fixation needed for ICC and ISH is often quite different, and these requirements make it difficult, in some cases, to combine the two techniques on the same preparation. Two kinds of fixative can be considered: chemical and physical fixatives.

Among the former, precipitating fixatives give a nice morphology, but are generally not well suited for RNA detection (see Reference 10). However, the Clarke and Carnoy fixatives are widely used for cytogenetic studies, and a modified Clarke solution proved to be a good alternative for ISH on amphibian oocytes (28). Non-precipitating aldehydic fixatives have been introduced in ISH studies in the 1980s, and 4% paraformaldehyde is now the most common fixative for both ICC and ISH and their combination (31). The fixation by the formaldehyde or its derivatives is largely reversible. Thus, the chemical properties of nucleic acids are not modified by the fixation and hybrids. The melting temperature (T_m) remains unchanged upon aldehyde fixation (6). Despite these favorable properties, it is preferable to use slight concentrations of formaldehyde on nervous tissue (12) to allow a better accessibility of the probes to the target. The glutaraldehyde is known to be of a poor interest in ISH studies since it creates a dense network of double bonds making the probe penetration difficult (see Reference 10). Finally, the use of picric acid should be banned since it modifies chemical properties of the probe and can alter it.

Although best suited for ICC, the fixation by perfusion often maintains, in the brain, some molecules that can react with the hybridization solution and induce a high background (Reference 34 and unpublished observations). Thus, immersion fixation or fixation on slides after sectioning, are usually the most appropriate ways to prepare tissues. However, perfusion is sometimes necessary for studies at the electron microscope level, and some mRNA species can hardly be detected at the ultrastructural level in such conditions. A possible requirement for a specific type of fixation, either for protein or for mRNA detection, is in fact the most critical restriction to the coupling between ICC and ISH.

Physical fixations have also been used and proved to be of great interest, at least for double ISH. Thus, drying is a very easy way to fix nervous tissues, and air-dried cryostat sections (8) or cell cultures (17) are more sensitive to mRNA detection than fixed material. However, without chemical fixation, ICC is unlikely to be performed and cannot be combined with ISH on the same preparation. Therefore, the use of such a protocol must be restricted to double ISH studies. Also, microwave heating has been reported to improve hybridization efficiency without altering tissue morphology (20).

Storing of Material

The nervous tissue preparations can be stored at -20°C, after or without fixation, in sealed boxes that may contain a chemical to absorb humidity. Importantly, the sections or cells must have been carefully air-dried before freezing. There is no significant decrease of the mRNA content for a period up to 1 year if slides are properly stored.

Compatibility of the Combined Reactions

A double labeling experiment needs that the two reactions are compatible in terms of fixation (see above), pretreatments, temperature of reaction, and detection. There is no major hindrance to the detection of two mRNAs. In contrast, a good preservation of the antigens (or mRNAs) in the conditions of the ISH (or ICC) reaction is a major condition for coupling ICC and ISH.

Pretreatments

No specific pretreatments are required for combining two detections, but standard pretreatments for ICC or ISH are still needed as long as they are consistent with the coupling. Preincubations aiming to avoid nonspecific labeling, that is prehybridization and/or the blocking step for ICC, are useful prior to each reaction and do not interfere with the coupled reaction. On the other hand, permeabilization steps for ISH studies, using high amounts of proteinase, are likely to denature tissue antigens. Proteinase seems to be useless on cryostat sections and can be replaced by a freeze-thaw protocol on vibratome sections or on cells.

Temperature of Reactions

The temperatures needed for the hybridization or posthybridization steps depend on the T_m of the probes and can reach 70°C for the riboprobes. Subsequently, double ISH experiments must be carried out either with oligoprobes only or with riboprobes only, but a mixture of the two types of probes appears almost impossible. Similarly, the temperatures usually range between 4° and 37°C for ICC, and it appears difficult to preserve antigenicity after performing ISH with riboprobes. On the contrary, temperatures needed for ISH with oligoprobes are compatible with ICC reactions.

The temperature used for posthybridization washings may be high (e.g., SSC 1×; 4 times for 15 minutes at 55°C followed by 30 minutes at room temperature when using oligoprobes) to efficiently eliminate possible background staining. In contrast, a lower temperature (e.g., twice in 2× SSC for 30 minutes each at room temperature; twice in 0.1× SSC for 30 minutes each at hybridization temperature when using oligoprobes) is preferable for combining with ICC or for electron microscope studies to ensure a better preservation of antigenicity or ultrastructural morphology, respectively.

Order of the Two Detections

When combining ICC and ISH, it has to be decided what reaction should be carried out first. Performing ICC prior to ISH brings up two major problems. First, the immunohistochemical reagents must remain devoid of RNase, and the detection of low amounts of mRNA make it necessary to use RNase inhibitors. Second, the detection system must not be altered or hindered by the subsequent ISH reaction and, conversely, must not prevent the accessibility of the probe to the target mRNA. In particular, nonspecific ISH labeling after immunoperoxidase reaction can be reduced either by acetylation (4), which prevents cation binding to diaminobenzidine reaction product, or by a prehybridization step (35) using a tRNA-enriched solution. If the antigen can stand up to the ISH conditions, it is then possible to invert the order of the two detections. Such a protocol makes it easier to preserve mRNAs. However, even if the hybridization step is carried out first, the development of ISH, either radioactive or nonradioactive, should be performed after ICC to avoid alteration of antigenicity (21,41).

Compatibility of the Detection Systems

First, the signals given by each technique should be readily distinguishable. Second, the detectable molecule, that is the reporter molecule introduced chemically or enzymatically, should not interfere with the hybridization reaction or the stability of the resulting hybrid. Moreover, it should remain accessible to the detection system used later on. The most common protocols of multiple labelings associate a radioactive to an enzymatic detection. Despite its lower sensitivity, the silver-enhanced gold labeling can also be used as an alternative to the radioactive reaction, since the gold particles can be clearly identified in brightfield or epi-illumination observation. If two enzymatic labelings have to be performed, the reaction products must be of sharply different colors (e.g., purple and brown) (see Protocol #5).

Recently, new protocols arose that allow multiple fluorescent detections (11; see also the section entitled New Trends). Beside classical fluorochromes, fluorescent dyes proved to be useful to detect some enzymes. As an example, the red fluorescence of the HNPP compound used in Protocol 2 is specifically detected with an exciter at the 568 nm line (Figure 2a). No green fluorescence is observed with an exciter at the 488 nm line (Figure 2c). The efficiency and specificity of this detection set expands the range of fluorescence-based techniques available for multiple labelings (see Protocol #2) (Figure 2, b, d, and f).

At the electron microscope level, an important feature is the resolution of the reporter molecules. Autoradiographic studies may require the use of ^{3}H instead of ^{35}S to sharpen the resolution of the signal. Nevertheless, beside the difficulties of such a technique, the resolution remains poor. Also, enzymatic products are known to diffuse throughout the cytoplasm. However, the resolution is fine enough to identify the positive subcellular compartments, and for ISH, enzymatic preembedding appears to be a good compromise between sensibility and resolution (39). In multiple labeling studies, an enzymatic reaction can be combined to silver-enhanced gold staining. The latter represents, to date, the marker ensuring the best resolution for pre-embedding ICC (1,2) and ISH (23,32,33,42) studies, provided that the duration and temperature of the silver enhancement procedure are well optimized.

A postfixation is usually advised after incubation with gold particle-conjugated antibody and may result in a possible masking of antigenic sites.

Sensitivity and Specificity

The sensitivity of the ISH reaction is not altered by coupling to another mRNA detection if both reactions use either oligo- (5) or riboprobes. When performing double ISH according to Protocol #1, we did not notice a decrease of the nonradioactive ISH sensitivity due to the use of DTT, even at high concentrations. In contrast, combining radioactive ISH and ICC results in a decrease of the ISH signal by 30% (unpublished data), probably because the incubations in stringent solutions are damaging to the immunocytochemical reaction.

Regarding the specificity of each combined reaction, a coating with 3% collodion in amyl acetate has proved to be necessary after double ISH and before dipping (Protocol #1) to avoid chemography inducing nonspecific labeling for both radioactive and enzymatic reactions (46). Otherwise, no artifacts are induced by the coupling if there is no cross reactivity between the detection systems.

Other Protocols

The protocols described should be considered as general protocols using oligoprobes for multiple labelings on cryostat or vibratome sections. They need few changes to be adapted to other material.

Cell cultures can be hybridized as cryostat sections after being air-dried and/or permeabilized by a freeze-thawing protocol or by using Triton® X-100 (18,19). For classical histological analysis, human samples are usually embedded in paraffin that has to be removed before an ICC or ISH reaction. Paraffin embedding provides a well preserved morphology, but the protocol of inclusion may be inconsistent with some ICC reactions, and it induces a decrease in the sensitivity of the ISH signal. Thus, the use of riboprobes appears highly recommended for such samples. As an alternative, some authors replaced paraffin by gelatin or polyethylene glycol (PEG), which allows the recovery of high sensitivity (7). Also, slices can be processed for multiple labeling after sectioning (see also the section entitled Compatibility with Other Techniques).

Riboprobes must be hybridized according to their specific protocol, but the hybrids can be visualized by the same devices as those used for oligoprobes.

Choice of the Most Suitable Protocol

The choice of the protocol mainly depends on the desired application. Also, the most suitable markers are different from one purpose to another. Thus, we will give examples of typical applications for each of the protocols previously listed.

Radioactive and Enzymatic Double ISH

This protocol has been broadly used and proved to be reliable and easy to perform either with riboprobes (37) or oligoprobes (5,24). Such a protocol allows one to locate a given mRNA synthesizing neuron among a chemically identified population. It can, for instance, give clues to the nature of a neuropeptide receptor, e.g., presynaptic or postsynaptic (5), and lead to distinguishing between the physiological roles of different receptor subtypes. Beside its high sensitivity, the use of radioactive probes gives the opportunity to quantify one signal (25,26) and, hence, to carry out functional studies. Furthermore, double ISH experiments, followed by the quantification of the radioactive signal, can demonstrate that the level of a gene expression may be regulated by some physiological or experimental stimulations only in a restricted population characterized by its neurochemical phenotype identified by the nonradioactive signal. In particular, this leads to quantitatively studying the expression of functional markers of cellular activity (transcription factors, second messengers) in a given neuronal population.

Double Fluorescent ISH

Since these detection procedures appear less sensitive, at least in the central nervous system, than the previous ones, their use is not well suited to mapping cell populations that express a given mRNA, especially if the mRNA expression is weak. In contrast, these techniques appear more appropriate for subcellular localization studies. Their major advantage lies in a high resolution of the fluorescent signals, especially when observed with a confocal microscope, thus making possible the study of intracellular localization of mRNAs. This protocol provided evidence for a differential subcellular compartmentalization of G proteins and vasopressin mRNAs within the same hypothalamic magnocellular neuron (39) or between two neuropeptides transcripts, namely galanin and vasopressin (Figure 2, b, d, and f), in the same model.

Double ISH at the Electron Microscope Level

Although double ISH at the electron

microscope level cannot be considered as a routine technique, it provides unsurpassed information on the relative distribution of two mRNA species at the subcellular level. However, appropriate controls are needed to ensure the specificity of the technique.

Double Labeling Combining ICC and ISH

This double labeling offers a wide range of applications, both at the optic and electron microscope levels depending on the chosen markers. For light microscopy, an enzymatic ICC can be combined with mRNA detection by using either the silver-enhanced gold particle protocol (see Protocol #4) or radioactive probes (see Protocol #5 modified by omitting the enzymatic ISH).

This coupling represents, first, an alternative to double ICC or double ISH to identify possible colocalizations between heterologous protein and mRNA (21,22, 41). Second, it allows functional studies when visualizing homologous protein and mRNA. It makes it possible to check the reality of the protein synthesis in messenger-containing neurons (Figure 1e). On the contrary, it may point out a possible lack of one gene expression product (12,41) (see also Figure 1e). If the peptide is not detectable, neurons can be seen as active cells, where translation products may be exported to nerve endings at a higher rate than they are synthesized. In contrast, a neuron in which a transcript is not visualized, although the corresponding peptide has been identified, may be considered as a resting cell, storing molecules that are likely to be released upon stimulation. Thus, combining ICC and ISH documents the functional heterogeneity of neuronal populations that were usually considered as a whole.

For these studies, the use of radioactively labeled probes is highly recommended because of their high sensitivity. Moreover, silver grains are easily distinguished from peroxidase precipitate (Figure 1c versus Figure 1, d and e). In some cases, it may be of importance to identify which subcellular compartment is actually accumulating the protein or the messenger (e.g., dendrites or axons). Then, ISH can be performed by using a fluoresent method or a digoxigenin-labeled probe detected by the silver-enhanced gold particle system. Whereas the sensitivity of the latter is lower than the autoradiographic procedure, its resolution is excellent (Figure 2e) and allows studies at the electron microscopic level (Figure 4a).

Triple Labeling Combining Enzymatic ICC and Radioactive and Enzymatic Double ISH

This protocol is a combination of others. One probe has to be detected with an enzymatic marker, and a good timing of the development of the two enzymatic reactions must be found. This triple labeling gives valuable information about the precise distribution of a peptide between the two populations identified upon the basis of their transcript content (13,22) (Figure 2, g and h). In summary, the main applications for which the protocols presented here can be divided are as follows:

1. Anatomical localization: multiple labelings allow one to describe the cellular or intracellular distribution of a given gene expression product in a chemically identified population.

2. Functional studies: given the nature of the target genes (e.g., receptors, second messenger molecules, inducible transcription factors) and the possible variations of their mRNA amounts, multiple labelings can document the roles of neurochemicals and their implications in functional regulations.

Controls

Usual control experiments, that aim to investigate possible excess or deficiency errors when performing ICC or ISH, are still necessary for multiple labelings. Besides these controls specific to each reaction, possible interferences must be also checked. An easy way is to compare the results of the multiple labelings to those obtained when performing each labeling separately. Another advantadge is to evaluate a possible decrease of sensitivity induced by the coupling procedure. When possible, it may be useful to invert the order of two detections to determine possible interactions between different steps. Changing the detection systems also ensures the specificity of the labelings.

Compatibility with Other Techniques

Other techniques can be coupled to ICC and/or ISH on the same preparation. Hodological tracing is one example of a neuroanatomical method that needs an in vivo injection. Such a treatment of the animal does not affect the detection of proteins and/or mRNAs.

The effects of exogenous drugs as well as physiological or pathological conditions can be analyzed by their consequences on the relative distribution and/or intensity of the combined stainings.

Also, binding studies can be combined with ICC and/or ISH. Of particular interest is the recent development of fluorescent ligands that allow studies at the cellular and subcellular levels (30) and make it possible to directly compare the synthesis rate of a given receptor with its localization and routing within living cells upon various stimulations.

New Trends

Improvements in the ability of multiple labelings to localize gene products lie mainly in amplification procedures. The recent introduction of the catalyzed reporter deposition (CARD) method in morphological studies represents a good example of new efforts aiming to increase the sensitivity of the techniques (3,19,43). The CARD method is based on the use of tyramides as described in Protocol #6. Our procedure can be applied either for ICC or ISH detection. The tyramide amplification sequence can be inserted within other protocols, and the specific reagents required for this detection (amplification diluent, blocking solution) do not affect the coupled reactions. The high signal generation capacity of the peroxidase tyramide reaction allows the detection of low amounts of target molecules. The diffusion of the deposited tyramide limits the resolution of the technique which remains suitable for electron microscope ISH studies (Figure 4b) (23) or for ICC localization of membrane-bound receptors (Figure 1f) (5). A dramatic amplification of the messengers might also be obtained by in situ reverse transcription polymerase chain reaction (RT-PCR) methods. However, with a few remarkable exceptions (see References 10, 27, and 29), technical problems arose which slowed down the development of this method. In particular, the good conditions for the amplification rounds are not easy to adjust, and the fixation must represent an acceptable compromise between the penetration of reagents and the retention of the amplification products. In situ RT-PCR methods are discussed in Chapter 9.

Together with the gain in sensitivity, the development of highly resolutive techniques allows one to further study the compartmentalization of gene products and especially the routing of proteins and mRNAs. Adequation between proteins and their corresponding mRNA in specific subcellular domains provided evidence for a local synthesis of proteins on their site of action, as it has been well documented for dendritic synthesis of bioactive molecules (33,38).

Neuroanatomical techniques are now broadly used in ontogenesis studies. To give functional clues, it is of importance to determine the period of gene expression throughout development. Thus, a time component must be taken into account in addition to the spatial localization of gene products. Thus, neuroanatomical methods led to describe hierarchies in regulatory interactions (15). Moreover, the codetection of a protein and its messenger is essential, and discrepancies identified between their relative distribution suggested a diffusion of the protein in tissues that do not express the transcript (9). Furthermore, the multiple labeling methods allow one to follow the regulation of a gene expression in differentiating neurons that are further identified by their neurochemical phenotype. Double ISH using riboprobes is thus likely to be a very useful technique for this approach, since it allows studies on whole mount embryos as well as on paraffin-embedded tissue.

A common feature of functional studies is now the use of genetically-modified systems. Multiple labelings represent useful tools to colocalize endogenous and exogenous genes and thus to identify cells or tissues that are expressing a tagged transgene.

In all these models, quantifying the gene expression has become an essential feature, and numerous devices are in development to quantitatively analyze radioactive or fluorescent signals by using the so-called beta imagers and imaging softwares.

A broad array of multiple labeling methods is now available and represent essential approaches not only for anatomical but also for functional studies. Their growing interest depends on their ability to answer an increasing number of questions related to cellular or developmental neurobiology.

ACKNOWLEDGMENTS

We are grateful to Prof. Tomas Hökfelt (Department of Neuroscience, Karolinska Institutet, Stockholm, Sweden) in whose laboratory some of the described protocols have been set up or improved. We thank the Association pour la Recherche Médicale en Aquitaine for providing financial support to the authors.

REFERENCES

1. Bernard, V., P. Somogyi, and J.P. Bolam. 1997. Cellular, subcellular, and subsynaptic distribution of AMPA-type glutamate receptor subunits in the neostriatum of the rat. J. Neurosci. *17*:819-833.
2. Bernard, V., O. Laribi, A.I. Levey, and B. Bloch. 1998. Subcellular redistribution of m2 muscarinic acetylcholine receptors in striatal interneurons *in vivo* after acute cholinergic stimulation. J. Neurosci. *18*:10207-10218.
3. Bobrow, M.N., T.D. Harris, K.J. Shaughnessy, and G.J. Litt. 1989. Catalyzed reporter deposition, a novel mmethod of signal amplification. Application to immunoassays. J. Immunol. Methods *125*:279-285.
4. Brahic, M., A.H. Haase, and E. Cash. 1984. Simultaneous *in situ* detection of viral RNA and antigens. Proc. Natl. Acad. Sci. USA *81*:5445-5448.
5. Broberger, C., M. Landry, H. Wong, J.N. Walsh, and T. Hökfelt. 1997. Subtypes Y1 and Y2 of the neuropeptide Y receptor are respectively expressed in proopiomelanocortin and neuropeptide Y-containing neurons of the rat hypothalamic arcuate nucleus. J. Neuroendocrinol. *66*:393-408.
6. Brudlag, D., C. Schlehuber, and J. Bonner. 1977. Properties of formaldehyde treated nucleohistone. Biochemistry *8*:3214.
7. Clayton, D.F. and A. Alvarez-Buylla. 1989. *In situ* hybridization using PEG-embedded tissue and riboprobes: increased cellular detail coupled with high sensitivity. J. Histochem. Cytochem. *3*:389-393.
8. Dagerlind, A., K. Friberg, A.J. Bean, and T. Hökfelt. 1992. Sensitive messenger RNA detection using unfixed tissue: combined radioactive and non-radioactive *in situ* hybridization histochemistry. Histochemistry *98*:39-49.
9. Driever, W. and C. Nusslein-Volhard. 1988. The bicoid protein determines position in the Drosophila embryo in a concentration-dependent manner. Cell *54*:95-104.
10. Fournier, J.G. 1994. Histologie Moléculaire. Technique and Documentation. Lavoisier, Paris.
11. Grino, M. and A.J. Zamora. 1998. An *in situ* hybridization histochemistry technique allowing simultaneous visualization by the use of confocal microscopy of three cellular mRNA species in individual neurons. J. Histochem. Cytochem. *46*:753-759.
12. Guitteny, A.F., P. Böhlen, and B. Bloch. 1988. Analysis of vasopressin gene expression by *in situ* hybridiza-

tion and immunohistochemistry in semi-thin sections. J. Histochem. Cytochem. *36*:1373-1378.
13. Hrabovszky, E., M.E. Vrontakis, and S.L. Pertersen. 1995. Triple-labeling method combining immunocytochemistry and in situ hybridization histochemistry: demonstration of overlap between fos-immunoreactive and galanin mRNA-expressing subpopulations of luteinizing hormone-releasing hormone neurons in female rats. J. Histochem. Cytochem. *4*:363-370.
14. Hunyady, B., K. Krempels, G. Harta, and E. Mezey. 1996. Immunohistochemical signal amplification by catalyzed reporter deposition and its application in double immunostaining. J. Histochem. Cytochem. *12*:1353-1362.
15. Ingham, P.W. 1988. The molecular genetics of embryonic pattern formation in Drosophila. Nature *335*:25-34.
16. Kagiyama, N., S. Fujita, M. Momiyama, H. Saito, H. Shirahama, and S.H. Hori. 1992. A fluorescent detection method for DNA, hybridization using 2-hydroxy-3 naphthoic acid-2′-phenylanilide phosphate as a subtrate for alkaline phosphatase. J. Histochem. Cytochem. *25*:467-471.
17. Kerekes, N., M. Landry, M. Rydh-Rinder, and T. Hökfelt. 1997. The effect of NGF, BDNF and bFGF on galanin message associated peptide in cultured rat dorsal root ganglia. Brain Res. *754*:131-141.
18. Kerekes, N., M. Landry, and T. Hökfelt. 1999. Leukemia inhibitory factor (LIF) regulates galanin/GMAP expression in cultured mouse dorsal root ganglia with a note on in situ hybridization technology. J. Neurosci. *89*:1123-1134.
19. Kerstens, H.M.J. and P.J. Poddighe. 1995. A novel in situ hybridization signal amplification method based on the deposition of biotinylated tyramide. J. Histochem. Cytochem. *4*:347-352.
20. Lan, H.Y., W. Mu, Y.Y. Ng, and D.J. Nikolic-Paterson. 1996. A simple, reliable, and sensitive method for nonradioactive in situ hybridization: use of microwave heating to improve hybridization efficiency and preserve tissue morphology. J. Histochem. Cytochem. *3*:281-287.
21. Landry, M., A. Trembleau, R. Arai, and A. Calas. 1991. Evidence for a colocalization of OT mRNA and galanin: a study combining in situ hybridization and immunohistochemistry. Mol. Brain Res. *10*:91-95.
22. Landry, M., D. Roche, E. Angelova, and A. Calas. 1997. Expression of galanin in gypothalamic magnocellular neurons of lactating rats. Coexistence with vasopressin and oxytocin. J. Endocrinol. *155*:467-481.
23. Landry, M. and T. Hökfelt. 1998. Subcellular localization of preprogalanin mRNA in perikarya and axons of hypothalamo-posthypophyseal magnocellular neurons. an in situ hybridization study at the light and electron microscope level. J. Neurosci. *84*:897-912.
24. Landry, M., K. Aman, and T. Hökfelt. 1998. The galanin-R1 receptor in anterior and mid-hypothalamus: distribution and regulation. J. Comp. Neurol. *399*:321-340.
25. Le Moine, C. and B. Bloch. 1995. D1 and D2 dopamine receptor gene expression in the rat striatum: sensitive cRNA probes demonstrate prominent segregation of D1 and D2 mRNAs in distinct neuronal populations of the dorsal and ventral striatum. J. Comp. Neurol. *355*:418-426.
26. Le Moine, C., V. Bernard, and B. Bloch. 1994. Quantitative in situ hybridization, using radioactive probes in the study of gene expression in heterocellular systems. In K.H.A. Choo (Ed.), Methods in Molecular Biology, In Situ Hybridization Protocols, Vol. 33. Humana Press, Totowa.
27. Martinez, A., M.J. Miller, K. Quinn, E.J. Unsworth, M. Ebina, and F. Cuttitta. 1995. Non-radioactive localization of nucleic acids by-direct in situ PCR and in situ RT-PCR in paraffin-embedded sections. J. Histochem. Cytochem. *8*:739-747.
28. Melton, D.A. 1987. Translocation of a localized maternal mRNA to the vegetal pole of Xenopus oocytes. Nature *328*:80-82.
29. Morel, G., M. Berger, B. Ronsin, S. Recher, S. Ricard-Blum, H.C. Mertani, and P.E. Lobie. 1998. In situ reverse transcription-polymerase chain reaction. Applications for light and electron microscopy. J. Biol. Cell *90*:137-154.
30. Nouel, D., M.P. Faure, J.A. St. Pierre, R. Alonso, R. Quirion, and A. Beaudet. 1997. Differential binding profile and internalization process of neurotensin via neuronal and glial receptors. J. Neurosci. *17*:1795-1803.
31. Pardue, M.L. 1987. In situ hybridization, p. 179. In B.D. Hames and S.J. Miggins (Eds.), Nucleic Acid Hybrydization. IRL Press, Oxford.
32. Prakash, N., S. Fehr, E. Mohr, and D. Richter. 1997. Dendritic localization of rat vasopressin mRNA: ultrastructural analysis and mapping of targeting elements. J. Neurosci. *9*:523-532.
33. Racca, C., A. Gardiol, and A. Triller. 1997. Dendritic and postsynaptic localizations of glycine receptor α subunit mRNAs. J. Neurosci. *17*:1691-1700.
34. Schachter, G.S. 1987. Studies of neuropeptide gene expression in brain pituitary, p. 111. In K.L. Valentino, J.H. Eberwin, and J.D. Barchas (Eds.), In Situ Hybridization. Applications to Neurobiology. Oxford University Press, New-York.
35. Shivers, B.D., R. Harlan, D.W. Pfaff, and B.S. Schachter. 1986. Combination of immunocytochemistry and in situ hybridization in the same tissue section of rat pituitary. J. Histochem. Cytochem. *34*:39-43.
36. Steinbusch, H.W., A.A. Verhofstad, and H.W. Joosten. 1978. Localization of serotonin in the central nervous system by immunohistochemistry: description of a specific and sensitive technique and some applications. Neuroscience *3*:811-819.
37. Svenningsson, P., C. Le Moine, B. Kull, R. Sunahara, B. Bloch, and B.B. Fredholm. 1997. Cellular expression of adenosine A_{2A} receptor messenger RNA in the rat central nervous system with special reference to dopamine innervated areas. J. Neurosci. *4*:1171-1185.
38. Torre, E.R. and O. Steward. 1996. Protein synthesis within dendrites: glycosylation of newly synthesized proteins in dendrites of hippocampal neurons in culture. J. Neurosci. *16*:5967-5978.
39. Trembleau, A. and F.E. Bloom. 1996. Spatial segregation of $G_{\alpha s}$ mRNA and vasopressin mRNA to distinct domains of the rough endoplasmic reticulum within secretory neurons of the rat hypothalamus. J. Mol.

Cell. Neurosci. *7*:17-28.
40. Trembleau, A., A. Calas, and M. Fevre-Montange. 1990. Ultrastructural localization of oxytocin mRNA in the rat hypothalamus by *in situ* hybridization using a synthetic oligonucleotide. Mol. Brain Res. *8*:37-45.
41. Trembleau, A., D. Roche, and A. Calas. 1993. Combination of non-radioactive *in situ* hybridization with immunohistochemistry: a new method allowing the simultaneous detection of two mRNAs and one antigen in the same section. J. Histochem. Cytochem. *41*:489-498.
42. Trembleau, A., M. Morales, and F.E. Bloom. 1996. Differential compartmentalization of vasopressin messenger RNA and neuropeptide within the rat hypothalamo-neurohypophysial axonal tracts: light and electron microscopic evidence. J. Neurosci. *1*:113-125.
43. Van Gijlswijk, R.P.M., H.J.M.A.A. Zijlmans, J. Wiegant, M.N. Bobrow, T.J. Erickson, K.E. Adler, H.J. Tanke, and A.K. Raap. 1997. Fluorochrome-labeled tyramides: use in immunocytochemistry and fluorescence *in situ* hybridization. J. Histochem. Cytochem. *45*:375-382.
44. Van Gijlswijk, R.P.M., D.J. van Gijlswijk-Janssen, A.K. Raap, M.R. Daha, and H.J. Tanke. 1996. Enzyme labelled antibody-avidin complexes, new flexible and sensitive immunochemical reagents. J. Immunol. Methods *189*:117-127.
45. Xu, Z.-Q., T.-J. Shi, M. Landry, and T. Hökfelt. 1996. Evidence for galanin receptors in primary sensory neurons and effect of axotomy and inflammation. Neuroreport *8*:237-242.
46. Young, W.S. III. 1989. Simultaneous use of digoxigenin and radiolabeled oligodeoxyribonucleotide probes for hybridization histochemistry. Neuropeptides *13*:271-275.

9
In Situ Reverse Transcription PCR for Detection of mRNA in the CNS

Helle Broholm[1] and Steen Gammeltoft[2]
[1]Rigshospitalet Copenhagen University Hospital, Department of Neuropathology, Copenhagen; and [2]Department of Clinical Biochemistry, Glostrup Hospital, Copenhagen University, Glostrup, Denmark, EU

OVERVIEW

The development of polymerase chain reaction (PCR) produced a technological breakthrough in nucleic acid detection by increasing molecular sensitivity capabilities. In situ PCR is a marriage of two established technologies: PCR and in situ hybridization based on the amplification within intact cells or tissue sections of specific DNA sequences, or mRNA species, to levels detectable by in situ hybridization and/or immunohistochemistry. Thus, PCR results can be correlated spatially with cell morphology.

This chapter focuses on the application of in situ PCR on detection of mRNA in rat brain sections. We provide a detailed one-day protocol for direct in situ reverse transcription PCR (RT-PCR) using incorporation of nonradioactive nucleotides (digoxigenin-11-dUTP) in the PCR product and detection with an antidigoxigenin antibody conjugated with alkaline phosphatase. We have used this approach for localization of 90 kDa ribosomal S6 kinase (RSK) mRNA in adult rat brain. Initially, we investigated the distribution of RSK1, RSK2, and RSK3 mRNA in rat brain by conventional in situ hybridization using radiolabeled oligonucleotides. The signals were, however, too weak for localization in brain section. Accordingly, we applied in situ RT-PCR to improve the sensitivity by PCR amplification of the target and by increasing the signal with immunocytochemistry. Using this protocol, RSK mRNA was successfully visualized in neuronal cell bodies throughout the adult rat brain (Figures 1 and 2).

BACKGROUND

In situ hybridization is an invaluable tool to identify, in tissue samples, the cell type that contains a specific nucleic acid. In situ hybridization techniques represent the histochemical counterparts to Northern and Southern blot analysis in molecular biology. These techniques enable investigators to delineate which cells in a mixed population express the mRNA of interest or harbor a particular viral DNA (28). In situ hybridization permits the localization of nuclear acid sequences at the cellular level with high specificity, but this is sometimes overshadowed by low detection sensitivity (19,25). The threshold for detection of mRNA are approximately 10 copies per cell, and most conventional nonisotopic protocols do not detect single-copy genes. Thus, constraints on the detection limits of the assay minimize its effectiveness in evaluating low level nucleic acid expression (see Chapter 8 for further discussion and for details on in situ hybridization protocols).

In situ PCR has been applied for the detection of low abundance viral DNA and mRNA in tissue and cellular samples. The first successful in situ PCR was described 10 years ago, where lentiviral DNA in infected cells was amplified in situ, and the amplification product was detected by in situ hybridization (6). The technique has primarily been applied for the detection of viral DNA. Several groups used in situ PCR to localize different human viruses in formalin-fixed paraffin-embedded tissues and blood cell suspensions (1,3,11,13,23,26). Some authors have used in situ PCR to localize mRNA in cells with low levels of expression (2,8,14,19,20,24,29). Many centers, however, have encountered problems with reaction failure or commonly seen false positive signals. Several reviews and books have dealt with these problems and published modifications and development of the in situ PCR protocols (5,7,15,19,22,25).

In Situ PCR Techniques

The term in situ PCR is used to describe the technique as a whole. Direct in situ PCR describes the technique whereby a label is incorporated directly into the amplicon during PCR of cellular DNA and subsequently detected in order to localize the amplified product. Indirect in situ PCR describes an alternative technique whereby the amplicon is produced by PCR of cellular DNA without label incorporation. The amplified product is then detected by standard in situ hybridization. This basic terminology can be extended to include amplification and detection of mRNA targets by direct or indirect in situ RT-PCR (Figure 3). Four essentially similar in situ amplification techniques have been described (15,25):

1. Direct in situ PCR: PCR amplification of cellular DNA sequences in tissue specimens using either a hapten-labeled nucleotide (dUTP) or hapten-labeled primer within the PCR

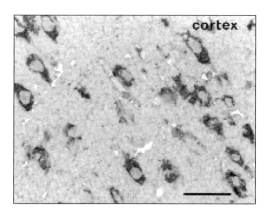

Figure 1. Localization of RSK2 mRNA in rat cerebral cortex by indirect in situ RT-PCR. Color staining is localized in the cytoplasm of neuronal cell bodies of interneurons in cortex. Glial cells are unstained.

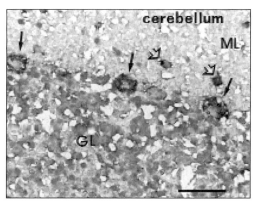

Figure 2. Localization of RSK2 mRNA in rat cerebellum by direct in situ RT-PCR. Color staining is localized in cytoplasm of the perikarya of Purkinje cells. Neurons in the granular cell layer are unstained.

mixture. Biotin, digoxigenin, or fluorescein is used to label the nucleotide or primer. The labeled PCR product is then detected in the tissue using standard detection techniques as for conventional in situ hybridization or immunocytochemistry.

2. Indirect in situ PCR: PCR amplification of cellular DNA sequences in tissue specimens using a standard PCR mixture. Following PCR, the localization of the amplified product is then detected by in situ hybridization using a labeled oligonucleotide or genomic probe. The labels used can be either isotopic (^{32}P, ^{35}S) or nonisotopic (e.g., biotin, digoxigenin, or fluorescein). To date, most studies have used nonisotopic labels.

3. Direct in situ RT-PCR: amplification of mRNA sequences in cells and tissue specimens by initially creating a complementary DNA (cDNA) template using reverse transcriptase and then amplifying the newly created cDNA template by PCR using labeled nucleotides as for direct in situ PCR. The labeled PCR product is detected by immunocytochemistry (Figure 4).

4. Indirect in situ RT-PCR: amplification of RNA sequences in cells and tissue specimens by creating a cDNA template using reverse transcriptase, amplifying the newly created DNA template by PCR, and then probing this DNA

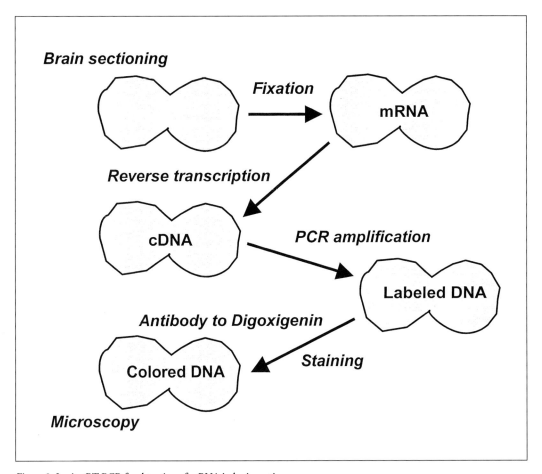

Figure 3. In situ RT-PCR for detection of mRNA in brain sections.

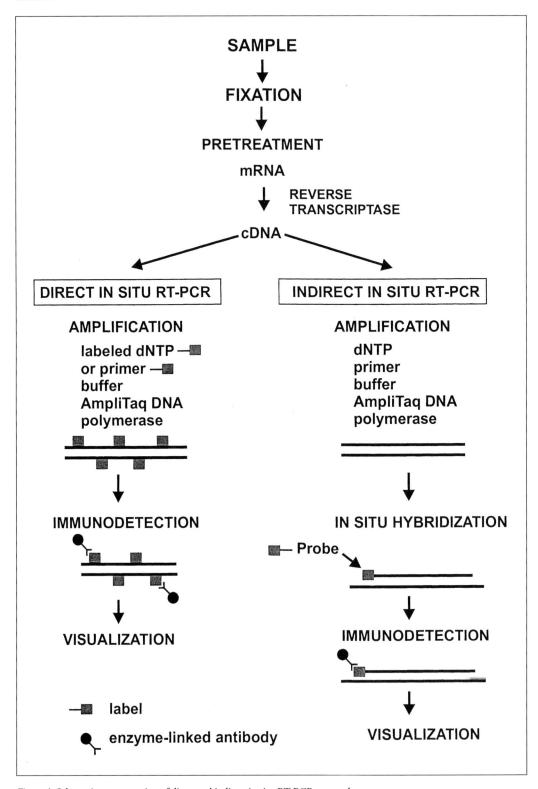

Figure 4. Schematic representation of direct and indirect in situ RT-PCR protocols.

with an internal oligonucletide probe (Figure 4).

We have used direct in situ RT-PCR for localization of 90 kDa RSK mRNA in adult rat brain (H. Broholm and S. Gammeltoft, unpublished data). RSK is involved in transcriptional regulation by growth factors and hormones via the mitogen-activated protein (MAP) kinase cascade in brain and other tissues (4,10). Three isoforms of RSK exist: RSK1, RSK2, and RSK3 with about 80% homology. The importance of RSK2 in brain function was recently underlined by the finding that inactivating mutations of human RSK2 gene are associated with a syndromic form of mental retardation, Coffin-Lowry syndrome (9,27). Thus, it is important to analyze the cellular distribution and function of RSK isoforms in the central nervous system (CNS).

We have evaluated two techniques of in situ RT-PCR for detection of mRNA in cells and tissues: *(i)* direct in situ RT-PCR with incorporation of a label directly into the PCR product and subsequent detection; and *(ii)* indirect in situ RT-PCR with amplification of cDNA and subsequent hybridization (Figures 1 and 2). We conclude that direct in situ RT-PCR is the method of preference, as it is simple, sensitive, specific, rapid, and reproducible. Indirect in situ RT-PCR may be more specific because the probe only hybridizes to the target-specific amplicon sequences produced during the PCR and not to DNA produced by nonspecific synthesis. On the other hand, the process of indirect in situ RT-PCR takes considerably longer than direct in situ PCR because of the additional hybridization and stringent washing steps required. Furthermore, manipulation of the tissue can result in detachment and loss of the sections. Finally, indirect in situ RT-PCR is generally found to be less sensitive than direct in situ RT-PCR.

During direct in situ RT-PCR, a label is incorporated directly into the amplicon throughout the PCR process. Two methods can be utilized for direct incorporation. In one method, a hapten-labeled nucleotide analog such as biotin-11-dUTP, digoxigenin-11-dUTP, or fluorescein-15-dATP is added to the PCR mixture. During amplification, the hapten becomes incorporated into the amplicon and can be detected by enzyme-conjugated antibodies and chromogenic substrates. Direct label incorporation results in the labeling of all nucleic acids synthesized during the PCR process with a high degree of sensitivity. With this approach, the detection of single-copy mRNA is made possible. The other method of direct incorporation is via hapten-labeled primers. This labeling method is less sensitive as less hapten is incorporated at each amplification step. Hapten-labeled oligonucleotide primers tend to be sticky, and the nonspecific binding can lead to high background. We used direct incorporation of digoxigenin-11-dUTP into the PCR product and detection by antidigoxigenin antibody conjugated to alkaline phosphatase and chromogenic substrate.

An important factor for the success of in situ PCR is the availability of instrumentation dedicated to the thermal cycling of PCR components on microscopic slides. We used GeneAmp® *In situ* PCR system 1000 manufactured by PE Biosystems (Foster City, CA, USA) (15). With this instrument, in situ PCR can be performed on nondisrupted cells and tissues immobilized onto the surface of microscopic slides. The capacity is 10 slides with two or three sections per slide. The slides are held firmly against a series of vertical slots comprising the block of the thermal cycler to ensure precise temperature control. The system also provides a unique containment system for maintaining PCR components over the sections during thermal cycling. A dedicated Assembly Tool is used to assemble AmpliCover™ Discs and AmpliCover Clips (all from PE Biosystems) onto 1.2-

mm-thick silane-coated slides to create a standardized reaction chamber of 35 to 50 µL volume over the immobilized sample. The stainless steel AmpliCover Clips hold the silicon-rubber AmpliCover Discs firmly in place to ensure that the contained reaction remains sealed throughout the cycling reaction.

We emphasize, however, that several problems were encountered with the technique leading to reaction failure or false positive signals. The implementation of in situ RT-PCR is labor-intensive and requires the optimization of each step involved in the procedure. No universally applicable technique is available, and a standard protocol for in situ RT-PCR does not exist. All variables should be carefully controlled for each application, and individual steps must be determined empirically. Critical steps include starting material, fixation of tissue, protease digestion, DNase digestion, primer design, and cycling conditions. Future users of in situ RT-PCR are recommended to read reviews that deal with technical problems and optimization of protocols (7,15–17,19,21,22,25).

PROTOCOL FOR DIRECT IN SITU RT-PCR

Materials and Reagents

- Tissue-Tek® O.C.T.-compound (Sakura Finetek, Amsterdam, The Nederlands).
- Multi-well culture plates (Corning Incorporates, Corning, NY, USA).
- 2-methylbutane (Sigma, St. Louis, MO, USA).
- In situ PCR glass slides (PE Biosystems).
- Humi-Cap dessicant capsules (United Desiccants, Pennsauken, NJ, USA).
- Paraformaldehyde (Sigma).
- Disposable syringe filters (Millipore, Bedford, MA, USA).
- Pepsin (Sigma).
- Proteinase K (Sigma).
- Diethyl-pyrocarbonate (DEPC; Sigma).
- Deoxyribonuclease I (DNase I; Roche Molecular Biochemicals, Mannhein, Germany).
- Thermanox cover slips (Miles Scientific, Naperville, IL, USA).
- Acetic anydride (Sigma).
- Triethanolamine (Sigma).
- Acetic acid (Sigma).
- GeneAmp RNA PCR Core Kit (PE Biosystems).
- GeneAmp In situ PCR System 1000.
- Albumin (Sigma).
- Antidigoxigenin Fab fragment conjugated with alkaline phosphatase (Roche Molecular Biochemicals).
- 4-nitroblue tetrazolium chloride (Sigma).
- 5-bromo-4-choloro-3-indolyl phosphate (BCIP; Sigma).

Buffers and Solutions

Buffered Paraformaldehyde 4%

Paraformaldehyde fixative is prepared by heating to 60°C a volume of water equal to slightly less that 2/3 the desired final volume of fixative. Weigh out a quantity of paraformaldehyde that will make a 4% solution and add it with a stir bar to the water. Cover. Transfer the solution to fume hood and maintain on heating plate at 60°C with stirring. Add 1 N NaOH drop-wise with a Pasteur pipet until the solution is clear. The solution will still have some fine particles that will not go away. Be careful not to overheat the solution. Remove from heat and add 1/3 volume 3× phosphate-buffered saline (PBS). Bring pH of solution to 7.2 with HCl, add water to final volume, and filter using a disposable Millipore filter. Cool to room temperature or to 4°C on ice.

Caution: Formaldehyde is a carcinogen and may cause allergic reactions.

3× and 1× Phosphate-Buffered Saline, pH 7.2

Prepare the following solutions: 390 mM NaCl/30 mM Na_2HPO_4 and 390 mM NaCl/30 mM NaH_2PO_4. Mix to obtain 3× PBS, pH 7.2 and autoclave or sterilize with a filter. Prepare 1× PBS by diluting 3× PBS 3-fold with water.

Diethyl-Pyrocarbonate-Water 0.1%

In all steps involving treatment of sections for in situ RT-PCR, it is recommended to use water with DEPC in order to inhibit RNase activity. Prepare 0.1% DEPC-H_2O by adding 1 mL 100% DEPC to 1000 mL MilliQ-purified water. Mix for 1 hour and incubate at 37°C. Autoclave the solution.

Pepsin 2 mg/mL

Prepare the solution with 10 mg pepsin in 4832 µL 0.1% DEPC-H_2O and 168 µL 3 M HCl. Freshly made for each experiment.

DNase Solution 50 U/mL

Prepare the solution by adding 1.0 µL DNase I (5 U), RNase-free to 100 µL 1× DNase buffer composed of Tris 40 mM, pH 7.6, $MgCl_2$ 6 mM, and $CaCl_2$ 2 mM.

Acetic Anhydride 0.3% in 0.1 M Triethanolamine

Triethanolamine (1.85 g) in 100 mL MilliQ water. Stir to ensure complete solution. Adjust pH to 8.0. Just before use, add 300 µL acetic anhydride. Keep everything in the hood.

Reverse Transcriptase Master Mixture (GeneAmp RNA PCR Core Kit)

Add the reagents in the order and proportions as follows: $MgCl_2$ 25 mM 4 µL, 10× PCR Buffer II 2 µL, DEPC-H_2O 9 µL, dNTP 10 mM 2 µL, RNase 20 U/µL 1 µL, Oligo(dT)$_{16}$ 1 µL, MuLV reverse transcriptase 50 U/µL 1 µL. Total volume 20 µL.

PCR Master Mixture (GeneAmp RNA PCR Core Kit)

The solution is prepared by adding the reagents in the order and proportions as follows: $MgCl_2$ 25 mM 3.2 µL, 10× PCR Buffer II 4.0 µL, DEPC-H_2O 26.9 µL, dNTP 10 mM 0.8 µL, digoxigenin-11-dUTP 1 mM 0.8 µL, *Taq* DNA polymerase 5 U/µL 0.3 µL, "Downstream" Primer 10 µM 2.0 µL and "Upstream" Primer 10 µM 2.0 µL. Total volume 40.0 µL.

Buffer 1 for Digoxigenin Detection

0.05 M Tris-HCl, pH 7.5, and 0.15 M NaCl.

Buffer 2 for Digoxigenin Detection

0.1 M Tris-HCl, pH 9.5, 0.05 M $MgCl_2$, and 0.1 M NaCl.

Blocking Buffer

Albumin 0.3% dissolved in Buffer 1.

Digoxigenin Antibody Solution

Antidigoxigenin Fab fragment conjugated with alkaline phosphatase is diluted 1:500 in Buffer 1.

4-Nitroblue Tetrazolium Chloride

10 mg/mL water.

5-Bromo-4-Chloro-3-Indoylphosphate

25 mg/0.5 mL dimethylformamide (DMF).

Protocol

General Concepts

Direct in situ RT-PCR technique represents the coming together of RT-PCR and in situ hybridization, allowing the amplification of specific mRNA sequences inside cells. First, cells or tissue sections are mounted on coated glass slides and fixed with a suitable fixative, usually neutral-buffered formaldehyde. For some applications, formaldehyde-fixed cells or tissue are permeabilized using proteolytic enzymes in order to permit access of PCR reagents into the cells and to target the nucleic acid. Before performing the reverse transcription procedure, degradation of the existing DNA is usually recommended to avoid amplification of the genomic DNA. Acetylation can be included during pretreatment to reduce nonspecific sticking of hapten-labeled oligonucleotide primers during indirect in situ RT-PCR. Reverse transcriptase reaction is performed to synthesize cDNA from mRNA, and PCR amplification in situ is performed in specifically designed in situ PCR thermocyclers. Digoxigenin-labeled nucleotides are incorporated during the PCR, and the amplified product is visualized by immunocytochemistry using antibody to digoxigenin conjugated with alkaline phosphatase. We emphasize that pretreatment with protease, DNase as well as acetylation, was abandoned in our protocol. This resulted in a simple and rapid 4-step protocol of direct in situ RT-PCR: *(i)* sample preparation; *(ii)* reverse transcriptase reaction; *(iii)* in situ PCR amplification; and *(iv)* immunodetection and visualization (Table 1).

Sample Preparation

Tissue Dissection

1. Kill the animal by decapitation without anesthesia.
2. Open the skull with scissors cutting through foramen magnum and the midline and gently remove the brain and place it on the bottom of a plastic petri dish on ice at 4°C.
3. Remove the meninges with a pincer. Cut the brain with a razor blade in 6 coronal sections of 4 to 5 mm and place them in wells of a 12-well cell culture plate on ice at 4°C.
4. Embed tissue in OCT and freeze it by immersion of the plate in liquid 2-methylbutane prechilled with dry ice. The frozen tissue is stored in sealed containers at -80°C.

Comment: The meninges are removed with a pincer to avoid interference with the coronal sectioning. Tissue should be oriented in the block appropriately for sectioning. Note the tissue number on the block directly and indicate which face of the block should be sectioned.

Sectioning and Fixation of Frozen Tissues

5. Cut 7-μm cryostat sections and thaw-mount them on 1.2-mm silane-coated in situ PCR glass slides.
6. Allow to air-dry and fix in 4% paraformaldehyde in 1× PBS, pH 7.2, at room temperature for 20 minutes. Two sections are placed on each slide, one for a positive test and one for a negative control (see below).
7. Wash slides once in 3× PBS for 5 minutes and twice in 1× PBS for 5 minutes.

9. In Situ Reverse Transcription PCR

Table 1. Protocol for Direct In Situ RT-PCR

Step	Objective	Technique	Comment
Tissue sectioning and adhesion	Disrupt tissue barriers and keep sections on the slide during incubations.	5–10-μm sections at -18°C on 1.2-mm silane-coated glass slides.	Sections should be less than 10 μm to permit diffusion of reagents to nucleic acid targets.
Fixation	Maintain tissue architecture and prevent diffusion of mRNA.	PBS-buffered 4% paraformaldehyde for 20 min.	Weak and brief fixation does not cause denaturation of proteins, mRNA, or DNA.
Protease digestion	Degrade protein–protein and protein–DNA networks.	Pepsin 2 mg/mL in 0.01 N HCl for 20–90 min at room temperature.	Omitted in our protocol.
DNase treatment	Degrade genomic DNA to prevent false positive reactions.	Deoxyribonuclease 1 U for 7–16 h.	Omitted in our protocol.
Acetylation	Prevent nonspecific sticking of hapten-labeled oligonucleotide primers used in indirect RT-PCR.	Acetic anhydride 0.3% in 0.1 M triethanolamine for 10 min and acetic acid 20% for 15 s.	Omitted in our protocol.
Reverse transcription	Synthesis of cDNA from mRNA.	MuLV reverse transcriptase 2.5 U with oligo(dT)$_{16}$ primer for 30 min at 42°C.	Standard reverse transcriptase procedure.
PCR	Amplification of cDNA.	Taq DNA polymerase 0.4 U with primer set and digoxigenin-11-dUTP in thermal cycler (45 cycles).	Standard PCR procedure.
Immunodetection	Increase signal and localize cellular mRNA (cDNA).	Antidigoxigenin antibody 1:500 for 60 min at 37°C.	Standard immunochemistry.
Visualization	Staining of cells for light microscopy.	Color development with substrate for alkaline phosphatase.	Standard histochemistry.

8. Dehydrate step-wise in 70%, 80%, 90%, and 100% ethanol for 2 minutes at each step and air-dry.
9. Slides are stored at -80°C in plastic slide boxes with Humi-Cap desiccant capsules.

Comment: The problem with keeping the tissue sections on the slide during the hybridization and washing procedures is reduced by the use of silane-coated glass slides and subsequent 4% paraformaldehyde fixation. mRNA is stable when stored as described above, and tissue sections may be used for up to 6 months after sectioning.

Pretreatment of Sections

Protease Digestion

Protease digestion is performed with 150 μL of pepsin 2 mg/mL in 0.01 N HCl per sample and incubated for 20 to 90 minutes at room temperature in a humid chamber. The slide is washed with DEPC water for 5 minutes. Others have used proteinase K 0.1 to 0.5 µg/mL for cells and 0.1 to 1.0 µg/mL for tissue sections before in situ RT-PCR of mRNA.

Comment: In our study (mild aldehyde fixation) protease digestion was omitted.

In the case of strong formaldehyde-fixed (10% for 16–24 h) and paraffin-embedded sections, however, protease digestion is generally recommended to reduce the network of protein–protein and protein–DNA cross-linking created by the fixative, and to allow diffusion of primers and *Taq* DNA polymerase to the nucleic acid targets. The duration of protease digestion depends on the type of fixative used and the extent of fixation. It is a critical parameter that must be determined empirically (22). Too little protease digestion will result in poor amplification efficiency, while over digestion will destroy tissue morphology and will permit diffusion of the amplicons out of the cells. In our experience, the protease digestion is difficult to optimize and should be avoided if possible. Using fresh frozen sections of rat brain, we tested whether protease digestion was needed for in situ RT-PCR of RSK mRNA. A comparison between protease-treated and untreated sections showed no difference in amplification and localization of the signal. In both cases, a strong signal was clearly localized to the cytoplasm of neuronal cell bodies in the brain.

Digestion with DNase

DNase treatment is performed by application of 20 µL of DNase solution (1 U/sample) to each section. Cover with Thermanox cover slips and incubate the slide at 37°C in a humid chamber for at least 7 hours. Digestion times shorter that 7 hours might not be adequate to completely degrade genomic DNA. After the digestion the slides are washed for 1 minute in ultra-pure water followed by 100% ethanol and air-drying.

Comment: In our study of RSK mRNA distribution in rat brain, DNase pretreatment of sections was omitted. Digestion with DNase may be used to degrade genomic DNA and prevent amplification of genomic sequences during in situ RT-PCR leading to false positive reactions. Although we omitted this step, we did not observe any nuclear localization of the digoxigenin label and found only staining of the cytoplasm indicating that amplification of genomic DNA was not a problem.

DNase treatment appears to be a problematic step, and on some occasions omission of this step is required to avoid nonspecific nuclear staining. A time-dependent increase in nonspecific nuclear staining is clearly seen after 2 and 24 hours of DNase treatment (19). This anomaly is probably caused by small fragments of nuclear DNA being used as nonspecific primers during the amplification step. Efficient amplification of genomic DNA can be avoided by using special primer designs such as a poly(dT) primer or primers flanking introns. Elimination of the need for DNase treatment has been successfully demonstrated in these instances (15).

Acetylation

Labeled reagents can stick nonspecifically to tissue sections because of static charge. One way of reducing static charge is to include an acetylation step during the pretreatment procedure. Acetylation is performed by incubation of sections with 0.3% acetic anhydride in 0.1 M triethanolamine for 10 minutes followed by 20% acetic acid for 15 seconds and then washing with water.

Comment: In our study, acetylation of brain sections before in situ RT-PCR was omitted. This step can be included to prevent nonspecific sticking of hapten-labeled oligonucleotide primer during indirect in situ RT-PCR (15). High background in problematic tissues (e.g., kidney, liver, and gastrointestinal tract) can be reduced significantly by inclusion of this step. This step is rarely used if hapten-labeled nucleotides are used for direct in situ RT-PCR detection of mRNA.

Reverse Transcriptase Reaction

10. Remove slides from the freezer and thaw for 5 minutes at room temperature.
11. Apply 20 μL reverse transcriptase master mixture on the sections. The reverse transcriptase reaction can be primed using random oligo(dT) primers or custom-designed forward (antisense) primers 3′ to the gene of interest, i.e., the downstream primer of the PCR amplification. We used oligo(dT)$_{16}$ to prime all mRNAs with a poly(A) tail for initiating first-strand cDNA synthesis.
12. Cover the sections with Thermanox cover slips.
13. Preincubate the slides at room temperature for 10 minutes to extend the oligo(dT)$_{16}$ by reverse transcriptase. The slides are incubated in a humid chamber at 42°C for 30 minutes. To avoid disrupting tissue material, carefully lift the cover slips without sliding them sideways.
14. Wash the slides in 1× PBS for 5 minutes at room temperature. Dehydrate the slides in 0.1% DEPC-H$_2$O with 50%, 70%, 90%, and 100% ethanol, 1 minute each, and air-dry.

Comment: Heat inactivation of reverse transcriptase, which is used for solution RT-PCR is not necessary for in situ RT-PCR as the enzyme is removed by washing the slides. All reagents and plast-ware should be RNase-free. The investigator should use gloves to avoid contamination of samples with skin bacteria and RNase. Negative controls are prepared by omitting MuLV reverse transcriptase from the reverse transcriptase master mixture.

PCR Amplification

15. Program the GeneAmp *In situ* PCR System 1000 as follows: a 2 minute hold at 95°C; 45 cycles of 95°C for 15 seconds and 59°C for 30 seconds; a 7 minute hold at 72°C; a 24 hour hold at 4°C. The hybridization temperature depends on the GC content and melting temperature (T$_m$) of the primers. Cycling conditions must be optimized for the target sequence and primer pair.
16. Set the thermal cycler at a 70°C hold.
17. Prepare 40 μL of PCR master mixture for each tissue section. For the application, the PCR master mixture components might have to be optimized.
18. Place a slide on the Assembly Tool and overlay the section with 40 μL of PCR solution. Be careful not to let the drop of solution spread along the glass.
19. Make an assembly using the AmpliCover Discs, Clips, and Assembly Tool contained in the GeneAmp *In situ* PCR System 1000. Continue with the other tissue spots.
20. As each slide is assembled, place it in the GeneAmp *In situ* PCR System 1000 set at a 70°C soak.
21. When all the slides are placed in the cycler, begin the thermal cycle program described above.
22. After cycling is complete, soak the slides at 4°C until disassembly and digoxigenin detection.

Comment: The protocol has been applied for in situ RT-PCR of RSK1, RSK2, and RSK3 cDNA in rat brain. The recommended primer length is between 20 and 30 nucleotides. In our study, the primer length varied between 21 and 23 bp. The GC content varied between 0.42 and 0.66. The PCR product length should be 100 to 600 bp. The length of the PCR product was 546 bp for RSK1, 412 bp for RSK2, and 328 bp for RSK3. To eliminate the possibility of generating PCR products from genomic DNA, it is important to design primers that bridge introns so as to distinguish the template source on the basis of product size. Before the in situ PCR experiments, all parameters for the PCR, including $MgCl_2$, pH, and annealing temperature must be optimized by solution PCR and agarose electrophoresis. Products should be cloned and sequenced to confirm identity. In our study, the T_m of the primers varied between 54° and 59°C, but a temperature of 59°C is used for a combined primer annealing and extension step for 30 seconds. The 7 minute hold at 72°C after the thermal cycling was used to allow primer extension and amplicon synthesis to be completed. We used 45 cycles to obtain the highest amplification of cDNA. Some authors recommend lower number of cycles to avoid nonspecific nuclear staining (19). In our study, however, no nuclear staining was observed. Negative controls are prepared by omitting primers or *Taq* DNA polymerase from the PCR master mixture.

Immunodetection and Visualization

23. Remove the AmpliCover Clips from each slide using the Disassembly Tool. Carefully lift the AmpliCover Disc without sliding it sideways so as not to disrupt the tissue material.
24. Rinse the slides briefly with Buffer 1 using a Pasteur pipet.
25. Wash the slides in a Coplin jar with Buffer 1 for 5 minutes followed by Blocking buffer for 20 minutes.
26. Drain the slides of Blocking buffer, and wipe a ring around each section with a hydrophobic pen.
27. Apply 50 µL of antidigoxigenin antibody solution.
28. Place the slides in a humid chamber and incubate at 37°C for 60 minutes.
29. In a Coplin jar, wash the slides in Buffer 2. Add 1 mL per slide of freshly prepared solution with appropriate substrates (4-nitroblue tetrazolium chloride and BCIP) and develop for 5 minutes to 1 hour at room temperature protected from light. Drain the substrate solution from the slides. Stop the reaction by rinsing in water.
30. The tissue is counterstained in Mayers Hematoxylin for 1 minute. Slides are mounted with Aquamount.

Comment: Do not let the sections dry out during the entire detection process. The length of development will depend on the initial copy number of the mRNA target sequences and the amplification efficiency. It has been estimated that after 30 cycles, amplification is of the order of 10- to 30-fold (3). Monitor the development of purple color on the slide every few minutes. In our study of in situ RT-PCR of RSK mRNA in rat brain, a 5 minute development was sufficient. A purple precipitate will appear in the location where digoxigenin is incorporated into DNA. For in situ RT-PCR, color development will appear over the cytoplasm, which is the cellular compartment where most mRNA is found. Over development will result in high levels of background staining.

DISCUSSION

In situ RT-PCR is a relatively new technique that extends the researcher's ability to localize specific mRNA targets in cells and tissue. It combines the specificity of in situ hybridization and the sensitivity of PCR and, with reliable multislide thermal cyclers available, can be a very reproducible technique. Two main protocols, direct and indirect, have been described for in situ RT-PCR. In the direct approach a labeled nucleotide is incorporated into the PCR products, whereas in the indirect method a subsequent in situ hybridization with a labeled probe is required to visualize the products (15,18,25). We chose the direct method because it is shorter, less expensive, and minimizes manipulation of the tissue, which can result in detachment and loss of sections. Some authors point out that direct methods can result in false positive results (12,15,18), but we find that use of proper controls will eliminate this problem.

Fixation and pretreatment of tissue or cells before in situ RT-PCR is of great significance for the successful result. Several groups have addressed these critical problems that may cause reaction failure or false positive results. During the development of our protocol, we evaluated the fixation and pretreatment carefully. The use of fixation of sections for in situ PCR has been discussed (25). Based on our experience with in situ RT-PCR, 4% paraformaldehyde in PBS buffer for 20 minutes is the best fixation to preserve integrity of tissue morphology and to retain mRNA and cDNA in the cells. Several authors recommend protease digestion of tissue sections that have been fixed in 10% paraformaldehyde for 16 to 24 hours and embedded in paraffin in order to degrade protein–protein and protein–DNA network and permit diffusion of PCR reagents to nucleic acid targets (22). On the other hand, protease overdigestion can result in loss of tissue architecture and migration of the PCR products from the cells (19). We find, however, that treatment of sections with pepsin is difficult to optimize and does not improve the results. Accordingly, we omitted protease digestion from our protocol.

The DNase treatment of sections for in situ RT-PCR is used to degrade genomic DNA in order to reduce amplification of DNA during PCR. This appears to be a problematic step that may result in nonspecific nuclear staining probably caused by small fragments of nuclear DNA being used as nonspecific primers during the amplification step (19). DNase digestion was omitted from our in situ RT-PCR protocol for amplification of RSK mRNA. Even after a high number of cycles (we used 45 cycles), we find that the staining is limited to the cytoplasm indicating that no amplification of genomic DNA occurs during the PCR amplification. Finally, acetylation of sections is recommended in order to reduce nonspecific sticking of hapten-labeled primers. This is not relevant for our procedure of direct in situ RT-PCR using hapten-labeled nucleotide for labeling of amplified DNA. In conclusion, we recommend a light and brief fixation of tissue sections, but avoid pretreatment.

The RT-PCRs were performed using a standard protocol for solution reverse transcriptase reaction and PCR from PE Biosystems. The same conditions were used for in situ RT-PCR and solution RT-PCR allowing control and analysis of the PCR products by gel electrophoresis and DNA sequencing. This improved the reliability of in situ RT-PCR. The PCR products ranged in size from 328 to 546 bp in our study. Others have analyzed the effect of size of the PCR product on sensitivity (19). Products smaller that 600 bp always yielded signal, whereas a longer product (1400 bp) did not. It is reasonable to conclude that in situ PCR matched the sensitivity of solution PCR. Immunodetection and visualization were standard immunohistochemistry procedures based on reagents from Roche Molecular Biochemicals. The signal obtained was of sufficient intensity, and no modifications were needed.

In conclusion, in situ RT-PCR is facile, versatile, can be done quickly, and does not require nucleic acid extraction and purification. Traditional in situ hybridization techniques are useful only if the target sequence copy number is greater than 10 per cell. Because targets can be exponentially amplified with PCR, the researcher can use in situ PCR to detect and localize low abundance nucleic acid sequences. In theory, a single DNA and mRNA molecule per cell should be detectable using in situ PCR and in situ RT-PCR, respectively.

ACKNOWLEDGMENTS

Birte Kofoed is gratefully acknowledged for expert and enduring technical assistance during implementation and application of in situ RT-PCR. The study was supported by grants from the Danish Research Center for Growth and Regeneration, Danish Cancer Society, Novo Nordisk Foundation, and the Foundation for Medical Research in Copenhagen County, Greenland and Faeroe Islands.

REFERENCES

1. Bagasra, O., S.P. Hauptman, H.W. Lischner, M. Sachs, and R.J. Pomerantz. 1992. Detection of human immunodeficiency virus type 1 provirus in mononuclear cells by *in situ* polymerase chain reaction. N. Engl. J. Med. *326*:1385-1391.
2. Chow, L.H., S. Subramanian, G.J. Nuovo, F. Miller, and E.P. Nord. 1995. Endothelin receptor mRNA expression in renal medulla identified by *in situ* RT-PCR. Am. J. Physiol. *269*:F449-F457.
3. Embretson, J., M. Zupancic, J. Beneke, M. Till, S. Wolinsky, J.L. Ribas, A. Burke, and A.T. Haase. 1993. Analysis of human immunodeficiency virus-infected tissues by amplification and *in situ* hybridization reveals latent and permissive infections at single-cell resolution. Proc. Natl. Acad. Sci. USA *90*:357-361.
4. Frodin, M. and S. Gammeltoft. 1999. Role and regulation of 90 kDa ribosomal S6 kinase (RSK) in signal transduction. Mol. Cell. Endocrinol. *151*:65-77.
5. Gu, J. 1995. In Situ Polymerase Chain Reaction and Related Technology. Eaton Publishing, Natick, MA.
6. Haase, A.T., E.F. Retzel, and K.A. Staskus. 1990. Amplification and detection of lentiviral DNA inside cells. Proc. Natl. Acad. Sci. USA *87*:4971-4975.
7. Herrington, C.S. and J.J. O'Leary. 1998. PCR In Situ Hybridization. Oxford University Press, New York.
8. Jacob, A., I.R. Hurley, L.O. Goodwin, G.W. Cooper, and S. Benoff. 2000. Molecular characterization of a voltage-gated potassium channel expressed in rat testis. Mol. Hum. Reprod. *6*:303-313.
9. Jacquot, S., K. Merienne, E. Trivier, M. Zeniou, S. Pannetier, and A. Hanauer. 1999. Coffin-Lowry syndrome: current status. Am. J. Med. Genet. *85*:214-215.
10. Jensen, C.J., M.B. Buch, T.O. Krag, B.A. Hemmings, S. Gammeltoft, and M. Frodin. 1999. 90-kDa ribosomal S6 kinase is phosphorylated and activated by 3-phosphoinositide-dependent protein kinase-1. J. Biol. Chem. *274*:27168-27176.
11. Koffron, A.J., M. Hummel, B.K. Patterson, S. Yan, D.B. Kaufman, J.P. Fryer, F.P. Stuart, and M.I. Abecassis. 1998. Cellular localization of latent murine cytomegalovirus. J. Virol. *72*:95-103.
12. Komminoth, P. and A.A. Long. 1993. In-situ polymerase chain reaction. An overview of methods, applications and limitations of a new molecular technique. Virchows Arch. B Cell Pathol. Incl. Mol. Pathol. *64*:67-73.
13. Komminoth, P., A.A. Long, R. Ray, and H.J. Wolfe. 1992. *In situ* polymerase chain reaction detection of viral DNA, single-copy genes, and gene rearrangements in cell suspensions and cytospins. Diagn. Mol. Pathol. *1*:85-97.
14. Kulaksiz, H., R. Arnold, B. Goke, E. Maronde, M. Meyer, F. Fahrenholz, W.G. Forssmann, and R. Eissele. 2000. Expression and cell-specific localization of the cholecystokinin B/gastrin receptor in the human stomach. Cell Tissue Res. *299*:289-298.
15. Lewis, F. 1996. An Approach to In situ PCR. PE Biosystems, Foster City, CA.
16. Long, A.A. 1998. *In-situ* polymerase chain reaction: foundation of the technology and today's options. Eur. J. Histochem. *42*:101-109.
17. Long, A.A. and P. Komminoth. 1997. *In situ* PCR. An overview. Methods Mol. Biol. *71*:141-161.
18. Long, A.A., P. Komminoth, E. Lee, and H.J. Wolfe. 1993. Comparison of indirect and direct in-situ polymerase chain reaction in cell preparations and tissue sections. Detection of viral DNA, gene rearrangements and chromosomal translocations. Histochemistry *99*:151-162.
19. Martinez, A., M.J. Miller, K. Quinn, E.J. Unsworth, M. Ebina, and F. Cuttitta. 1995. Non-radioactive localization of nucleic acids by direct *in situ* PCR and *in situ* RT-PCR in paraffin-embedded sections. J. Histochem. Cytochem. *43*:739-747.
20. Nakai, M., T. Kawamata, T. Taniguchi, K. Maeda, and C. Tanaka. 1996. Expression of apolipoprotein E mRNA in rat microglia. Neurosci. Lett. *211*:41-44.
21. Nuovo, G.J. 2000. *In situ* localization of PCR-amplified DNA and cDNA. Methods Mol. Biol. *123*:217-238.
22. Nuovo, G.J. 1994. PCR *In Situ* Hybridization. Raven Press, New York.
23. Nuovo, G.J., F. Gallery, P. MacConnell, J. Becker, and W. Bloch. 1991. An improved technique for the *in situ* detection of DNA after polymerase chain reaction amplification. Am. J. Pathol. *139*:1239-1244.
24. Ohtaka-Maruyama, C., F. Hanaoka, and A.B. Chepelinsky. 1998. A novel alternative spliced variant of the transcription factor AP2alpha is expressed in the murine ocular lens. Dev. Biol. *202*:125-135.
25. O'Leary, J.J., R. Chetty, A.K. Graham, and J.O. McGee. 1996. *In situ* PCR: pathologist's dream or nightmare? J. Pathol. *178*:11-20.
26. Patterson, B.K., M. Till, P. Otto, C. Goolsby, M.R. Furtado, L.J. McBride, and S.M. Wolinsky. 1993. Detection of HIV-1 DNA and messenger RNA in individual cells by PCR-driven *in situ* hybridization

and flow cytometry. Science *260*:976-979.
27. Trivier, E., D. De Cesare, S. Jacquot, S. Pannetier, E. Zackai, I. Young, J.L. Mandel, P. Sassone-Corsi, and A. Hanauer. 1996. Mutations in the kinase Rsk-2 associated with Coffin-Lowry syndrome. Nature *384*:567-570.
28. Wilcox, J.N. 1993. Fundamental principles of *in situ* hybridization. J. Histochem. Cytochem. *41*:1725-1733.
29. Wolfahrt, S., B. Kleine, and W.G. Rossmanith. 1998. Detection of gonadotrophin releasing hormone and its receptor mRNA in human placental trophoblasts using in-situ reverse transcription-polymerase chain reaction. Mol. Hum. Reprod. *4*:999-1006.

10 | Immunocytochemical Labeling Methods and Related Techniques for Ultrastructural Analysis of Neuronal Connectivity

Patrizia Aimar[1], Laura Lossi[1], and Adalberto Merighi[1,2]
[1]Department of Veterinary Morphophysiology—Neuroscience Research Group; and [2]Rita Levi Montalcini Center for Brain Repair, Università degli Studi di Torino, Torino, Italy, EU

OVERVIEW

We describe a series of techniques based on the use of immunocytochemical labeling methods for the study of the neurochemistry and connectivity of cells within the central nervous system (CNS). According to our current view, neurons communicate to each other mainly at the level of synapses. Although it is now widely accepted that nonsynaptic neuron-to-neuron communication and neuron-to-glia crosstalk also occur, ultrastructural localization of transmitters–modulators at synapses and analysis of neuronal connections still remain major challenges for a correct understanding of the way in which neuronal networks are organized and operate.

This chapter is mainly devoted to the description of a number of immunocytochemical procedures that are performed on material that, as far as possible, has been prepared according to the protocols for conventional electron microscopy, particularly regarding the use of osmium tetroxide as a fixative for the achievement of optimal ultrastructure. We will also briefly consider several techniques which can be used in combinations with the above methods and may be useful to further dissect the complex interactions between the cells of the CNS.

In situ hybridization techniques at the ultrastructural level are considered in Chapter 8. A detailed description of the methods used for combined ultrastructural and electrophysiological analysis of central neurons is given in Chapter 11, and examples of applications of gold labeling procedures to the study of cell proliferation and apoptosis are considered in Chapter 14.

BACKGROUND

Immunocytochemical Labeling Methods at the Electron Microscope Level

Immunocytochemical labeling methods rely on the possibility to successfully identify the molecule(s) of biological interest directly on tissue sections by using an antigen–antibody reaction that is then visualized in the light and/or electron microscope by adequate means. Historically, the first immunocytochemical reactions at the electron microscopic level were performed in the early 1970s (30). Since then, the original methods have been widely modified and perfected, and in the 1980s, they started to be more extensively used for analysis of central (and peripheral) neurons.

The study of the neurochemistry and connectivity of the CNS poses unique problems due to a series of inherent difficulties for the achievement of satisfactory immunolabeling. The main factors to be considered for this purpose may be summarized as follows: *(i)* retention of antigens within the tissue and the preservation of their antigenicity; *(ii)* preservation of adequate ultrastructure; and *(iii)* the absence of a barrier likely to prevent penetration of antibodies into the tissue and interaction with their respective antigens. The achievement of the optimal balance of these three factors (particularly for the cells of the CNS) is a rather demanding task. Nevertheless, one has to keep in mind that it is obviously pointless to obtain labeling in the absence of good ultrastructure, and vice versa.

In general terms, immunocytochemical labeling methods at the ultrastructural level fall into two main categories: the pre- and postembedding methods. In the former, as it appears from the flow chart diagram of Figure 1, the immunostaining reaction is carried out before osmium postfixation and the embedding procedure, while in the latter, the immunostaining is performed directly on ultrathin sections of tissues which have previously been embedded in plastic resins. Under several aspects, these two intrinsically different approaches have proved to be complementary, since the advantages of the pre-embedding protocols roughly correspond to the disadvantages of the postembedding methods, and vice versa. As we will discuss in one of the following sections, the use of small-sized colloidal gold particles and/or gold clusters in conjunction with silver intensification procedures in pre-embedding methods represents an appealing alternative for the localization of particularly labile antigens. The main advantages and disadvantages of these approaches have been summarized in Figure 2.

Irrespective of the protocol, primary fixation is an insuppressible preliminary step to all immunocytochemical procedures, although ultrathin frozen sections of unfixed material or cryosubstitution techniques might represent possible alternatives (19,24–27,32,59,79). Nonetheless, these methods are very expensive, show many technical difficulties, and their general practicability is still limited. As an additional alternative, some of the problems related to the embedding procedures can be avoided by cutting ultrathin frozen sections of fixed tissue, although this procedure is also very demanding in terms of costs and technical skill (46,52,95). Readers are referred to the existing literature for an additional discussion of the general principles of primary fixation and pre- and postembedding procedures (see for example References 7,8,60,62,65,80–82,91).

Pre-Embedding Immunolabeling

The protocols, which have been most widely employed for pre-embedding ultrastructural studies of the nervous tissue, are the peroxidase–antiperoxidase (PAP) and the avidin–biotin–peroxidase complex (ABC) methods. Historically, the PAP method (90) was used first, and then the more sensitive ABC procedure (43). Both these methods are of the immunoenzymatic type. Nonetheless, immunoenzymatic techniques are not the only available alternative for pre-embedding immunocytochemistry, since other methods rely on radioimmunological procedures (39,71,74,78). We will consider here the protocols based on use of horseradish peroxidase (HRP) and gold particles as reporter molecules.

The protocols based on HRP are generally associated to the use of 3,3′diaminobenzidine (DAB) as a chromogen to reveal the site(s) of positive reaction. As mentioned above, they have a series of advantages–disadvantages that have been summarized in Figure 2, together with some explicative comments. These methods basically consist of a PAP or ABC reaction which is similar to that employed for light microscopy. However, the mandatory need for adequate ultrastructural preservation is obviously linked with several problems that are encountered during fixation, tissue sectioning, and immunostaining. Perhaps the most serious drawbacks associated with the use of these methods are related to the generally poor tissue penetration of immunoreactants and the appearance of nonspecific staining often due to the multiple steps involved in the immunolabeling procedure (see Reference 78). We have considered some of these issues in Table 1.

Very small (about 1 nm) colloidal gold particles or gold clusters (Nanogold™) can also be used for pre- (and post-) embedding immunocytochemical labeling. The use of silver-intensified 1 nm colloidal gold particles or Nanogold in pre-embedding procedures has been introduced in the recent past to overcome some of the problems that are inherent to the "classic" pre- and postembedding immunogold labeling methods (see Figure 2 and Table 1). We will describe and discuss this approach in the following sections of this chapter.

Post-Embedding Immunolabeling

When choosing to use immunogold post-embedding staining methods, it is nec-

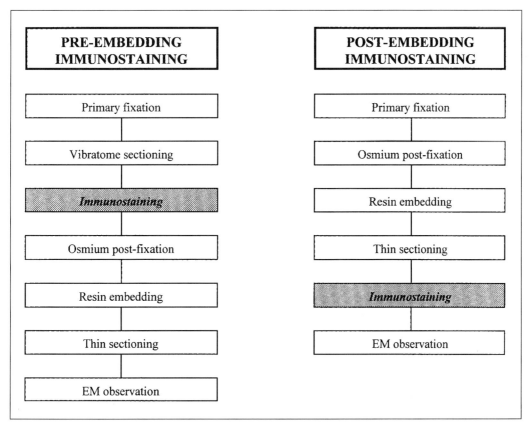

Figure 1. Flow chart diagram of principal steps in pre- and post-embedding immunocytochemical staining of the nervous tissue.

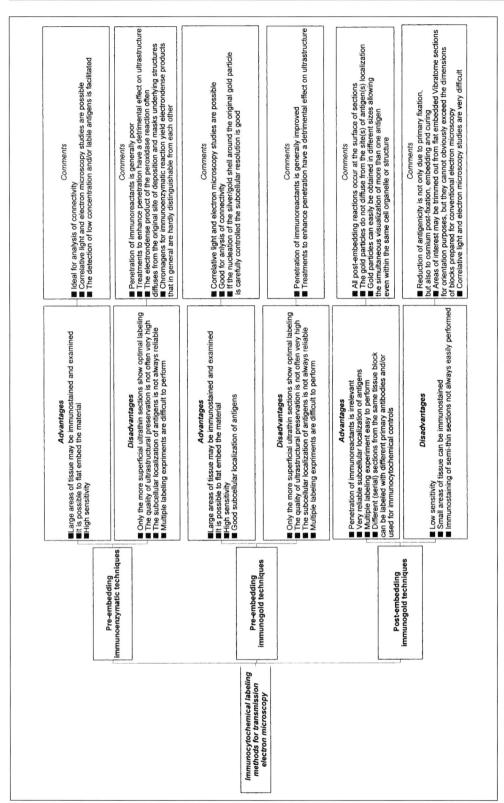

Figure 2. Block diagram showing the main advantages–disadvantages of the most widely employed methods for immunocytochemical labeling in transmission electron microscopy.

Table 1. Practical Hints in Pre-Embedding Procedures

Step in the Preparative Procedure	Available Alternatives	Suggested Guidelines and/or Comments
Fixation	In general: 4% paraformaldehyde + 0.02%–0.5% glutaraldehyde in 0.1 M Sörensen buffer, pH 7.4 Possible additions: 0.2% picric acid (for carbohydrates residues) 1% acrolein (to increase the rate of fixation)	In pre-embedding procedures, primary fixation is carried out with minimal concentrations of glutaraldehyde to minimize the detrimental effects on tissue antigenicity. Large protein antigens (transmitter-synthesizing enzymes, receptors) are often sensitive to minimal amounts of glutaraldehyde and thus require a pre-embedding approach for their visualization.
Tissue Sectioning	Vibrating microtome (Vibratome) Tissue chopper Cryostat or freezing microtome	For CNS analysis, the vibrating microtome represents the best available choice. The tissue chopper often produces sensible tissue damage. Cryostat sections generally show poor ultrastructure.
Penetration of Immunoreactants	0.25% Triton (for 15 min prior to immunostaining) Freeze-thawing Use of high ionic strength solutions Use of Fab' fragments Use of ultra-small gold	Even very low concentration Triton has detrimental effects on ultrastructure. It is advisable not to exceed the concentration indicated. Freeze-thawing is performed on tissue blocks which should not exceed 5 mm thickness before Vibratome sectioning. Tissue is infiltrated in 30% sucrose in 0.1 M phosphate buffer as a cryoprotectant, then immersed in liquid nitrogen for 1–2 minutes and finally thawed in phosphate buffer at room temperature.
Immunostaining with DAB	Conventional staining: Incubation in 0.06% DAB in PBS for 10 minutes at room temperature. Addition of 0.01% H_2O_2 and incubation for further 15 minutes. Intensification protocol: Intensification of the DAB product can be obtained by adding cobalt chloride and nickel ammonium sulphate at the final concentrations of 0.05% and 0.04% before incubation with H_2O_2	Intensification of DAB reaction offers higher sensitivity and results in black end-product which is easier to photograph for correlative light and electron microscopy studies. Prepare the medium by adding dropwise the correct amount of 1% cobalt chloride and 1% nickel ammonium sulphate to the final concentration indicated.
Flat-Embedding	Plastic coverslips Acetate foils (blank EM negatives)	Avoid the use of glass slides and hydrophobic repellent coating, since it is difficult to detach samples for re-embedding.

essary to be aware that problems will be encountered which are not only related to fixation and the labeling procedure, but also to all the preparative stages of postfixation, dehydration, resin infiltration, and curing. All these aspects have been extensively reviewed in previous publications from our laboratory (2,62). For the sake of brevity, we have summarized some relevant aspects in Table 2. Nonetheless, we intend to consider below certain issues of particular importance and mention some of the more recent findings which may represent significant ameliorations of the existing protocols.

Embedding can be carried out using both hydrophobic epoxy resins and hydrophilic acrylic embedding media (18). For most tissue, the quality of tissue preservation is definitely superior after epoxy resin embedding and osmium postfixation, although embedding in hydrophilic media results in higher labeling efficiency. This latter is a consequence of less detrimental effects on tissue antigenicity (17). A method of Epon preparation has recently been described to obtain approximately the same immunogold labeling for epoxy sections as for acrylic sections without any etching (16).

Osmium postfixation represents a fundamental step in post-embedding immunocytochemistry of epoxy sections from the brain and/or spinal cord. In our opinion, the use of osmium postfixation is mandatory for adequate preservation of the delicate membrane structures of different synapses and subcellular organelles within the CNS. The discussion of the mechanism of action of osmium as a fixative for electron microscopy (EM) is beyond the purpose of this review. It is sufficient to remember here that (membrane) lipids remain soluble in organic solvents such as alcohols or acetone (40,69) and are extracted during embedding procedures unless secondary fixation with osmium tetroxide is applied or aldehyde-fixed tissue is embedded at low temperatures using hydrophilic resins (17,18,96). Recently, an osmium-free method of Epon embedment has been introduced that preserves both ultrastructure and antigenicity for post-embedding immunocytochemistry of the CNS (70). The method is based on primary fixation with high concentration glutaraldehyde and replacement of osmium postfixation with tannic acid followed by other heavy metals and p-phenylenediamine. Pretreatments with strong oxidizers (section etching) are usually required to restore antigenicity of osmicated material. Different chemicals have been employed to this purpose, but sodium metaperiodate (10) so far gives the most satisfactory results (see Reference 62 for further discussion). It seems of relevance the finding that, by using of a sodium citrate-buffered sodium metaperiodate solution at high temperature, a significantly higher gold labeling density can be achieved (93).

If those mentioned above may be considered among the most serious problems to be solved to achieve a successful post-embedding immunolabeling of the nervous tissue, perhaps the more relevant advantages in the use of this approach are the possibilities to obtain a reliable subcellular localization of molecules under study, the easiness to perform multiple labelings, and the possibility to quantify the immunocytochemical signal (see Figure 2 and Results and Discussion).

PROTOCOLS

Protocol for Pre-Embedding Immunostaining with Ultra-Small Gold Clusters (Nanogold)

Materials and Reagents

- Formaldehyde EM grade (Electron Microscopy Science, Fort Washington, PA, USA).
- Glutaraldehye EM grade (Electron

Table 2. Practical Hints in Post-Embedding Procedures

Step in the Preparative Procedure	Available Alternatives	Suggested Guidelines and/or Comments
Fixation	In general: 4% paraformaldehyde + 2.5% glutaraldehyde in 0.1 M Sörensen buffer, pH 7.4 or 1% paraformaldehyde + 2% glutaraldehyde in 0.1 M Sörensen buffer, pH 7.4	In post-embedding procedures, fixation is usually carried out with high concentration glutaraldehyde that is necessary to achieve optimal structural preservation. Up to 5% glutaraldehyde is used for immunocytochemical labeling of transmitter amino acids (see for example Reference 15).
Postfixation	1% osmium tetroxide in water or 1% osmium ferrocyanide	Potassium permanganate can be used as an alternative to osmium tetroxide when embedding is carried out in hydrophilic media that need to be cured at low temperature by UV light (62).
Dehydration	Ethanol or Acetone	The total length of dehydration should not exceed 45 minutes to avoid extraction of tissue components and reduction of antigenicity.
Embedding	Araldite or Epon or Epon-Araldite	Embedding in Araldite is usually ideal for subsequent etching with sodium metaperiodate. Epon or Epon-Araldite require longer etching to restore tissue antigenicity (62). Keep the embedding steps as short as possible to avoid extraction of antigens by resin monomer. Embedding with hydrophilic resins (LR White, LR Gold, Lowicryl K4M) is usually unsatisfactory for CNS analysis (62).
Sectioning and Counterstaining	Collect sections onto nickel or gold grids Counterstain very shortly in uranyl acetate and lead citrate	Nickel (gold) grids avoid interaction of the metal with oxidizing solutions. It is preferable to use uncoated grids unless strictly necessary (single slot grids for reconstruction studies or in certain double labeling protocols) (62).

Microscopy Science).
- Bovine Serum Albumin, Fraction V (BSA; Sigma, St. Louis, MO, USA).
- Triton® X-100 (Sigma).
- Normal Goat Serum (NGS; Sigma).
- Nanogold antirabbit (Nanoprobes, Stony Brook, NY, USA).
- Nanogold antimouse (Nanoprobes).
- Goldenhance-EM™ (Nanoprobes).

Procedure

1. Perfuse the animal with a fixative consisting of 4% paraformaldehyde, 0.02% to 0.05% glutaraldehyde and 0.2% picric acid in Sörensen buffer 0.1 M, pH 7.6.
2. Remove the areas of interest and trim them to small blocks (not exceeding 5 mm in thickness). Postfix for 2 additional hours in the same fixative.
3. Cryoprotect blocks by infiltration with increasing concentrations of sucrose in phosphate-buffered saline (PBS). Infiltration is carried out as follows: 10% sucrose, 1 hour; 20% sucrose, 2 hours; 30% sucrose, overnight (or as long as blocks sink at the bottom of vials).
4. Quickly freeze blocks in liquid nitrogen for 1 to 2 minutes and then allow them to thaw in PBS at room temperature.
5. Thoroughly wash blocks in PBS and cut them with a Vibratome at a thickness of 30 to 50 μm.
6. Carefully wash sections in PBS. Optional: incubate in 0.2% Triton X-100 for 15 minutes prior to immunostaining.
7. Incubate in NGS (diluted 1:30 in PBS containing 0.2% BSA) for 60 minutes at room temperature.
8. Incubate in primary antibodies at optimal titer (in PBS containing 0.2% BSA) overnight preferably under continuous agitation. The length and temperature of incubation in the primary antibody must be varied upon necessity. Incubation as long as 72 hours at 4°C can be required for low affinity antibodies.
9. Thoroughly wash in 0.5 M Tris-buffered saline (TBS), pH 7.0 to 7.4 containing 1% BSA (TBS-BSA).
10. Incubate overnight at room temperature in the appropriate Nanogold reagent diluted 1:100 in TBS-BSA.
11. Carefully wash sections as follows: PBS 2 × 5 minutes; PBS + 0.1% gelatin for 5 minutes; distilled water for 5 minutes.
12. Prepare the Goldenhance-EM reagent just before use following the manufacturer's instructions: mix well 1 volume of reagent A (enhancer) and 1 volume of reagent B (activator); wait for 20 minutes and add 1 volume of reagent C (initiator).
13. Check staining intensity at the light microscope. Usually the reaction is completed after 10 minutes, but the intensification time should be adjusted from case to case.
14. Wash quickly in distilled water and then transfer sections for a few seconds in a solution of photographic fixer. Thoroughly wash in distilled water.
15. Optional: postfix sections in 2% glutaraldehyde in Sörensen buffer 0.1 M, pH 7.4, for 30 minutes at room temperature.
16. Thoroughly wash in Sörensen buffer, postfix in osmium tetroxide, dehydrate, and infiltrate in resin as described below in steps 5 through 11 of the Protocol for Preparation of the Nervous Tissue for Post-embedding Immunocytochemistry.
17. Flat-embed sections between 2 blank EM negatives.

10. Ultrastructural Analysis of Neuronal Connectivity

18. Areas of interest may be photographed in the light microscope, trimmed to be directly glued at the top of an empty resin block, and cut with the ultramicrotome.

Protocol for Preparation of the Nervous Tissue for Post-Embedding Immunocytochemistry

Materials and Reagents

- Formaldehyde EM grade.
- Glutaraldehye EM grade.
- BSA.
- Triton X-100.
- Sodium metaperiodate (Sigma).
- NGS.
- Osmium tetroxide (Electron Microscopy Science).
- Uranyl acetate (Electron Microscopy Science).

Procedure

1. Perfuse the animal with a fixative consisting of 1% paraformaldehyde and 2% glutaraldehyde in Sörensen buffer 0.1 M, pH 7.6.
2. Remove the areas of interest and trim them to small blocks (not exceeding 5 mm in thickness). Postfix for 2 additional hours in the same fixative.
3. Reduce blocks to small (1 mm^3) cubes. Optional: it is advisable to cut Vibratome sections (200 µm) of the areas of interest to be eventually flat-embedded for the purpose of better orientation or correlative light microscopy studies.
4. Thoroughly wash in Sörensen buffer 0.1 M, pH 7.6.
5. Postfix Vibratome sections (or blocks) in osmium ferrocyanide for 60 minutes at 4°C. The osmium ferrocyanide is prepared just prior to use by adding one volume of 2% aqueous osmium tetroxide to one volume of 3% potassium ferrocyanide.
6. Wash in maleate buffer, pH 5.1, 4 × 5 minutes.
7. Stain for 60 minutes at 4°C with 1% uranyl acetate in maleate buffer, pH 6.0.
8. Wash in maleate buffer, pH 5.1, 4 × 5 minutes.
9. Dehydrate as follows: 50% ethanol for 5 minutes; 80% ethanol for 5 minutes; 95% ethanol for 5 minutes; 100% ethanol 3 × 10 minutes.
10. Transfer sections in propylene oxide 2 × 10 minutes.
11. Infiltrate in resin as follows: propylene oxide–resin without accelerator (vol/vol): 3:1 for 15 minutes; 2:1 for 15 minutes; 1:1 for 2 hours (not longer). Embed in resin plus accelerator and cure for 24 hours.

Protocol for Post-Embedding Single Immunogold Staining

Materials and Reagents

- Syringe filters—pore size 0.22 µm (Whatman, Maidstone, England).
- 200 to 300 mesh nickel grids (Electron Microscopy Science).
- BSA.
- Triton X-100.
- NGS.
- Egg albumin (Sigma).
- Goat antirabbit (mouse) IgG gold conjugated (10 or 20 nm) (British Bio-Cell International, Cardiff, Wales, UK).
  An alternative: protein A-gold complex (10 or 20 nm) (British BioCell International).

Procedure

All solutions must be filtered with disposable filters. All steps are performed directly on grid. Grids are incubated on drops (30–50 µL) of solutions.

1. Collect sections onto precleaned uncoated grids.
2. Float grids on drops of a saturated aqueous solution of sodium metaperiodate for 5 to 15 minutes at room temperature (see also Table 2).
3. Rinse in 0.5 M TBS, pH 7.0 to 7.6, containing 1% Triton X-100.
4. Incubate for 60 minutes at room temperature in TBS containing 1% BSA (TBS-BSA) to which 10% normal serum from the donor species of the gold-conjugated IgGs has been added (generally NGS). If protein A-gold complexes are used, incubate in TBS-BSA containing 1% egg albumin.
5. Transfer onto drops of primary antisera at optimal titer for 24 to 72 hours at room temperature.
6. Thorough wash in TBS-BSA and incubate for 60 minutes at 37°C with the appropriate gold conjugate diluted 1:15 in TBS-BSA.
7. Extensively rinse in TBS-BSA, and postfix for 10 minutes at room temperature in 2.5% glutaraldehyde in Sörensen or cacodylate buffer.
8. Wash grids in double-distilled water and allow to dry protected from dust.
9. Counterstain with Reynold's lead citrate and uranyl acetate.

Protocol for Post-Embedding Double Immunogold Staining Using Primary Antibodies Raised in Different Species

Materials and Reagents

See section on single immunogold staining.

Procedure

1–4. See section on single immunogold staining.
5. Incubate in a mixture of the first primary antibodies at optimal titers, i.e., rabbit anti-A and mouse anti-B.
6. Thoroughly wash in TBS-BSA and incubate for 60 minutes at 37°C with the appropriate mixture of gold conjugates diluted 1:15 in TBS-BSA, i.e., goat antirabbit 10 nm gold and goat antimouse 20 nm gold.
7–9. See section on single immunogold staining.

RESULTS AND DISCUSSION

Pre-Embedding Versus Post-Embedding Immunolabeling Methods

The comparison of Figures 3 and 4 is useful to the purpose of describing the main advantages and disadvantages in the use of pre- and post-embedding procedures for the localization of neural antigens in the CNS. In this case, we have used a mouse monoclonal antibody against the glial fibrillary acidic protein (GFAP) (Roche Molecular Biochemicals, Mannheim, Germany) to localize this intermediate filament protein in astrocytes from the hippocampus and spinal cord. We will briefly consider the following main issues: *(i)* ultrastructural preservation, *(ii)* sensitivity of the methods, *(iii)* subcellular localization of antigen(s) under study, *(iv)* possibility to quantify the results of immunocytochemical labeling, and *(v)* to perform multiple stainings.

<u>Ultrastructural Preservation</u>

As mentioned in the Background section, the need for optimal ultrastructural preservation is mandatory, particularly in the study of neural connectivity and neurochemical characterization of different synapses in the neuropil of the CNS. Although the use of pre-embedding immunolabeling with ultra-small gold offers several advantages over the more classical immunoenzymatic methods (as previously discussed), it is clear that the ultrastructural preservation, which can be achieved by this means, is generally inferior to that obtained when post-embedding immunogold labeling is performed on osmicated material in epoxy resins. The detrimental effect(s) on ultrastructure are mainly linked to the (general) use of milder aldehyde fixation and the treatments that are necessary to improve tissue penetration. In particular, treatment with detergents (usually Triton X-100) even for a short period of time is sufficient to dissolve part of the cell membranes (see Reference 78 for further discussion). Under this aspect, the use of freeze-thawing may be less detrimental. However, in both cases, there is a

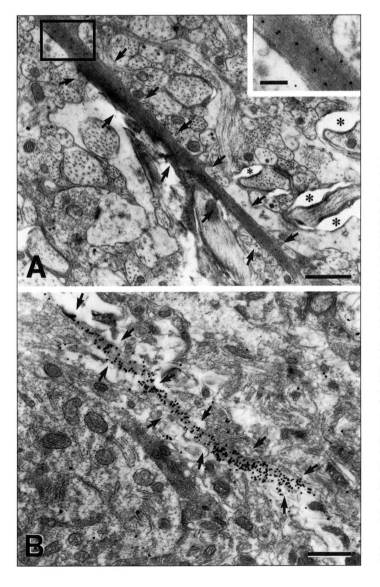

Figure 3. Ultrastructural visualization of GFAP immunoreactivity in the rat hippocampus using post-embedding immunogold (A) or pre-embedding immunogold (B) labeling techniques. Post-embedding immunolabeling (A) has been carried out directly on ultrathin sections from osmicated tissue using a 10 nm IgG gold conjugate. An astrocytic process is indicated by the arrows. Gold particles are seen over the glial filaments (insert). Note that they are quite limited in number and show a "patchy" distribution although filaments run all along the glial process. Note also that cell membranes are nicely preserved, and that some holes–cleft (asterisks) can be observed as a consequence of section etching with sodium metaperiodate. Pre-embedding immunolabeling (B) has been carried out using an ultra-small gold probe (Nanogold) followed by silver intensification. Silver-intensified gold particles are easily seen at low power and almost completely fill an astrocytic process which is indicated by the arrows. Note the intensity of the immunocytochemical signal, but also the less satisfactory preservation of cell membranes due to the use of Triton X-100 pretreatment. Scale bars = 1 µm; insert = 0.25 µm.

significative variability in the degree of preservation not only within the same tissue block, but also among areas of the same section. Therefore, a certain degree of unpredictability of the final result in terms of ultrastructural preservation has always to be expected after pre-embedding immunolabeling. It is worthy to mention here that although the use of ultra-small gold offers some ameliorations with respect to the DAB immunoenzymatic procedure, it still enhances tissue penetration of primary antibodies. To improve tissue morphology and yet allow for good antibody and reagent penetration, it was recently proposed to use high glutaraldehyde concentration (3%) and sodium metabilsulfite (31) as a fixative for pre-embedding with ultra-small gold.

Sensitivity of the Methods

The relatively poor sensitivity of the post-embedding immunolabeling methods

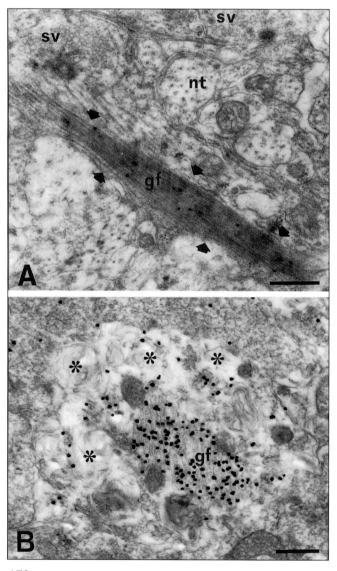

Figure 4. Ultrastructural visualization of GFAP immunoreactivity in the rat hippocampus using post-embedding immunogold (A) or pre-embedding immunogold (B) labeling techniques. High power views clearly show the differences in staining intensity of glial filaments (gf) and the extent of preservation of subcellular structures such as synaptic vesicles (sv) or neurotubules (nt). Also, at this magnification, the detrimental effect of tissue permeabilization is clearly apparent (asterisks). Scale bars = 0.5 μm.

perhaps represents the most limiting factor in the use this approach for the localization of neural antigens within the CNS. Clearly, a much stronger signal is generally achieved after pre-embedding immunolabeling. Again, the comparison of Figures 3 and 4 is self explanatory. Besides the already mentioned chemical parameters to be taken into consideration, i.e., primary fixation, osmium postfixation, dehydration, embedding, and curing, which altogether decrease the sensitivity of the post-embedding methods by "killing" tissue antigenicity, two physical parameters need also to be considered: *(i)* a proper surface exposition of the antigen, and *(ii)* its (relative) spatial distribution. The post-embedding reactions on plastic sections are surface reactions, and labeling is restricted to epitopes that are exposed at the cutting planes of the ultrathin sections. This might explain why post-embedding immunogold labeling is, in our example, restricted to certain areas of the astrocyte processes which, on the other hand, appear to contain glial filaments all along their course (Figure 3). The spatial distribution of the antigen(s) may also affect the actual possibility of a successful localization with the pre- or post-embedding approaches. If one considers two antigens which, in absolute terms, have the same concentration inside the cell, but a different cellular localization, such as a cytosolic transmitter synthesizing enzyme and a biologically active peptide packed in large dense cored vesicles (LGVs), the difficulties in localizing the former by a post-embedding approach are obvious, since a low number of gold particles scattered over the general cytoplasm are difficult to ascribe to specific staining rather than to unwanted background. This example and Figure 5 help to clarify why the post-embedding methods are ideal for localizing antigens concentrated (stored) in synaptic vesicles or other subcellular organelles rather than evenly distributed inside the cell. On the other hand, the fact that immunoenzymatic pre-embedding reactions with DAB are self-enhancing reactions is often associated with high sensitivity (but poor cellular resolution, since the chromogen deposition usually diffuses from the actual sites of the antigen–antibody interaction) and makes these methods ideal for studies of molecules which have a rather general distribution within the cytosol, although at relatively low concentrations.

Subcellular Localization of Immunolabeling

The problem of reliable subcellular localization of immunolabeling is directly correlated to the so-called spatial resolution of the techniques. This latter becomes a critical factor which must be carefully considered to determine the actual site in which the antigen(s) under study is (are) localized. The spatial resolution of an immunocytochemical technique is defined as the distance between the (particulate) marker employed for visualization and the epitope recognized by the primary antibody. In the case of the immunoenzymatic techniques, this is rather difficult to estimate due to the nature of the electrondense marker and the possibility of its diffusion from the original site(s) of deposition. This is the reason why the use of immunogold labeling techniques gives the possibility of localizing antigens to different subcellular compartments more precisely than after any other ultrastructural immunocytochemical methods. The minimum value of the spatial resolution that can theoretically be achieved with the protein A-gold technique is about 16 nm, using a colloidal gold particle of 3 nm (80). Using the IgG-gold technique the spatial resolution is slightly less satisfactory, since IgGs are bigger than protein A (68). For example, using IgG-gold conjugates with gold particles of 10 nm, the resolution is approximately 21 nm. Therefore, the bigger the gold particle, the less satisfactory is the spatial resolution of the technique. Gold clusters of small size covalently attached to Fab′ fragments (Nanogold) offer a better resolution since Fab′ fragments are about 1/3 the size of IgGs (38,39). In this latter case, however, one has to take into consideration the need for silver enhancement of the ultra-small size gold probe. The discussion of this aspect is not merely academic, since it might be necessary, for example, to localize certain molecules within the synaptic vesicles (mean diameter 40 nm) or on their membranes, so that a resolution in the magnitude of 20 nm clearly becomes a critical factor for a correct interpretation of the final result(s) obtained after immunogold labeling.

Quantification of Immunostaining

In epoxy resin-embedded material, gold

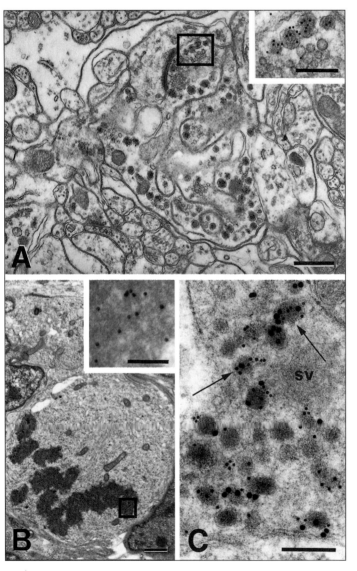

Figure 5. Post-embedding immunogold labeling of neural antigens in osmicated tissue. (A) Localization of calcitonin gene-related peptide (CGRP) immunoreactivity (10 nm gold particles) over LGVs in terminals within the superficial dorsal horn of the spinal cord. In the insert, the subcellular site of positive reaction is clearly restricted to LGVs. (B) Incorporation of bromo-deoxyuridine (BrdU) in a mitotic granule cell precursor of the postnatal cerebellar cortex. Exogenously administered BrdU has been visualized using an indirect immunogold method with 15 nm colloidal gold. Gold particles are scattered over mitotic chromosomes. (C) Double immunogold labeling of substance P (10 nm gold, long arrows) and CGRP (20 nm gold) over the LGVs in a nerve terminal of the superficial dorsal horn. Staining has been obtained using the simultaneous labeling method. sv = synaptic vesicles. Scale bars = 0.2 μm (A and C), 1 μm (B).

labeling is restricted to antigens exposed to the surface of the ultrathin sections (7,9). Since immunoreagents do not penetrate sections, tissue permeability is not influencing the distribution of labeling as after pre-embedding immunocytochemistry. Therefore, it is possible to quantify the gold particle label over different subcellular compartments. Quantitative measurement of gold labeling in ultrathin frozen sections requires a 3-dimensional analysis, since in this case, immunoreactants do penetrate tissues (52,72,73).

Results are usually expressed in terms of number of gold particles per area and should be corrected for background labeling over empty resin (28,29,36,42,63,80, 86,88,91). Computerized systems and dedicated software have been developed to this purpose (6,11,28,76).

Multiple Labelings

The availability of gold probes of different sizes makes the use of post-embedding immunogold methods as the methods of choice for multiple labeling experiments. Several alternatives are available to this purpose, and these include the use of the simultaneous labeling procedure when the two (or more) primary antibodies available are raised in different species (Figure 5), the double face methods, the double protein A (protein G) procedure, and a multiple labeling protocol based on the use of paraformaldehyde vapors to block unreacted primary immunoglobulin binding sites. The advantages and disadvantages of these methods have been already reviewed in detail (62). Finally, it should be mentioned here that silver intensification of ultra-small gold probes can be used not only as a single step procedure after immunogold labeling to facilitate the visualization of gold particles at low magnification, but also for producing different sized gold particles in multiple labeling procedures (44,49,50,92).

Use of Combined Methods for the Analysis of Neuronal Connectivity

As thoroughly discussed in the previous sections, several labeling procedures at the ultrastructural level can often be performed after high glutaraldehyde, osmium, and epoxy embedment, i.e., using a material that was subjected to a quasi-conventional preparative procedure. This has opened the way not only to combine together the different immunolabeling protocols, but also to use these methods in conjunction with several other procedures particularly relevant to the study of neuronal connections. These combined approaches have a wide range of potential applications and can be theoretically employed to solve a number of problems in different fields of basic and applied neurobiology research.

Combination Pre- and Post-embedding Procedures

As discussed above, pre- and post-embedding methods have to be considered as fully complementary to each others under several aspects. The combined use of pre- and post-embedding immunocytochemical labeling has been particularly useful for the study of the connectivity and synaptic interactions of central neurons (see for example References 20,21,34,35, and 67). In general terms, a pre-embedding labeling procedure is performed first, after low or mild aldehyde fixation, using a primary antibody directed towards the more labile antigen to be detected and a suitable chromagen (usually DAB). Then stronger aldehyde fixation, osmication, and embedding, follow. The second, more resistant, antigen under study is eventually visualized using a post-embedding staining protocol. The entire procedure is technically very demanding and not always so straight. The choice of such an approach mainly depends on the impossibility to visualize the more labile antigen with a post-embedding protocol. However, besides the resistance of antigen(s) to strong fixation, there are several other variables to be considered, as widely discussed in the previous sections, among which the membrane or cytosolic rather than particulate distribution of antigen(s) under analysis appears to be particularly significant.

Combination with Enzyme Histochemical Techniques

Again, this approach combines a pre- and a post-embedding procedure, but the first pre-embedding labeling is a histochemical reaction, which is used to reveal endogenous enzyme activities. In this case, primary fixation must be compatible with retention of the activity of enzyme(s) under study. Moreover, a proper chromagen for ultrastructural visualization of the enzyme –substrate reaction is needed. Fortunately, a number of enzymes, such as for example acetylcholinesterase (12,87), alkaline phosphatase (33,37,89), β-galactosidase (85), and nicotinamide adenine dinucleotide phosphate-diaphorase (NADPH-d) (1,13, 94), retain their activities in tissues fixed for ultrastructural examination. The possibility to label nitric oxide (NO)-producing cells by means of a NADPH-d reaction modified for ultrastructural use in combination with post-embedding immunogold (3,13) seems to be of particular interest. An additional field of further exploitation of post-embedding gold labeling methods relies on the possibility to combine them with the ultrastructural visualization of enzymes such as β-galactosidase and human placental alkaline phosphatase, which have been engineered as reporter genes in transgenic animals (37,85). For further discussion of this type of approach see Chapter 7.

Combination with Neuroanatomical Tract-Tracing Methods

The combination of pre- and post-embedding immunogold labeling and neuroanatomical tract-tracing methods is an useful tool in the study of neuronal connections. There are several possibilities to trace neuronal connections at the light level. Any of the methods so far developed for light microscopy has inherent advantages and disadvantages. In general terms, any of the tract-tracing methods to be successfully employed in conjunction with post-embedding ultrastructural immunocytochemistry has to be compatible with a good preservation of cell organelles and membranes and with the obvious need to use an electrondense tracer. Successful tracing of neuronal pathways at the ultrastructural level has been achieved using: *(i)* Golgi impregnation (87); *(ii)* retrograde labeling with either free HRP or HRP conjugated to plant lectins or other electron dense markers such as colloidal gold particles (5,23,53–56,58,75,78); *(iii)* anterograde labeling with free HRP (14,15,22, 45,51) or biotinylated dextran amine (84); and *(iv)* HRP microiontophoretic filling of single neurons or fibers, following electrophysiological characterization (4,57,77). There are numerous examples in the literature in which these methods have been employed in conjunction with post-embedding immunogold labeling (61,66,83,84). Other approaches such as: *(i)* anterograde tracing using radiolabeled amino acids (41); *(ii)* degeneration methods (47,48); and *(iii)* microiontophoretic filling of single neurons under visual control (64) can also be potentially combined with post-embedding immunogold labeling.

As a general protocol, all these procedures involved the administration of the tracer in vivo or in a slice preparation, followed by primary fixation in low concentration glutaraldehyde and the pre-embedding visualization of the tracer by enzyme histochemistry. If a pre-embedding immunogold procedure is then chosen, tissue is processed as described in the Protocols section. If a post-embedding approach is preferred, tissue is then postfixed in higher glutaraldehyde concentration, osmicated, and embedded. Finally a "classical" immunogold labeling procedure is employed to detect directly on tissue sections the antigen(s) of interest (see Protocols). Since the entire procedure is generally carried out on vibratome slices, correlative light and electronic studies are made possible.

ACKNOWLEDGMENTS

Much of the work described here was supported by grants from the University of Torino, the Italian Ministero dell'Istruzione dell'Università e della Ricerca (MIUR - Cofin 2000, Cofin 2001), and Consiglio Nazionale delle Ricerche (CNR). We are greatly indebted to Dr. Jim Hainfeld, Nanoprobes, Inc., for his generous gift of Nanogold probes and Goldenhance-EM reagent.

REFERENCES

1. Aimar, P., I. Barajon, and A. Merighi. 1998. Ultrastructural features and synaptic connections of NADPH-diaphorase positive neurones in the rat spinal cord. Eur. J. Anat. 2:27-34.
2. Aimar, P., L. Lossi, and A. Merighi. 1997. Immunogold labeling for transmission electron microscopy: exploring new frontiers. Cell Vision 4:394-407.
3. Aimar, P., L. Pasti, G. Carmignoto, and A. Merighi. 1998. Nitric oxide-producing islet cells modulate the release of sensory neuropeptides in the rat substantia gelatinosa. J. Neurosci. 18:10375-10388.
4. Alvarez, F.J., A.M. Kavookjian, and R. Light. 1992. Synaptic interactions between GABA-immunoreactive profiles and the terminals of functionally defined myelinated nociceptors in the monkey and cat spinal cord. J. Neurosci. 12:2901-2917.
5. Basbaum, A.I. 1989. A rapid and simple silver enhancement procedure for ultrastructural localization of the retrograde tracer WGAapoHRP-Au and its use in double-label studies with post-embedding immunocytochemistry. J. Histochem. Cytochem. 37:1811-1815.
6. Beier, K. and H.D. Fahimi. 1985. Automatic determination of labeling density in protein A-gold immunocytochemical preparations using an image analyzer. Application to peroxisomal enzymes. Histochemistry 82:99-100.
7. Bendayan, M. 1984. Protein A-gold electron microscopic immunocytochemistry: methods, applications, and limitations. J. Electron Microsc. Tech. 1:243-270.
8. Bendayan, M. 1987. Introduction of the protein G-gold complex for high-resolution immunocytochemistry. J. Electron Microsc. Tech. 6:7-13.
9. Bendayan, M., A. Nanci, and F.W.K. Kan. 1987. Effect of tissue processing on colloidal gold cytochemistry. J. Histochem. Cytochem. 35:983-996.
10. Bendayan, M. and M. Zollinger. 1983. Ultrastructural localization of antigenic sites on osmium-fixed tissues appliying the protein A-gold technique. J. Histochem. Cytochem. 31:101-109.
11. Blackstad, T.W., T. Karagulle, and O.P. Ottersen. 1990. Morforel, a computer program for two-dimensional analysis of micrographs of biological specimens, with emphasis on immunogold preparations. Comput. Biol. Med. 20:15-34.
12. Bolam, J.P., C.A. Ingham, and A.D. Smith. 1984. The section-Golgi-impregnation procedure-3. Combination of Golgi-impregnation with enzyme histochemistry and electron microscopy to characterize acetylcholinesterase-containing neurons in the rat neostriatum. Neuroscience 12:711-718.
13. Bouwens, L. and G. Kloppel. 1994. Cytochemical localization of NADPH-diaphorase in the four types of pancreatic islet cell. Histochemistry 101:209-214.
14. Broman, J. and F. Ådahl. 1994. Evidence for vesicular storage of glutamate in primary afferent terminals. Neuroreport 5:1801-1804.
15. Broman, J., J. Westman, and O.P. Ottersen. 1990. Ascending afferents to lateral cervical nucleus are enriched in glutamate-like immunoreactivity: a combined anterograde transport-immunogold study in the cat. Brain Res. 520:178-191.
16. Brorson, S.H. and F. Skjorten. 1996a. Improved technique for immunoelectron microscopy. How to prepare epoxy resin to obtain approximately the same immunogold labeling for epoxy sections as for acrylic sections without any etching. Micron 27:211-217.
17. Brorson, S.H. and F. Skjorten. 1996b. The theoretical relationship of immunogold labeling on acrylic sections and epoxy sections. Micron 27:193-201.
18. Causton, B.E. 1984. The choice of resins for electron immunocytochemistry, p. 29-36. In J.M. Polak and I.M. Varndell (Eds.), Immunolabelling for Electron Microscopy. Elsevier Publishing, Amsterdam.
19. Chang, J.P. and M. Yokama. 1970. A modified section freeze substitution technique. J. Histochem. Cytochem. 18:683-684.
20. Cheng, P.Y., L.Y. Liu Chen, C. Chen, and V.M. Pickel. 1996. Immunolabeling of Mu opioid receptors in the rat nucleus of the solitary tract: extrasynaptic plasmalemmal localization and association with Leu5-enkephalin. J. Comp. Neurol. 371:522-536.
21. Cheng, P.Y., A. Moriwaki, J.B. Wang, G.R. Uhl, and V.M. Pickel. 1996. Ultrastructural localization of mu-

opioid receptors in the superficial layers of the rat cervical spinal cord: extrasynaptic localization and proximity to Leu5-enkephalin. Brain Res. 731:141-154.
22. Cruz, F., D. Lima, and A. Coimbra. 1987. Several morphological types of terminal arborizations of primary afferents in laminae I-II of the rat spinal cord, as shown after HRP labeling and Golgi impregnation. J. Comp. Neurol. 261:221-236.
23. De Zeeuw, C.I., J.C. Holstege, F. Calkoen, T.J.H. Ruigrok, and J. Voogd. 1988. A new combination of WGA-HRP anterograde tracing and GABA immunocytochemistry applied to afferents of the cat inferior olive at the ultrastructural level. Brain Res. 447:369-375.
24. Dudek, R.W. and A.F. Boyne. 1986. An excursion through the ultrastructural world of quick-fozen pancreatic islets. Am. J. Anat. 175:217-243.
25. Dudek, R.W., A.F. Boyne, and M. Freinkel. 1981. Quick-freeze fixation and freeze-drying of isolated rat pancreatic islets: application to the ultrastructural localization of inorganic phosphate in the pancreatic beta cell. J. Histochem. Cytochem. 29:321-325.
26. Dudek, R.W., G.V. Childs, and A.F. Boyne. 1982. Quick-freezing and freeze drying in preparation for high quality morphology and immunocytochemistry at the ultrastructural level: application to pancreatic beta cells. J. Histochem. Cytochem. 30:129-138.
27. Dudek, R.W., I.M. Varndell, and J.M. Polak. 1984. Quick-freeze fixation and freeze drying for electron microscopic immunocytochemistry, p. 235-248. In J.M. Polak and I.M. Varndell (Eds.), Immunolabelling for Electron Microscopy. Elsevier Publishers, Amsterdam.
28. Eneström, S. and B. Kniola. 1990. Quantitiative ultrastructural immunocytochemistry using a computerized image analysis system. Stain Technol. 65:263-278.
29. Eneström, S. and B. Kniola. 1995. Resin embedding for quantitative immunoelectron microscopy. A comparative computerized image analysis. Biotech. Histochem. 70:135-146.
30. Faulk, W.P. and G.M. Taylor. 1971. An immunocolloid method for the electron microscope. Immunochemistry 8:1081-1083.
31. Gilerovitch, H.G., G.A. Bishop, J.S. King, and R.W. Burry. 1995. The use of electron microscopic immunocytochemistry with silver-enhanced 1.4-nm gold particles to localize GAD in the cerebellar nuclei. J. Histochem. Cytochem. 43:337-343.
32. Gingras, D. and M. Bendayan. 1994. Compartmentalization of secretory proteins in pancreatic zymogen granules as revealed by immunolabeling on cryo-fixed and molecular distillation processed tissue. Biol. Cell 81:153-163.
33. Gomez, S. and A. Boyde. 1994. Correlated alkaline phosphatase histochemistry and quantitative backscattered electron imaging in the study of rat incisor ameloblasts and enamel mineralization. Microsc. Res. Tech. 29:29-36.
34. Gong, L.W., Y.-Q. Ding, D. Wang, H.X. Zheng, B.Z. Qin, J.S. Li, T. Kaneko, and N. Mizuno. 1997. GABAergic synapses on mu-opioid receptor-expressing neurons in the superficial dorsal horn: an eletron microscopy study in the cat spinal cord. Neurosci. Lett. 227:33-36.
35. Gracy, K.N. and V.M. Pickel. 1996. Ultrastructural immunocytochemical localization of the N-methyl-D-aspartate receptor and tyrosine hydroxylase in the shell of the rat nucleus accumbens. Brain Res. 739:169-181.
36. Gundersen, V., O. Shupliakov, L. Brodin, O.P. Ottersen, and J. Storm-Mathisen. 1995. Quantification of excitatory amino acid uptake at intact glutamatergic synapses by immunocytochemistry of exogenous D-aspartate. J. Neurosci. 15:4417-4428.
37. Gustinich, S., A. Feigenspan, D.K. Wu, L.J. Koopman, and E. Raviola. 1997. Control of dopamine release in the retina: a transgenic approach to neural networks. Neuron 18:723-726.
38. Hainfeld, J.F. and F.R. Furuya. 1992. A 1.4 nm gold gluster covalently attached to antibodies improves immunolabeling. J. Histochem. Cytochem. 40:177-184.
39. Hainfeld, J.F. and R.D. Powell. 1997. Nanogold technology: new frontiers in gold labeling. Cell Vision 4:408-432.
40. Hayat, M.A. 1977. Principles and techniques of electron microscopy. Vol. 1. Biological Applications. Edward Arnold, New York.
41. Holstege, J.C. and J.M. Vrensen. 1988. Anterograde tracing in the brain using autoradiography and HRP-histochemistry. A comparison at the ultrastructural level. J. Microsc. 150:233-243.
42. Howell, K.E., U. Reuter-Carlson, E. Devaney, J.P. Luzio, and S.D. Fuller. 1987. One antigen one gold? A quantitative analysis of immunogold labelling of plasma membrane 5′-nucleotidase in frozen thin sections. Eur. J. Cell Biol. 44:318-326.
43. Hsu, S.M., L. Raine, and H. Fanger. 1981. Use of avidin-biotin-peroxidase complex (ABC) in immunoperoxidase techniques. A comparison between ABC and unlabeled antibody (PAP) procedures. J. Histochem. Cytochem. 29:577-587.
44. Huiz, J.M., M.L. Campos, R.H. Helfert, and R.A. Altschuler. 1996. Silver intensification of immunocolloidal gold on ultrathin plastic sections applied to the study of the neuronal distribution of GABA and glycine. J. Hirnforsch. 37:51-56.
45. Kechagias, S. and J. Broman. 1994. Compartmentation of glutamate and glutamine in the lateral cervical nucleus: further evidence for glutamate as a spinocervical tract neurotransmitter. J. Comp. Neurol. 340:531-540.
46. Keller, G.A., K.T. Tokuyasu, A.H. Dutton, and S.J. Singer. 1984. An improved procedure for immunoelectronmicroscopy: ultrathin plastic embedding of immunolabelled ultrathin frozen sections. Proc. Natl. Acad. Sci. USA 81:5744-5747.
47. Knyihar-Csillik, E., B. Csillik, and A.B. Oestreicher. 1992. Light and electron microscopic localization of B-50 (GAP43) in the rat spinal cord during transganglionic degenerative atrophy and regeneration. J. Neurosci. Res. 32:93-109.
48. Knyihar-Csillik, E., B. Csillik, and P. Rakic. 1982. Ultrastructure of normal and degenerating glomerular terminals of dorsal root axons in the substantia gelatinosa of the Rhesus monkey. J. Comp. Neurol. 210:357-375.

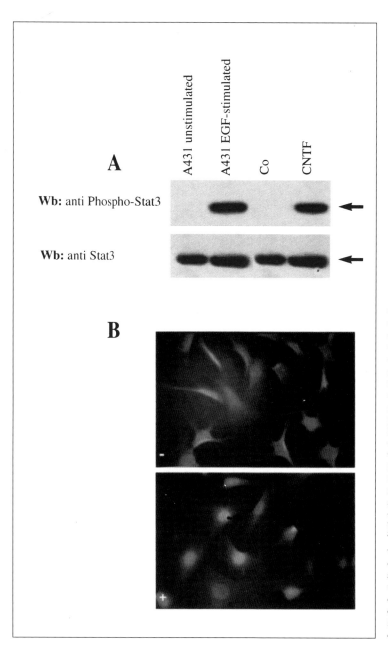

Chapter 1, Figure 2. STAT3 activation in CNS progenitor cells following CNTF treatment. (A) Cell lysates obtained from untreated and CNTF-treated primary neuronal cultures generated from the E14 rat striatum primordia were immunoblotted with anti-phospho-STAT3 antibody (upper panel). A tyrosine phosphorylated STAT3 band is visible in response to CNTF. The same membrane was stripped and reacted with anti-STAT3 antibody (lower panel). (B) STAT3 translocates into the nucleus of ST14A cells upon cytokine stimulation. The cells were incubated in the absence or presence of ligand for 15 minutes. The cellular distribution of STAT3 was examined by immunofluorescence. Untreated ST14A cells show a diffuse STAT3 distribution. On the other hand, STAT3 is detected exclusively in the nuclei of the treated cells, where a strong immunofluorescence signal is clearly visible (arrows).

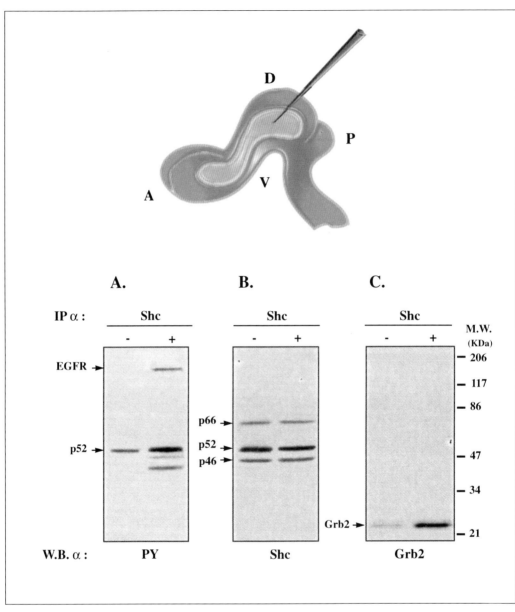

Chapter 1, Figure 3. p52ShcA phosphorylation and interaction with Grb2 in embryonic telencephalic vesicles following intraventricular injection of EGF. Upper panel: Injection of growth factors into the telencephalic vesicles of E15 (embryonic day 15) embryos. Injection was performed by a procedure we have developed for embryonic transplantation of CNS progenitor cells (2,9). Ten microliters of EGF (10 ng/μL) were placed intraventricularly into E15 embryos. The figure shows a schematic drawing of the rat embryonic neural tube and of the ventricular system where the growth factors were delivered. At E15, the cells lining the neural tube are still immature and proliferating actively. Intraventricular injection of growth factors at this early stage of brain maturation can therefore target this particular population of CNS progenitor cells. (A, anterior; P, posterior; D, dorsal; V, ventral). (Lower panel) (A) Phosphotyrosine immunoblot of ShcA immunoprecipitates after in vivo EGF treatment. Ten minutes after injection, the embryos were removed, and the telencephalic vesicles were isolated and subjected to immunoprecipitation with anti-ShcA antiserum followed by immunodecoration with 4G10 antiphosphotyrosine antibodies. Phosphorylated p52^{ShcA} is indicated. Control embryos were injected with vehicle. A 170-kDa phosphorylated band (arrow) corresponding to the coprecipitated EGFR is also visible in the treated group. (B) Anti-ShcA immunoblot of the filter in panel A. The membrane was stripped and reacted with ShcA monoclonal antibody. As shown, ShcA proteins were immunoprecipitated to the same extent in control and EGF-treated groups. (C) Anti-Grb2 immunoblot of the same immunoprecipitates as in panel A. The arrow indicates the 23 kDa Grb2 protein, which coprecipitates with ShcA more abundantly in the treated group.

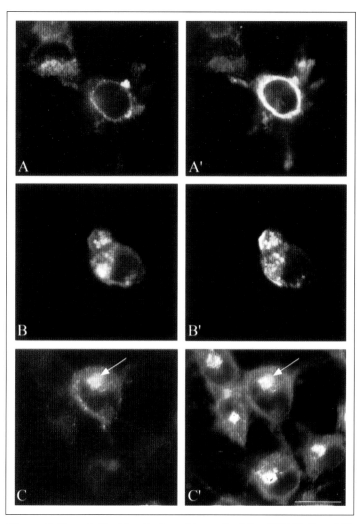

Chapter 2, Figure 2. Dual localization of internalized fluo-SRIF (A, B, and C) and of sst$_{2A}$ immunoreactivity (A′ and B′) or of the TGN marker syntaxin 6 (C′) in COS-7 cells transfected with cDNA encoding the sst$_{2A}$ receptor. Cells were incubated with 20 nM fluo-SRIF at 37°C for 5 to 45 minutes, fixed, and immunocytochemically reacted with either sst$_{2A}$ or syntaxin 6 antibodies. (A and A′) After 5 minutes of incubation, there is complete overlap between the ligand (A) and sst$_{2A}$-immunoreactivity (A′) at the periphery of the cell. (B and B′) After 30 minutes of incubation, fluo-SRIF (B) and sst$_{2A}$ immunoreactivity (B′) are both distributed more centrally within the cells and are partially dissociated. (C and C′) At 45 minutes, fluo-NT (C) is concentrated next to the nucleus, where it colocalizes extensively (arrows) with the TGN marker syntaxin 6 (C′). Abbreviation: N, nucleus. Scale bar: 10 μm.

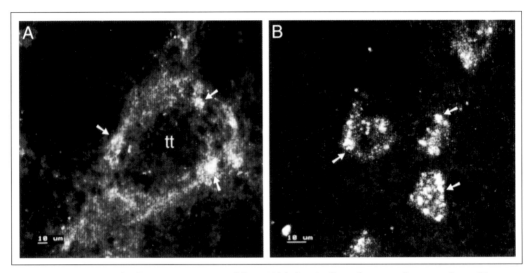

Chapter 2, Figure 3. Confocal microscopic images of fluo-NT labeling in slices of rat ventral tegmental area. Slices were pulse-labeled for 3 minutes with 10 nM fluo-NT, and sections were scanned 10 minutes (A) and 30 minutes (B) after washout with Ringer buffer. At 10 minutes (A), labeling is evident over both perikarya (arrows) and neuropil. At 30 minutes (B), nerve cell bodies are still intensely labeled, but the neuropil labeling is markedly reduced. Note that at 30 minutes, the labeling is detected in the form of small puntate fluorescent granules that pervade the perikaryal cytoplasm (arrows). Images were reconstructed from a stack of 25 serial optical sections separated by 0.12 μm steps and scanned at 32 scans per frame. Scale bars: 10 μm.

Chapter 8 Figure 1. Double labelings on sections of peripheral (a) and central (c–f) rat nervous system and on primary cell cultures (b). (a) Dorsal root ganglia sections have been hybridized with digoxigenin- and radioactively-labeled probes complementary to calcitonin gene related peptide (CGRP) (purple precipitate) and galanin receptor 1 (silver grains) mRNA, respectively (Protocol #1). Double labeled (arrows) and neurons containing CGRP only are visualized in normal rats. (b) Primary culture of adult dorsal root ganglia double stained for neuropeptide tyrosine (enzymatic precipitate) and its Y1 receptor (radioactive signal) by using the same protocol. Large size neurons are double labeled (arrow), whereas others display only the radioactive signal (open arrowhead). Thus, the Y1 receptor can be considered as both a pre- and postsynaptic receptor. (c) Double detection of heterologous gene products: double labeling combining two enzymatic detections (modified from Protocol #5). Galanin peptide is visualized by immunoperoxidase, whereas the development of alkaline phosphatase activity leads to the detection of oxytocin mRNA in the supraoptic nucleus of a lactating female rat hypothalamus. Upon such a stimulation, some neurons appear doubly labeled beside two distinct populations of cells singly labeled either for galanin (small arrowhead) or oxytocin mRNA (large arrowhead). (d) Double detection of heterologous gene products: double labeling combining radioactive ISH detection of vasopressin mRNA to immunoperoxidase for tyrosine hydroxylase (TH) in hypothalamic supraoptic nucleus of a dehydrated rat (modified from Protocol #5). Most neurons are double labeled (open arrows), although some cells remain single labeled either for TH (small arrowhead) or vasopressin mRNA (open arrowhead). Note that double labeled cells are easier to distinguish than when using a double enzymatic detection. (e) Double detection of homologous gene products: double labeling combining radioactive ISH detection of vasopressin mRNA to immunoperoxidase for vasopressin peptide in the rat hypothalamic supraoptic nucleus (modified from Protocol #5). Most neurons are double labeled (open arrows). However, rare cells remain single labeled for vasopressin and can be considered as resting neurons storing peptide (small arrowhead). (f) Double fluorescent immunolabeling using the CARD method to detect neuropeptide tyrosine Y1 receptor containing neurons (green) and neuropeptide tyrosine containing afferent fibers (red) in the rat arcuate nucleus (see Protocol #6). Yellow color corresponds to the superimposition of both fluorescences and indicates double labelings, that is contacts between fibers and neurons (arrows). The neuropeptide tyrosine containing fibers establish contact on soma and processes of Y1 receptor expressing neurons. Scale bars: 25 μm.

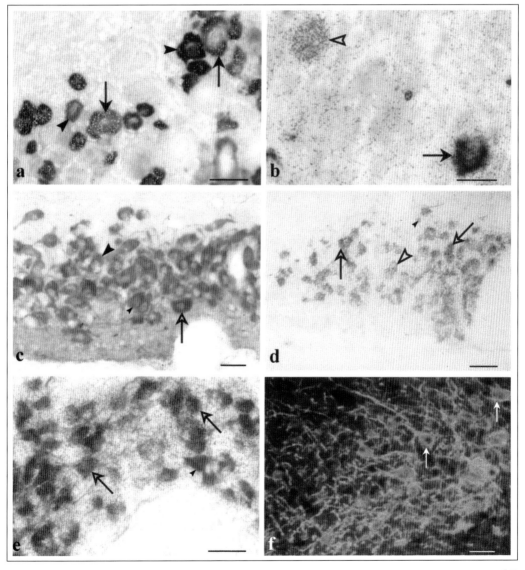

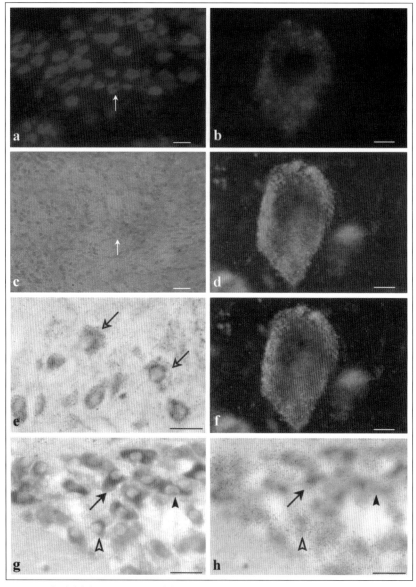

Chapter 8. Figure 2. Multiple labelings on sections of the rat peripheral (a and c) and central (b and d–h) nervous system. (a) The HNPP detection set gives a red fluorescence with exciter at 568 nm line leading to identify galanin mRNA containing neurons (arrow) in the axotomized superior cervical ganglia. (c) No green fluorescence is detected on the same section with exciter at 488 nm line. Such a specificity allows the combination of fluorescent ISH to galanin mRNA visualized by the HNPP detection set (b) with a fluorescein isothiocyanate (FITC)-conjugated–streptavidin visualization of biotin-labeled probes complementary to vasopressin mRNA (d) in hypothalamic magnocellular neurons of a salt-loaded rat supraoptic nucleus. The analysis at the confocal microscope and the merge image (f) lead to identify specific subcellular compartments involved in the synthesis of either galanin (perinuclear localization, red fluorescence) or vasopressin (green fluorescence, peripheral localization). The coupling between ICC with an immunoperoxidase protocol and ISH for vasopressin mRNA (e) or oxytocin mRNA (g) can be performed by using a silver-enhanced gold particle method (e) (Protocol #4) or a double enzymatic detection (g) (Protocol #5) identical to that of Figure 1c. Double labeled cells (arrows) are easily visualized with the former protocol. The latter needs to calibrate the development of the enzyme activity to be able to distinguish both colors. Triple labeling is shown in panels g and h (Protocol #5) and combine ICC for galanin (brown precipitate) to double radioactive and enzymatic ISH for vasopressin and oxytocin mRNAs, respectively. Both micrographs correspond to the same field, but different focuses are necessary to identify the three markers that are not in the same plane on the preparation. Some neurons are triple labeled (arrow). Besides, neurons containing oxytocin only (filled arrowhead), cells double labeled for vasopressin mRNA and galanin (open arrowhead) are seen. Scale bars: 50 μm (panels a, c, e, f, and g); 5 μm (panels b, d, and f).

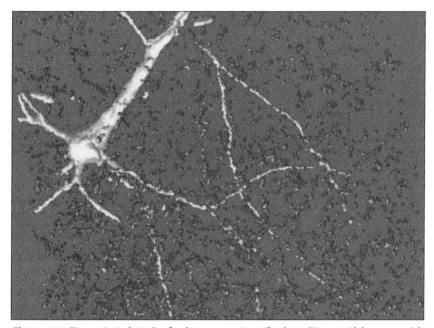

Chapter 11. Figure 5. (color) Confocal reconstruction of a layer V pyramidal neuron with surrounding VAChT-IR fibers illustrating the pattern of innervation of the cell. Note the abundance of immuno-reactive boutons in the area of the basal dendrites. This 3D reconstruction was based on ten 1-μm-thick serial optical sections. The reconstruction was performed with the aid of the MCID-M4 image analysis system. Scale bar = 50 μm.

Chapter 11. Figure 6. (Color) Confocal reconstruction of a dendrite from a layer V cortical pyramidal neuron illustrating the close association with surrounding VAChT-IR terminals. This 3D reconstruction was based on four 1-µm-thick serial optical sections. The reconstruction was performed using an image analysis system, as for Figure 5. The progressive 3D rotation from the top to the bottom of the panel illustrates how this approach can allow us to identify close association (but not synapses) of boutons with postsynaptic dendrites of the cell.

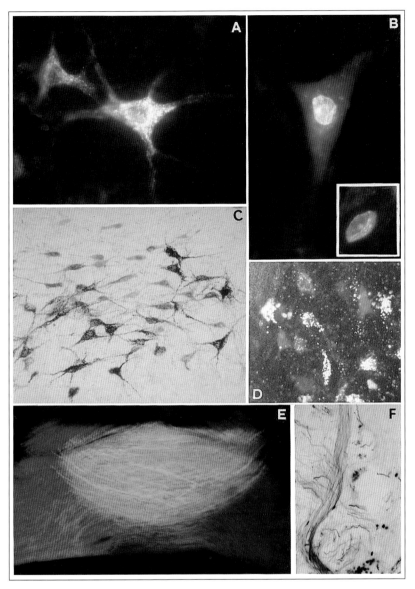

Chapter 12. Figure 2. Photomicrographs that illustrate different retrograde (A–D), anterograde (E,F) tracers, as well double retrograde (B,D), double anterograde (E) labeling, and retrograde labeling combined with an endogenous protein immunoreactivity. Panels A, B, and E show fluorescent dyes, panels C and F are in bright-field illumination, and panel D illustrates a combination of fluorescence and dark-field illumination. Panels A through D and F were derived from experiments in rats and panel E in the frog. (A) FB retrogradely labeled reticulospinal neurons; note the labeling of the cytoplasm and proximal dendrites. (B) Neuron of the brainstem reticular core double labeled after FB injection in the spinal cord and DY injection in the thalamus; FB labeling is revealed by blue fluorescence in the cytoplasm and DY labeling by gold fluorescence in the nucleus; the inset illustrates a single DY-labeled neuron. (C) Neurons of the substantia nigra retrogradely labeled by WGA-apoHRP-gold injection in the striatum (black granular labeling of the cytoplasm revealed by silver intensification) and combined with calbindin immunohistochemistry (brown diffuse cell immunostaining resulting from the use of DAB as chromogen). (D) Thalamic neurons double labeled after FB injection in the cortex and WGA-apoHRP-gold injection in the thalamic reticular nucleus; FB labeling is revealed by blue fluorescence and the gold-conjugated tracer by the granular labeling (resulting from silver intensification) that assumes a gold color in dark-field illumination (courtesy of Dr. Roberto Spreafico, Neurological Institute Besta, Milan, Italy). (E) Double anterograde labeling of the optic chiasm with fluorescent dextran amines: Texas Red® conjugated dextran amine (red fluorescent) in the left optic tract and fluorescein-conjugated dextran amine (green fluorescent) on the right side; the superimposition of the two colors (double exposure photograph) results in the yellowish labeling at the fiber crossing (courtesy of Dr. Bernd Fritzsch, Creighton University, Omaha, NE, USA). (F) Biotin–dextran amine fiber labeling.

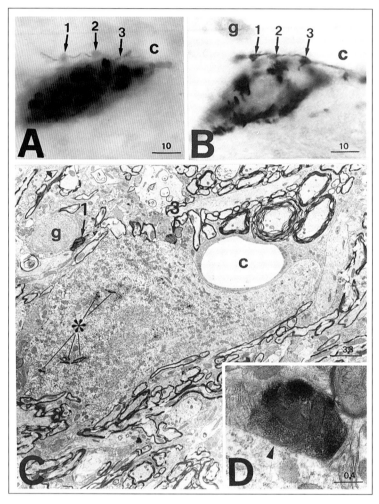

Chapter 13. Figure 2. Correlated light (A and B) and electron (C and D) analysis of a neuron in the pontine A_5 noradrenergic cell group. A capillary (c) and a glial cell (g) are used as landmarks. The neuron is retrogradely labeled with CTb (chromagen TMB; black granules in panels A and B and electron dense crystals marked with an asterisk in panel C) from the cervical spinal cord dorsal horn. Its noradrenergic nature is demonstrated in panel A by the diffuse blue staining due to immunoreactivity for DBH (chromagen Fast Blue removed in panel B during processing for ultrastructural analysis). The neuron receives appositions from axonal varicosities (chromogen DAB: brown staining in panel A, black product in panel B, and electrondense mark in panels C and D) labeled with BDA from the VLM (1–3 in the thick sections in panels A and B; 1 and 3 in the ultrathin section in panel C). In panel D, bouton 3 is shown at a higher magnification. A synapse cut obliquely is established between the bouton and the perikarya (arrowhead) of the retrogradely labeled neuron. Scale bars are expressed in micrometers.

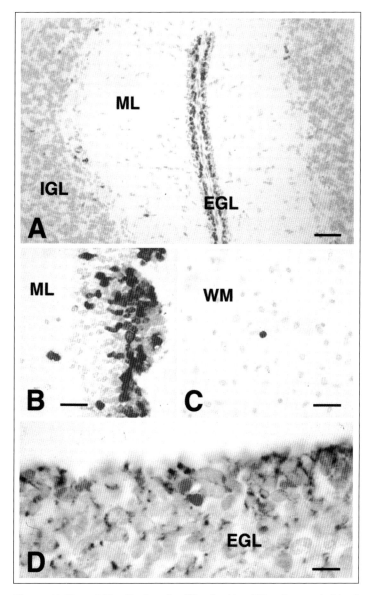

Chapter 14. Figure 2. Visualization of proliferating (A and B) and apoptotic (C and D) cells in the postnatal cerebellum. Proliferating cells have been detected by immunocytochemical labeling for the ki-67 nuclear antigen. (A) Rabbit cerebellum at P5. Note that the pattern of staining is remarkably similar to that shown in Figure 1B. (B) Human cerebellum at P30. Dual color immunocytochemistry for ki-67 (peroxidase, brown) and bcl-2 protein (alkaline phosphatase, blue), shows numerous proliferating cells with nuclear ki-67 positivity in the outer (proliferative) part of the EGL expressing in their cytoplasm the bcl-2 protein, a negative modulator of apoptosis. (C) An apoptotic cell in the cerebellar white matter of a P5 rabbit. TUNEL method. (D) Combined visualization of apoptotic cells (TUNEL alkaline phosphatase—new fuchsin = red) and vimentin (a marker of undifferentiated neural precursors) immunoreactivity (nickel intensified peroxidase = black) in the EGL. Abbreviations: EGL, external granular layer; IGL, internal granular layer; ML, molecular layer; P5, postnatal day 5; P30, postnatal day 30; WM, white matter. Scale bars: panel A = 100 μm; panels B and C = 25 μm; panel D = 10 μm.

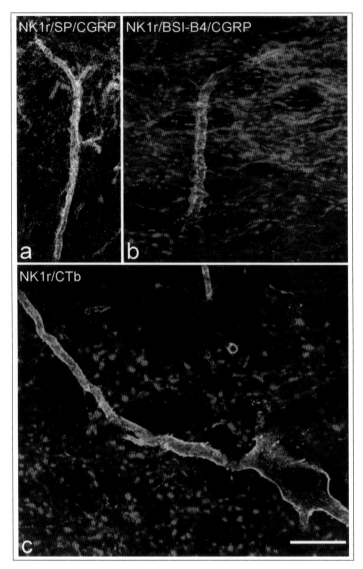

Chapter 15. Figure 1. The association between large NK1 receptor–immunoreactive lamina III/IV neurons with long dorsal dendrites and three different types of primary afferent in the dorsal horn of the rat spinal cord, seen in transverse (a) or parasagittal (b and c) sections. Each image contains a dendrite belonging to a NK1 receptor–immunoreactive neuron (green), and in panel c, the cell body of one of these neurons is also seen. (a) A section also reacted with a monoclonal antibody to substance P (blue) and an antiserum against CGRP (red). Most of the immunoreactive axons contain both peptides and therefore appear pink, and where these overlap the NK1 receptor–immunoreactive dendrite, they appear white. The dendrite is associated with many varicosities which contain both substance P and CGRP (i.e., substance P-containing primary afferent terminals). (b) The section has been immunostained with CGRP antiserum (blue) and incubated in biotinylated BSI-B4, which was revealed with avidin–rhodamine (red). Many of the CGRP-immunoreactive axons have also bound the lectin and appear pink or purple. Numerous red axons are present in the lower half of the field. These have bound the lectin but are not CGRP-immunoreactive and, therefore, belong to non-peptidergic C fibers. Very few of these axons make contact with the NK1-immunoreactive dendrite. (c) Shown is the cell body and proximal dendrites of a lamina III NK1 receptor–immunoreactive neuron from a rat in which CTb had been injected into the sciatic nerve 3 days previously. CTb (red) has been taken up and transported by myelinated afferents belonging to the sciatic nerve, which terminate extensively in lamina III. Although there are many CTb-immunoreactive axons surrounding the cell, very few make contact with it. Panel a was generated from 8 optical sections at 1 μm separation, panel b from 9 optical sections at 0.5 μm separation, while panel c is from a single optical section. Scale bar = 20 μm for each image. Panel b is reprinted with permission from Neuroscience 94, Sakamoto, H., R.C. Spike, and A.J. Todd, Neurons in laminae III and IV of the rat spinal cord with the neurokinin-1 receptor receive few contacts from unmyelinated primary afferents which do not contain substance P, p. 903-908, Copyright (1999), with permission from Elsevier Science. Panel c is reproduced with permission from Reference 15.

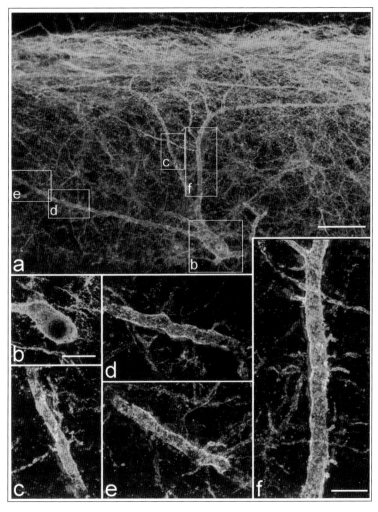

Chapter 15. Figure 2. Contacts formed by substance P-containing axons onto a lamina III NK1 receptor–immunoreactive spinothalamic tract neuron. (a) Shown is the cell body and dendrites of the neuron in a parasagittal section scanned to reveal only NK1 receptor-immunostaining (green). Boxes indicate the regions shown in the remaining parts of the figure. (b) The cell body of the neuron contains CTb (blue), which had been injected into the thalamus 3 days previously, thus allowing identification of this as a spinothalamic neuron. (c–f) Different parts of the dendritic tree receive numerous contacts from substance P-immunoreactive axons, which are orientated along the lengths of the dendrites (red). Panel a was obtained from 9 optical sections 1.5 µm apart, panels c and e through f from 4 optical sections, and panel d from 7 optical sections, each 0.5 µm apart. Scale bars: panel a = 50 µm, panel b = 20 µm, panels c through f = 10 µm. Reproduced with permission from Reference 14.

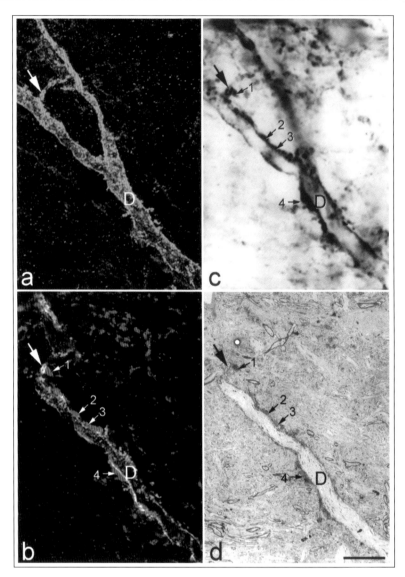

Chapter 15. Figure 3. Combined confocal microscopy and EM of a dorsal dendrite belonging to a lamina III NK1 receptor–immunoreactive neuron from a parasagittal section that had been reacted with NK1 receptor, substance P, and CGRP. (a) NK1 receptor immunoreactivity in a series of 7 optical sections at 0.5 μm intervals. The arrow shows a small branch given off from the main dendritic shaft (D). (b) A single confocal section from the series used to generate panel b. In this case, NK1 receptor appears green, substance P is blue, and CGRP is red. Boutons, which contain both peptides, appear pink. Several substance P-immunoreactive varicosities are in contact with the NK1 receptor–immunoreactive dendrite, and 4 of these are indicated with the small numbered arrows. The varicosity numbered 2 is only substance P-immunoreactive, whereas those numbered 1, 3, and 4 have both substance P and CGRP immunoreactivity. (c) The corresponding region seen with light microscopy after immunoperoxidase reaction and at a focal depth approximately equivalent to the confocal image in panel b. The main dendritic shaft (D) and part of its small branch (arrow) are clearly seen, whereas most of the right-hand branch is out of focus. Many substance P-immunoreactive varicosities are visible, including the ones indicated with arrows in panel b. (d) Low magnification electron micrograph of the corresponding region. The section is at a depth nearly equivalent to the confocal image in panel b. Scale bar = 10 μm. Modified and reproduced with permission from Reference 14.

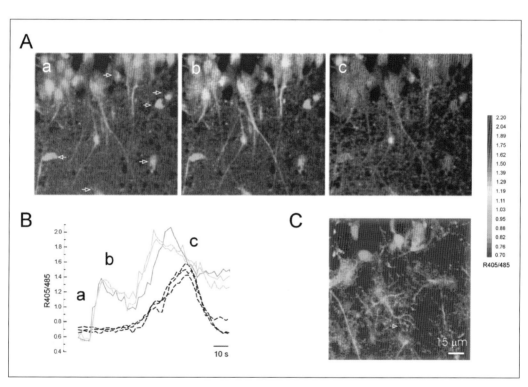

Chapter 16. Figure 1. Effects of antioxidant agents in loading of cells from acute brain slices with Ca^{2+} fluorescent dyes.
(A) Indo-1 loaded cells from the CA1 hippocampal region of a young rat at postnatal day 10 following slicing and incubation procedures in the presence of antioxidant agents. Most of the neuronal dendrites resulted well loaded. The time series of pseudocolor images illustrates the $[Ca^{2+}]_i$ changes occurring in these neurons following perfusion of the slice with 60 mM KCl. The sequence shows the $[Ca^{2+}]_i$ increase in pyramidal neurons occurring several seconds before that in astrocytes (open arrows). The R405/485 is displayed as a pseudocolor scale. Sampling rate 2 seconds. The stimulation with high K^+ extracellular solution was obtained by iso-osmotic replacement of Na^+ with K^+. (B) Kinetics of the $[Ca^{2+}]_i$ changes in the neurons (continuous lines) and astrocytes (dotted lines), indicated by arrows in panel A, following KCl stimulation, as expressed by the ratio between indo-1 emission wavelength at 405 and 485 nm. Letters a through c correspond to images a through c in panel A. (C) Indo-1 loaded cells from the CA1 hippocampal region of a young rat at postnatal day 10 following slicing and incubation procedures in the absence of antioxidant agents. Under these conditions, dendrites are less loaded with indo-1 while astrocyte processes (arrow) are, in general, more loaded with the dye. Furthermore, several dying neurons can be observed scattered in the CA1 pyramidal layer.

49. Krenacs, T. and L. Dux. 1994. Silver-enhanced immunogold labeling of calcium-ATPase in sarcoplasmic reticulum of skeletal muscle [letter, comment]. J. Histochem. Cytochem. *42*:967-968.
50. Krenacs, T., M. van Dartel, E. Lindhout, and M. Rosendaal. 1997. Direct cell/cell communication in the lymphoid germinal center: connexin43 gap junctions functionally couple follicular dendritic cells to each other and to B lymphocytes. Eur. J. Immunol. *27*:1489-1497.
51. Light, A.R. and A.M. Kavookjian. 1988. Electron microscopic localization of peptide-like immunoreactivity in labelled dorsal root terminals in the spinal substantia gelatinosa of the monkey, p. 57-69. *In* F. Cervero, G.J. Bennett, and P.M. Headley (Eds.), Processing of Sensory Information in the Superficial Dorsal Horn of the Spinal Cord. Plenum Press, New York.
52. Liou, W., H.J. Geuze, and J.W. Slot. 1996. Improving structural integrity of cryosections for immunogold labeling. Histochem. Cell Biol. *106*:41-58.
53. Lue, J.H., Y.F. Jiang Shieh, J.Y. Shieh, and C.Y. Wen. 1996. The synaptic interrelationships between primary afferent terminals, cuneothalamic relay neurons and GABA-immunoreactive boutons in the rat cuneate nucleus. Neurosci. Res. *24*:363-371.
54. Maxwell, D.J., W.H. Christie, A.D. Short, J. Störm-Mathisen, and O.P. Ottersen. 1990. Central boutons of glomeruli in the spinal cord of the cat are enriched with L-glutamate-like immunoreactivity. Neuroscience *36*:83-104.
55. Maxwell, D.J., W.M. Christie, O.P. Ottersen, and J. Störm-Mathisen. 1990. Terminals of group Ia primary afferent fibers in Clarke's column are enriched with L-glutamate-like immunoreactivity. Brain Res. *510*:346-350.
56. Maxwell, D.J., W.M. Christie, A.D. Short, and A.G. Brown. 1990. Direct observations of synapses between GABA-immunoreactive boutons and muscle afferent terminals in lamina VI of cat spinal cord. Brain Res. *530*:215-222.
57. Maxwell, D.J., R. Kerr, E. Jankowska, and J.S. Riddell. 1997. Synaptic connections of dorsal horn group II spinal interneurons: synapses formed with the interneurons and by their axon collaterals. J. Comp. Neurol. *380*:51-69.
58. Maxwell, D.J., O.P. Ottersen, and J. Storm-Mathisen. 1995. Synaptic organization of excitatory and inhibitory boutons associated with spinal neurons which project through the dorsal columns of the cat. Brain Res. *676*:103-112.
59. Menco, B.P., A.M. Cunningham, P. Qasba, N. Levy, and R.R. Reed. 1997. Putative odour receptors localize in cilia of olfactory receptor cells in rat and mouse: a freeze-substitution ultrastructural study. J. Neurocytol. *26*:297-312.
60. Merighi, A. 1992. Post-embedding electron microscopic immunocytochemistry, p. 51-87. *In* J.M. Polak and J.V. Priestly (Eds.), Electron Microscopic Immunocytochemistry. Oxford University Press, London.
61. Merighi, A., F. Cruz, and A. Coimbra. 1992. Immunocytochemical staining of neuropeptides in terminal arborization of primary afferent fibers anterogradely labeled and identified at light and electron microscopic levels. J. Neurosci. Methods *42*:105-113.
62. Merighi, A. and J.M. Polak. 1993. Post-embedding immunogold staining, p. 229-264. *In* A.C. Cuello (Ed.), Immunohistochemistry II. John Wiley & Sons, London.
63. Merighi, A., J.M. Polak, and D.T. Theodosis. 1991. Ultrastructural visualization of glutamate and aspartate immunoreactivities in the rat dorsal horn with special reference to the co-localization of glutamate, substance P and calcitonin gene-related peptide. Neuroscience *40*:67-80.
64. Merighi, A., E. Raviola, and R.F. Dacheux. 1996. Connections of two types of flat cone bipolars in the rabbit retina. J. Comp. Neurol. *371*:164-178.
65. Morris, J.F., D.V. Pow, and F.D. Shaw. 1989. Strategies for the ultrastructural study of peptide containing neurons, p. 83-124. *In* G. Fink and A.J. Harmar (Eds.), Neuropeptides: A Methodology. John Wiley & Sons, Chichester.
66. Negyessy, L., J. Takacs, I. Divac, and J. Hamori. 1994. A combined Golgi and post-embedding GABA and glutamate electron microscopic study of the nucleus dorsomedialis thalami of the rat. Neurobiology *2*:325-341.
67. Nirenberg, M.J., J. Chan, R.A. Vaughan, G.R. Uhl, M.J. Kuhar, and V.M. Pickel. 1997. Immunogold localization of the dopamine transporter: an ultrastructural study of the rat ventral tegmental area. J. Neurosci. *17*:5255-5262.
68. Ottersen, O.P. and C.R. Bramham. 1988. Quantitative electron microscopic immunocytochemistry of excitatory amino acids, p. 93-100. *In* E.A. Cavalheiro, J. Lehmann and L. Turski (Eds.), Frontiers in Excitatory Amino Acids Research. Alan R. Liss, New York.
69. Pease, D.C. 1964. Histological Techniques for Electron Microscopy. Academic Press, New York.
70. Phend, K.D., A. Rustioni, and R.J. Weinberg. 1995. An osmium-free method of epon embedment that preserves both ultrastructure and antigenicity for post-embedding immunocytochemistry. J. Histochem. Cytochem. *43*:283-292.
71. Pickel, V.M., J. Chan, and C. Aoki. 1993. Electron microscopic immunocytochemical labelling of endogenous and/or transported antigens in rat brain using silver intensified one-nanometre colloidal gold, p. 265-280. *In* A.C. Cuello (Ed.), Immunohistochemistry II. John Wiley & Sons, Chichester.
72. Posthuma, G., J.W. Slot, and H.J. Geuze. 1984. Immunocytochemical assays of amylase and chymotrypsinogen in rat pancreas secretory granules. Efficacy of using immunogold-labeled ultrathin cryosections to estimate relative protein concentrations. J. Histochem. Cytochem. *32*:1028-1034.
73. Posthuma, G., J.W. Slot, and H.J. Geuze. 1987. Usefulness of the immunogold technique in quantitation of a soluble protein in ultra-thin sections. J. Histochem. Cytochem. *35*:405-411.
74. Priestley, J.V. 1984. Pre-embedding ultrastructural immunocytochemistry: immunoenzyme techniques, p. 37-52. *In* J.M. Polak and I.M. Varndell (Eds.), Immunolabelling for Electron Microscopy. Elsevier Publishers, Amsterdam.

75. Priestley, J.V. and A.C. Cuello. 1989. Ultrastructural and neurochemical analysis of synaptic input to trigemino-thalamic projection neurones in lamina I of the rat: a combined immunocytochemical and retrograde labelling study. J. Comp. Neurol. 285:467-486.
76. Pulczynski, S. and O. Myhre Jensen. 1994. Quantitation of immunogold with an interactive image analysis system: a new, practical approach to antibody-induced modulation, internalization and intracellular transport of surface antigens in viable hematopoietic cells. Anal. Quant. Cytol. Histol. 16:393-399.
77. Rèthelyi, M., A.R. Light, and E.R. Perl. 1989. Synaptic ultrastructure of functionally and morphologically characterized neurons of the superficial spinal dorsal horn of cat. J. Neurosci. 9:1846-1863.
78. Ribeiro-Da-Silva, A., J.V. Priestley, and A.C. Cuello. 1993. Pre-embedding ultrastructural immunocytochemistry, p. 181-228. In A.C. Cuello (Ed.), Immunohistochemistry II. John Wiley & Sons, Chichester.
79. Robards, A.W. and U.B. Sleytr. 1985. Low temperature methods in biological electron microscopy, p. 1-551. In A.M. Glauert (Ed.), Practical Methods in Electron Microscopy. Elsevier Publishers, Amsterdam.
80. Roth, J. 1982. The protein A-gold (pAg) technique—a qualitative and quantitative approach for antigen localization on thin sections, p. 107-134. In G.R. Bullock and P. Petrusz (Eds.), Techniques in Immunohistochemistry (Vol. I). Academic Press, London.
81. Roth, J. 1984. The protein A-gold technique for antigen localization in tissue sections by light and electron microscopy, p. 113-122. In J.M. Polak and I.M. Varndell (Eds.), Immunolabelling for Electron Microscopy. Elsevier Publishers, Amsterdam.
82. Roth, J. 1996. The silver anniversary of gold: 25 years of the colloidal gold marker system for immunocytochemistry and histochemistry. Histochem. Cell. Biol. 106:1-8.
83. Saha, S., T.F. Batten, and P.N. McWilliam. 1995. Glutamate-immunoreactivity in identified vagal afferent terminals of the cat: a study combining horseradish peroxidase tracing and post-embedding electron microscopic immunogold staining. Exp. Physiol. 80:193-202.
84. Schwarz, C. and Y. Schmitz. 1997. Projection from the cerebellar lateral nucleus to precerebellar nuclei in the mossy fiber pathway is glutamatergic: a study combining anterograde tracing with immunogold labeling in the rat. J. Comp. Neurol. 381:320-334.
85. Sekerkova, G., Z. Katarova, F. Joo, J.R. Wolff, S. Prodan, and G. Szabo. 1997. Visualization of beta-galactosidase by enzyme and immunohistochemistry in the olfactory bulb of transgenic mice carrying the LacZ transgene. J. Histochem. Cytochem. 45:1147-1155.
86. Slot, J.W., G. Posthuma, L.Y. Chang, J.D. Crapo, and H.J. Geuze. 1989. Quantitative aspects of immunogold labeling in embedded and non-embedded sections. Am. J. Anat. 185:271-281.
87. Somogyi, P. 1990. Synaptic connections of neurones identified by golgi impregnation: characterisation by immunocytochemical, enzyme histochemical and degeneration methods. J. Electron Microsc. Tech. 15:332-351.
88. Somogyi, P., K. Halasy, J. Somogyi, J. Störm-Mathisen, and O.P. Ottersen. 1986. Quantification of immunogold labelling reveals enrichment of glutamate in mossy and parallel fibre terminals in cat cerebellum. Neuroscience 19:1045-1050.
89. Song, Z.M., S.J. Brookes, and M. Costa. 1994. Characterization of alkaline phosphatase-reactive neurons in the guinea-pig small intestine. Neuroscience 63:1153-1167.
90. Sternberger, L.A., P.J.J. Hardy, J.J. Cucculis, and H.G. Meyer. 1970. The unlabeled antibody-enzyme method of immunohisto-chemistry. Preparation and properties of soluble antigen-antibody complex (horseradish peroxidase-anti-horseradish peroxidase) and its use in identification of spirochetes. J. Histochem. Cytochem. 18:315-333.
91. Stirling, J.W. 1990. Immuno- and affinity probes for electron microscopy: a review of labeling and preparation techniques. J. Histochem. Cytochem. 38:145-158.
92. Stirling, J.W. 1993. Use of tannic acid and silver enhancer to improve staining for electron microscopy and immunogold labeling. J. Histochem. Cytochem. 41:643-648.
93. Stirling, J.W. and P.S. Graff. 1995. Antigen unmasking for immunoelectron microscopy: labeling is improved by treating with sodium ethoxide or sodium metaperiodate, then heating on retrieval medium. J. Histochem. Cytochem. 43:115-123.
94. Tang, F.R., C.K. Tan, and E.A. Ling. 1995. The distribution of NADPH-d in the central grey region (lamina X) of rat upper thoracic spinal cord. J. Neurocytol. 24:735-743.
95. Tokuyasu, K.T. 1984. Immuno-cryoultramicrotomy, p. 71-82. In J.M. Polak and I.M. Varndell (Eds.), Immunolabelling for Electron Microscopy. Elsevier Publishers, Amsterdam.
96. van Lookeren Campagne, M., A.B. Oestreicher, T.P. Van Der Krift, W.H. Gispen, and A.J. Verkleij. 1991. Freeze-substitution and lowicryl HM20 embedding of fixed rat brain: suitability for immunogold ultrastructural localization of neural antigens. J. Histochem. Cytochem. 39:1267-1279.

11

Combined Electrophysiological and Morphological Analyses of CNS Neurons

Alfredo Ribeiro-da-Silva[1,2] and Yves De Koninck[1,3]

Departments of [1]Pharmacology and Therapeutics and [2]Anatomy and Cell Biology, McGill University, Montreal; [3]Neurobiologie Cellulaire, Centre de Recherche Université Laval–Robert Gittard, Beauport, QC, Canada

OVERVIEW

Immunocytochemical and electrophysiological techniques have each independently yielded important information regarding the chemical anatomy of and function in the central nervous system. Yet it is now apparent that the combined application of these two techniques can yield information of considerably greater value than the simple addition of the data derived from the isolated use of each of the two techniques. In fact, such a combined approach is the only way in which it is possible to know how the morphological properties of one specific neuron correlate with the physiological properties of that same cell. With the application of this type of approach, it becomes possible to draw a correlation between the physiological and morphological properties of intracellularly labeled neurons with their neurotransmitter-specific innervation and with the transmitter and receptor content of the neurons themselves.

In this chapter, we will provide details of the protocols for two approaches used to achieve the objectives described above. One involves the use of intracellular recording from central nervous system (CNS) neurons in the whole animal. The other involves whole cell patch clamp recording from live slices. Each approach has its advantages and limitations and should be selected based on the type of issue one seeks to study.

BACKGROUND

The approach combining intracellular electrophysiological recording in whole animals with light and electron microscopic immunocytochemistry of CNS neurons has been described by us in detail in previous publications (9,10,19,20). The main advantages of this approach include: (*i*) the electrophysiological experiments are performed in vivo, in the anesthetized animal, so the synaptic circuitry and physiological responses are maintained as close to normal as possible; (*ii*) one can study responses elicited by natural stimuli; and (*iii*) as the animal is perfused with histological fixatives at the end of the experiment, the morphological preservation obtained at the ultrastructural level is very good to excellent. The main disadvantages of the in vivo approach are (*i*) the low success rate, as only about 25% of the experiments yield cells that have properly been characterized electrophysiologically and successfully labeled and recovered with good morphological preservation for electron microscopy; and (*ii*) the difficulty in obtaining prolonged stable recordings and controlling the extracellular milieu. Over the years, we improved success rates by using a more powerful intracellular amplifier (17) for iontophoretic injection of the marker and modifying the protocol to freeze-thaw the material prior to histochemistry, to avoid cracking of the tissue. We have successfully applied this approach to the study of the peptidergic synaptic input to physiologically characterized neurons of the cat dorsal horn (see Reference 9).

A very interesting complementary approach is the method of juxtacellular labeling with biocytin or neurobiotin during in vivo extracellular recording (29). This approach has the advantage that extracellular recording is less invasive (e.g., avoiding disruption of the integrity of the cell) and allows prolonged stable recording of undisturbed cellular activity. On the other hand, extracellular recording does not allow resolution of subliminal activity, measurement of conductance changes, and control of membrane potential.

Another complementary approach that is becoming increasingly popular is to perform recordings in live slices of CNS tissue. Slice preparations are particularly useful to obtain stable recordings, for placement of multiple electrodes in the vicinity of a single neuron and to control the extracellular milieu for detailed pharmacological studies of synaptic events. In particular, with the increasing use of tight seal whole cell patch clamp recording in CNS slices, the level of resolution obtained allows the study of elementary components of synaptic transmission such as miniature (action potential independent) synaptic currents that are thought to result from the release of transmitter contained in single vesicles from terminals in synaptic contact with the neuron from which the recording was obtained. Thus, the use of the whole cell patch clamp technique allows the application of quantitative analyses to study in detail synaptic events and the properties of single channels underlying them. The high resolution advantages provided by the tight seal whole cell recording configuration can also be exploited in vivo (18,23,25). This latter approach remains, however, difficult and limited to recording from neurons located very near the surface of the brain or spinal cord. In summary, advantages of the whole cell recording in slices include: (*i*) stable intracellular recording conditions; (*ii*) ability to control and manipulate the extracellular milieu; (*iii*) ability to restrict drug application to subcellular compartments of the neurons; (*iv*) option to perform recordings in voltage clamp mode; (*v*) the cell can be visualized with a fluorescent dye during or immediately at the end of the recording, providing an immediate identification of the cell type and assessment of the quality of its morphology (or allowing imaging with ion-sensitive dyes); (*vi*) simple diffusion of the label from the pipet into the cell occurs in the course of the recording and does not require active iontophoresis as with conventional sharp electrodes; and (*vii*) the success rate is higher than with whole animal approaches. The main challenge with this approach, however, is to obtain morphological preservation of sufficient quality for electron microscopy (EM) studies. With proper optimization, we have been able to obtain excellent preservation for light microscopy, and quite acceptable preservation for EM studies. Yet, certain substances remain difficult to detect by immunocytochemistry, because they are depleted during the incubation of the slices. This is especially true for amino acid neurotransmitters, such as GABA, glycine, and glutamate. However, in the case of GABA for example, such inconvenience can be avoided with the use of antibodies against glutamic acid decarboxylase (GAD), as the enzyme is not depleted. In this chapter, we will show images from the adult rat neocortex in which pyramidal neurons were recorded from and filled with a marker and then processed for immunocytochemistry at either the light or electron microscopic levels.

In summary, each of the approaches described above has its advantages and limitations with respect to the type of objective sought. The ultimate choice of recording conditions and tissue preparation should thus be mainly guided by the type of question being asked. For the purpose of this chapter, we will contrast our in vivo intracellular recording approach with whole cell recording and labeling in slices.

PROTOCOLS

Protocols for the Whole Animal (In Vivo) Approach

As an example, we provide the details of the protocol as applied to the lumbar region of the cat spinal cord. The protocol can obviously be applied to other CNS regions and other species with minor modifications. A diagrammatic representation of this approach is shown in Figure 1.

Materials and Reagents

- α-Chloralose; halothane (Somnothane; Hoechst, Frankfurt, Germany).
- Lidocaine hydrochloride (Xylocane; Astra Scientific, Pleasanton, CA, USA).
- Pancuronium bromide (Pavulon, Organon Teknika, Durham, NC, USA).
- Borosilicate glass capillaries with inner filament (WPI, Sarasota, FL, USA).
- Horseradish peroxidase (HRP—Type VI; Sigma, St. Louis, MO, USA).
- Potassium chloride or acetate; Tris-HCl or Tris-acetate buffers.
- Heparin, sodium salt (Sigma).
- Sodium nitrite (Fisher Scientific, Pittsburgh, PA, USA).
- Paraformaldehyde (BDH Chemicals, Poole, England, UK).
- Glutaraldehyde (25% solution; EM grade; Mecalab, Montreal, QC, Canada).
- Sucrose and glycerol (Fisher Scientific).
- Liquid nitrogen; isopentane (Fisher Scientific).
- Tissue culture plates of 12 wells (Falcon, Becton Dickinson, Franklin Lakes, NJ, USA).
- Sodium borohydride (Sigma).
- Monoclonal primary antibodies (e.g., mouse antienkephalin or rat antisubstance P; Medicorp, Montreal, QC, Canada or PharMingen, San Diego, CA, USA).
- Monoclonal antiperoxidase antibody (mouse or rat; Seralab, Sussex, England, UK).
- Polyclonal secondary antibodies (antimouse or antirat IgG, respectively; (American Qualex, San Clemente, CA, USA).
- Sodium phosphate buffer (PB), pH 7.4; phosphate-buffered saline (PBS), pH 7.4.
- 3,3′-diamininobenzidine tetrahydrochloride (DAB; Sigma).
- H_2O_2 (30%; American Chemicals Ltd., Montreal, QC, Canada).
- Cobalt chloride and nickel ammonium sulphate (Fisher Scientific).
- Osmium tetroxide (4% solution; Mecalab).

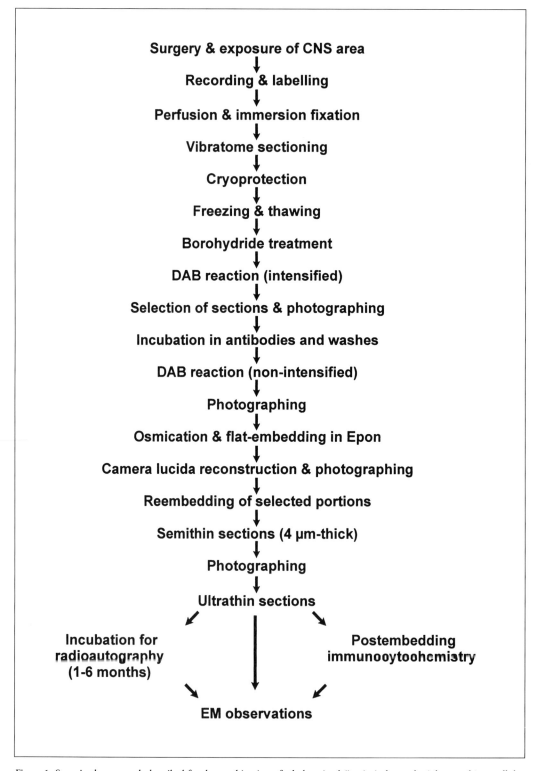

Figure 1. Steps in the approach described for the combination of whole animal (in vivo) electrophysiology and intracellular injection of HRP with the ultrastructural detection of multiple antigenic sites.

- Absolute ethanol (analytical grade; Sigma).
- Propylene oxide (BDH Chemicals).
- Epon (prepared from Epon 812 replacement, DDSA, NMA, and DMP30 as described in Reference 31; Mecalab).
- 3-aminopropyletriethoxysilane (APES; Sigma).
- Plastic coverslips (Fisher Scientific); thick acetate foil (available from any stationary store).
- Standard plastic EM capsules (Mecalab).

Procedure

Preparation of the Animals

1. Cats are anaesthetised with α-chloralose (60 mg/kg, i.v.) after induction with halothane.
2. The carotid artery and the jugular vein are cannulated for blood pressure monitoring and injection of drugs (Intramedic PE 160 polyethylene tubing).
3. A tracheal glass cannula is inserted via a tracheostomy for artificial ventilation of the animal.
4. Spinal segments L5 to L7 are exposed for recording. If the identification of the neurons by antidromic activation from supraspinal levels is not needed, the spinal cord can be transected at the L1 vertebral level; this allows the elimination of supraspinal influences and reduces movement of the lumbar cord due to intrathoracic pressure changes. Just prior to the transection, the L1 segment should be injected with lidocaine hydrochloride (xylocaine 1%; 0.1 mL) to minimize spinal shock. The exposed part of the spinal cord is covered with a pool of warm mineral oil to prevent cooling and drying.
5. To reduce movement of the spinal cord due to respiratory movements, the cats should be paralyzed with pancuronium bromide (1 mg/kg i.v.), given pneumothorax bilaterally and ventilated artificially through the tracheal cannula. End tidal CO_2 should be monitored and maintained at around 4% and body temperature at 38°C.
6. Firm clamping of the vertebra to a spinal frame is important to avoid movement during recording. In addition, while the dura mater is cut dorsally to expose the lumbar spinal segments (L5–L7), it can also be cut transversely underneath the rostral segment (e.g., L5) to eliminate tension in the dura mater and transmission of movement via this route.
7. For natural sensory stimulation, the fur on the hind leg ipsilateral to the side of recording can be cut unevenly to leave some hairs but also to expose the skin in places for thermal and touch stimulation. Electrical stimulation can also be useful to serve as search stimuli or for measurement of conduction velocity. For this, we dissect several nerves on the foot and at the level of the ankle (e.g., the superficial peroneal, the tibial and the sural nerves). For further details, the reader is referred to Reference 10.

Preparation of Electrodes

1. Electrodes are prepared from borosilicate glass capillaries (with an inner filament; 1.5 mm OD; 1.1 mm ID) pulled into fine-tipped micropipet (sharp electrodes) using a vertical (e.g., Narishige PN-3) or a horizontal (e.g., Brown & Flaming P-87) puller.
2. The pipets are filled with 4% to 8% HRP in 0.5 M KCl or KOAc, in 0.05 M Tris-HCl or Tris-acetate buffer.

Avoiding the Cl⁻ salts may be important to prevent reversal of Cl⁻-mediated inhibitory postsynaptic potentials (IPSPs). At the ranges of pH typically used (7.4–8.6), HRP has a net positive charge and can therefore be injected by iontophoresis through the recording electrode using positive (depolarizing) current injection. To avoid the formation of bubbles near the tip of the micropipet, we prefill the tip of the electrode with distilled water and fill the remainder of the electrode with the solution containing HRP. As it is important that enough time is given for the HRP to diffuse to the tip of the pipet, we fill the pipets a few hours prior to the recordings. The impedance of these electrodes ranges from 40 to 120 MΩ depending on tip size, given the type of cell targeted.

HRP Injection

HRP is injected after the intracellular physiological characterization of the cell is complete. To ensure sufficient iontophoretic injection of HRP, we use an amplifier–current source that has an input voltage range of ± 150 V. The high voltage source assures constant current ejection through the electrode even when the resistance of the electrode increases during ejection, for example as a result of plugging of the electrode tip with HRP (17).

1. We inject HRP using 500 to 700 ms positive current pulses of 1 to 5 nA at a frequency of 1 Hz for a duration of 5 to 15 minutes; these parameters are selected depending on the quality of the recording and the type of neuron studied. For large neurons (40–80 μm) the equivalent of an injection of 40 nA for 1 minute is used as a guideline; much less (as little as 1–5 nA for 1 min) can be used for smaller cells. Pulse injections are preferred to continuous current injections because the quality of the membrane potential and the size of the action potential or synaptically elicited responses can be tested between the current pulses to ensure that the electrode remains within the cell and to stop the injection as soon as the cell is lost.

2. After the end of the injection, the animals are kept anaesthetized for a couple of hours to allow the diffusion of HRP in the cell. However, we have obtained excellent results with a waiting time of just 30 minutes or even less.

Animal Perfusion

After the end of the electrophysiological part of the study, the animal is moved to a perfusion table for histological fixation. It is important that artificial ventilation is maintained up to the moment in which the perfusion is started, to avoid CNS ischaemia, leading to deterioration of ultrastructure. We usually connect the tracheal tube to a bottle with a mixture of 95% O_2/5% CO_2 and manually ventilate the lungs with the help of a Y tube.

1. The thorax of the animal is opened to expose the heart. The pericardium is removed. Slowly, 1 mL heparin (10 USP U/mL) and 3 mL of 1% sodium nitrite are injected into the left ventricle. The sodium nitrate solution should be injected particularly slowly to avoid cardiac fibrillation.

2. The animal is perfused through the ascending aorta with 2000 mL of a mixture of 4% paraformaldehyde/0.5% glutaraldehyde in 0.1 M PB, pH 7.4, after a brief washout of the vascular system (20 s maximum) with perfusion buffer (for composition see Reference 6). A high flow rate should be used at the beginning (for about 5

min) then reduced so that total perfusion time is about 30 minutes.

3. At the end of the perfusion, the relevant tissue is removed from the animal and immersed in the same fixative that was used for the perfusion at 4°C. Immersion postfixation should last 2 hours, but we have also used overnight postfixation with no loss of antigenicity, for the detection of the neuropeptides substance P and enkephalin. At the end of the fixation, the tissue is placed in 10 % sucrose in PB at 4°C until used.

4. For antigens that are highly sensitive to glutaraldehyde, we advise perfusion with 2000 mL a mixture of 4% paraformaldehyde/0.1% glutaraldehyde/15% (vol/vol) saturated picric acid in PB for 30 minutes followed by 2000 mL of the same mixture (but devoid of glutaraldehyde) for a further 30 minutes. Then, the animal should be perfused with 2000 mL of 10% sucrose in PB for 30 minutes. Subsequently, tissue is placed in 10% sucrose in PB in the refrigerator until use.

Tissue Processing for the Demonstration of the Cell

After removal from the animals, the tissue is processed first for the demonstration of the cell and then for immunocytochemistry.

1. The tissue should be trimmed to the proper size (block of about $1 \times 0.5 \times 0.5$ cm), ensuring that the area containing the labeled cell is not cut off. Cut 50-μm-thick sections on a Vibratome (Technical Products International [TPI], St. Louis, MO, USA), using cooled PBS or PB in the bath. We recommend that the sections are collected serially into a tissue culture plate with 12 wells (5 sections per well maximum).

2. The sections are cryoprotected by immersion in a mixture of 30% sucrose and 10% glycerol followed by 30% sucrose and 20% glycerol, in PB (1 h in each). Subsequently, the sections are snap-frozen by immersion in liquid nitrogen-cooled isopentane and thawed at room temperature. The procedure should be repeated 3 times. For the freeze-thaw procedure, all sections from each well should be placed in a small cup-like container made up of fine metal mesh and returned to the original well of the tissue culture plate at the end of the procedure.

3. The sections are rinsed twice in PBS (10 min each) and treated with 1% sodium borohydride in PBS for 30 minutes. Then, the sections are washed extensively in PBS until all bubbles disappear (usually 5 times over a period of 60 min is sufficient).

4. The sections are incubated in 0.05 % solution of DAB in PBS with cobalt chloride and nickel ammonium sulfate as described in detail elsewhere (31). At the end of 10 minutes, H_2O_2 is added to a final concentration of 0.01% (dilute 500 μL of 30% H_2O_2 in 14.5 mL of distilled water and add 5 μL of this solution to each 500 μL of DAB). After about 10 minutes, the reaction should be stopped by replacing the DAB by PBS.

5. The sections are rinsed for 2×15 minutes in PBS and mounted temporarily with PBS between a glass slide and a coverslip. Examination with a light microscope should be used to identify the sections that contain the cell. The sections that contain parts of the cell should be photographed at this stage, preferentially with a digital camera. The cell appears black or dark grey because of the metal intensification of the DAB reaction product.

Immunocytochemistry

Subsequently, the sections that contain the cell should be processed for ultrastructural immunocytochemistry. As it is the most likely approach to be used, we describe below a pre-embedding protocol for the detection of a single antigenic signal using a monoclonal primary antibody and DAB.

1. After washing in PBS, the sections should be incubated in the primary antibody and diluted to the proper working dilution in PBS overnight at 4°C. We found that, when using monoclonal antibodies, preincubation in normal serum of the species of the secondary antibody is not required. However, we strongly recommend shaking during incubations.

2. The following steps are all performed at room temperature. After washing 2 × 15 minutes in PBS, incubate the sections for 1½ to 2 hours in an antibody generated against the IgG of the species of the primary antibody (e.g., rabbit or goat antimouse IgG). Usually, dilutions vary from 1:30 to 1:50 in PBS, as the antibody has to be used in excess.

3. After 2 washes, the sections are incubated in a monoclonal antiperoxidase antibody, diluted in PBS, of the same species as the primary antibody, for 2 hours. Subsequently, after a 15-minute wash in PBS, the tissue is incubated in 5 µg/mL of HRP in PBS for 1 to 2 hours. After washing 3 × 10 minutes in PBS, the sections are incubated in 0.06% DAB in PBS for 10 minutes, then H_2O_2 is added to a final concentration of 0.01%. The reaction should be monitored under a stereomicroscope and stopped by replacing the DAB solution with PBS. Wash 3 × 10 minutes in PBS. At this stage, we recommend that the sections be temporarily mounted on glass slides using PBS and a glass coverslip and photographed in color. The cell (in back) will be easy to distinguish from the brown color of the immunostaining (nonintensified DAB reaction).

Osmication and Epon Embedding

1. Contents of each well are moved to glass scintillation vials, and after a short wash in PB, the material is osmicated in 1% OsO_4 in PB for 1½ hours at 4°C. After use, OsO_4 should be discarded into a bottle with corn oil. After 2 washes in PB, tissue is dehydrated in ascending alcohol concentrations (50%, 70%, 90%, and 95% for 5 min each; 2 × 10 min in 100%) and propylene oxide (2 × 10 min), followed by a 1:1 propylene oxide–Epon mixture (overnight or for 2 h), 2 hours in propylene oxide–Epon 1:2, and 1 to 2 hours in pure Epon.

2. Finally, each section is flat-embedded in Epon between acetate foil and a plastic coverslip. It is important that all sections be flat-embedded with the same side facing the plastic of the coverslip. The sections are cured in an oven overnight at 55°C.

Observation by Light Microscopy

The protocol described below allows the observation of the same fields by light microscopy and EM.

1. The Epon should be separated from the acetate foil so that the Epon embedded sections remain attached to the coverslips. The coverslips should be labeled and stored in small cardboard boxes.

2. Sections are examined with a microscope equipped with a camera lucida. The parts of the cell present in each

section are drawn. Blood vessels and other tissue landmarks can be used to properly identify the location of each segment of the cell and how they connect. Figure 2 shows an example of a reconstructed neuron. It facilitates the subsequent identification of the parts of the cell present in each Vibratome section if the sections are numbered and the parts of the cell present in each section are identified by the section number in a copy of the camera lucida reconstruction.

3. At this stage, all sections that contain parts of the cell should be photographed, preferentially with a digital camera.

Ultrastructural Observation

1. After photography, the flat-embedded sections should be re-embedded in Epon (see Reference 31). After curing in the oven at 55°–60°C, pyramids should be trimmed to the largest surface compatible with semithin sections. Serial 4-μm-thick sections are obtained using an ultramicrotome and a diamond knife specialized for semithin sections. Sections should be attached to APES-subbed glass slides, prepared in advance as described elsewhere (10,21). Subsequently, the 4-μm-thick sections are photographed and re-embedded by placing on top of each section a drop of Epon. A previously polymerized Epon block (cylinder with flat bases) with a block label inside (prepared 24 h before) is placed on top of each section. The glass slide with the Epon cylinders over each section is cured in an oven at 55°C for 24 to 48 hours. The blocks with sections attached are easy to separate from the glass when warm.

2. After trimming the blocks, ultrathin sections are cut with a diamond knife and collected onto formvar-coated 1-slot copper grids. They are counterstained lightly with uranyl acetate and lead citrate and observed under the electron microscope.

3. If a second antigenic signal is to be detected by postembedding immunogold, the ultrathin sections should be collected on nickel mesh grids. Details on how to process the sections for postembedding immunogold are given in several publications (see e.g., Reference 22). We recommend that etching with sodium metaperiodate be reduced to the minimum or abolished if possible (1,20). See also Chapter 10.

4. The micrographs of the semithin sections can be used as a guide to identify the parts of the cell when doing the EM observations.

Protocols for the Slice Approach

We have been using in recent years a combination of tight seal whole cell patch clamp electrophysiological recording with light and electron microscopic immunocytochemistry. The protocols described below have been optimized to obtain satisfactory recordings from cells in slices from adult (> 30 day old) or even aged (> 28 month old) animals and to allow the best possible preservation for confocal and electron microscopic examination. The patch clamp set-up that we have been using for these studies includes a fixed-stage upright light microscope equipped for infrared interference contrast (differential interference contrast or gradient contrast) (11,12), videomicroscopy, and fluorescence imaging. This approach allows the cell to be injected with a fluorescent dye during recording, it is possible to visualize the neuron being labeled and therefore get immediate direct information on the morphological type of cell being studied. As an

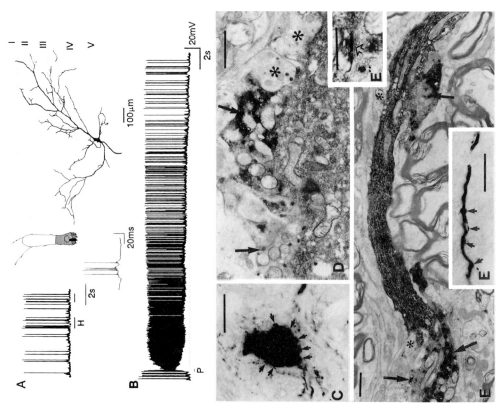

Figure 2. Dorsal horn neuron responding with a slow prolonged excitatory postsynaptic potential (EPSP) after noxious cutaneous stimulation. (A) Right: Camera lucida reconstruction of the cell body, located in lamina V, that was associated with a discrete area of substance P-immunoreactive (IR) fibers (see panel C). Left: Response of the neuron to low threshold mechanical stimulation (hair, H). The period of application of each of two stimuli is represented by the horizontal bars below the trace. Inset: Enlargement, on a faster time scale, of a typical burst of action potentials riding on EPSPs during the stimulus. Center: The cutaneous receptive field of the neuron is represented on the schematic diagram; darkened area represents the low threshold touch receptive field, and the larger hatched area represents the high threshold pinch receptive field. (B) Response of the neuron to high threshold mechanical stimulation (pinch, P). The period of the stimulus is represented by the horizontal bar below the trace. Note the marked depolarization during the noxious stimulus associated with action potential inactivation. Note also the prolonged depolarization after the end of the stimulus associated with increased firing frequency. (C) Light microscopic photograph of a parasagittal, 4-µm-thick, Epon flat-embedded section showing part of the neuronal cell body. Note the numerous substance P-IR profiles apposed to the cell body (arrows). The cell body was located within a region of intense immunoreactivity corresponding to one of the clusters of substance P-IR fibers commonly found in lamina V. Scale bar = 20 µm. (D) Electron microscopic photographs of an ultrathin section taken from the 4-µm-thick section shown in panel C. Note the 2 substance P-containing varicosities (arrows) apposed to the cell body. Synapses between these varicosities and the cell body could be demonstrated in adjacent sections (data not shown). Asterisks indicate non-IR axonal varicosities. Scale bar = 1 µm. (E) Electron micrographs of an ultrathin section taken from the 4-µm-thick section in panel E'. The portion of the dendrite shown in panel E corresponds to the curved portion of the dendrite shown in panel E' (the 2 arrows on the left of E' indicate the 2 boutons pointed to by 2 of the arrows in E). The immunoreactive profiles (arrows) belong to an axon that appears to wrap in a spiral-like fashion around the dendrite (arrows in panel E') of the cell and make several contacts with the dendrite. Asterisks indicate non-immunoreactive varicosities apposed to the dendrite. Scale bar = 1 µm. (E'') Details of the synapse (open arrow) established by the profile on the right in panel E (right arrow in panel E) obtained from an adjacent ultrathin section. Scale bars in panel E' = 20 µm; in panel E'' = 0.5 µm. Reproduced from De Koninck et al. (9).

example of this type of approach, we present below protocols for recording from and intracellular labeling of pyramidal neurons of the rat neocortex, followed by immunocytochemical processing for the demonstration of their cholinergic innervation. A diagrammatic representation of the protocols for confocal and electron microscopy is shown in Figure 3.

Materials and Reagents

- Sodium pentobarbital; sucrose; glycerol.
- NaCl, KCl, $CaCl_2$, $MgCl_2$, glucose, $NaHCO_3$, NaH_2PO_4.
- Cyanoacrylate cement.
- Borosilicate glass capillaries with inner filament.
- Cs-gluconate, CsCl, 4-(2-hydroxyethyl)-1-piperazine-ethanesulfonic acid (HEPES), BAPTA, ATP, GTP.
- Lucifer Yellow (Sigma).
- Biocytin (Calbiochem-Novabiochem, La Jolla, CA, USA) or neurobiotin (Vector Laboratories, Burlingame, CA, USA).
- Paraformaldehyde; glutaraldehyde (EM grade); picric acid (Fisher Scientific).
- Gelatin (Sigma).
- Tissue culture plates of 12 wells.
- Sodium borohydride; PB, pH 7.4; PBS, pH 7.4; Triton® X-100; H_2O_2.
- Normal donkey serum (Sigma).
- Antiserum against the vesicular acetylcholine transporter (VAChT; generous offer of Dr. R.H. Edwards, University of California, San Francisco).
- Rhodamine-tagged donkey antirabbit IgG (Jackson Immuno Research Laboratories, West Grove, PA, USA).
- Gelatin-subbed microscope glass slides.
- Glass coverslips (Fisher Scientific).
- Mounting medium: krystalon (Electron Microscopy Science, Gibbstown, NJ, USA).
- Avidin–biotin complex (Vector Laboratories).
- Antirabbit IgG ABC kit (Vectastain Elite Kit; Vector Laboratories).
- Cobalt chloride.
- Nickel ammonium sulfate.
- Osmium tetroxide (4% solution).
- Absolute ethanol (analytical grade).
- Propylene oxide.
- Epon (prepared from Epon 812 replacement, DDSA, NMA and DMP30 as described in Reference 13).
- APES.
- Plastic coverslips.
- Thick acetate foil (available from any stationary store).
- Standard plastic EM capsules.

Procedures

Slice Preparation

1. Adult rats (1–2 months old) are anaesthetized with Na^+-pentobarbital (30 mg/kg), and perfused intracardially for 15 to 20 seconds with ice-cold oxygenated (95% O_2, 5% CO_2) sucrose-substituted ACSF (S-ACSF) containing (in mM): 252 sucrose, 2.5 KCl, 2 $CaCl_2$, 2 $MgCl_2$, 10 glucose, 26 $NaHCO_3$, 1.25 NaH_2PO_4, (pH 7.35; 340–350 mOsm).

2. After decapitation, the brain is rapidly removed and immersed in ice-cold oxygenated (95% O2, 5% CO_2) S-ACSF for approximately 1 minute.

3. Slices are obtained using a Vibratome 1000 with the specimen holder modified to have a fluid chamber and a brass platform. The brain is glued, caudal side down, to the brass platform with

cyanoacrylate cement. The chamber is then filled with oxygenated ice-cold S-ACSF and 400-μm-thick slices are cut.

4. The freshly cut slices are incubated in oxygenated S-ACSF for 30 minutes at room temperature. The slices are then transferred to a storage chamber filled with oxygenated normal ACSF (126 mM NaCl instead of sucrose, 300–310 mOsm) and incubated at room temperature or 33°C for at least 1 hour before transferring to a recording chamber.

Whole Cell Patch Clamp Recording

1. Whole cell patch pipets are pulled from borosilicate glass capillaries and filled with different combinations of electrolytes depending on the type of membrane mechanisms one wants to focus on. In the present example, we used a Cs-gluconate-based-solution. The Cs was used to block K conductances because we focused on synaptically elicited responses (Figure 4). The anion used was gluconate to maintain

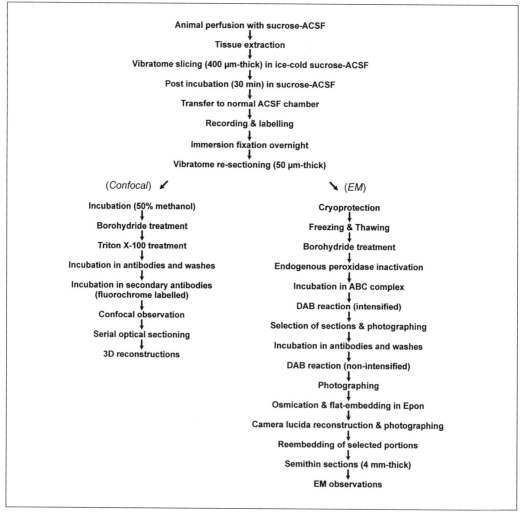

Figure 3. Steps in the approaches described for the combination of tight seal whole cell patch clamp electrophysiology in slices and intracellular injection of biocytin (or neurobiotin) and/or Lucifer yellow with the immunocytochemical detection of antigenic sites.

the normal chloride gradient. When focusing exclusively on inhibitory synaptic events (IPSPs/IPSCs [inhibitory postsynaptic currents]), a CsCl-filled solution has the advantage of increasing the driving force for Cl$^-$, which can help amplify the currents. As an example, the intracellular solution we used was composed of (in mM): 100 Cs gluconate, 5 CsCl, 10 HEPES, 2 MgCl$_2$, 1 CaCl$_2$, 11 BAPTA, 4 ATP, 0.4 GTP, 0.5% Lucifer Yellow, and/or 0.2% biocytin (a biotin-lysine complex) (pH adjusted to 7.2; osmolarity 275–280 mOsm).

2. Recordings are obtained by lowering the patch electrode onto the surface of visually identified neurons (we do not apply positive pressure to the pipet to avoid extracellular leakage of biocytin). While monitoring current responses to 5 mV pulses, a brief suction is applied to form greater than 5 GΩ seals. The membrane patch is then ruptured by further gentle suction to the electrode, establishing a whole cell recording configuration. The pressure is maintained to atmospheric throughout the recording.

3. Tight seal whole cell recordings in voltage clamp mode were performed using an Axopatch 200B amplifier (Axon Instruments, Foster City, CA, USA) with greater than 80% series resistance compensation for the recording. The access resistance is monitored throughout each experiment, and only recordings with stable access of 5 to15 MΩ are used for analysis of synaptic activity.

4. At the beginning of each recording, a set of 200 millisecond long, 5 mV hyperpolarizing pulses are used to measure the input resistance and membrane time constant of each neuron.

5. Recordings of greater than 10 to 15 minutes are sufficient to label the neurons. Simple diffusion of the dye included in the pipet into the cell during the course of the recording is sufficient to obtain complete labeling.

6. The recordings are stored on videotape using a digital data recorder (e.g., VR-10B; Instrutech, Great Neck, NY, USA). Stored recordings can then be played back off-line and sampled on a computer using the appropriate software.

Processing of Slices for Confocal Microscopy

1. At the end of the recording, the slices are fixed by immersion in 4% paraformaldehyde and 0.1% to 0.5% glutaraldehyde in 0.1 M PB, pH 7.4, for 2 hours at room temperature and postfixed overnight in 4% paraformaldehyde in PB at 4°C.

2. After fixation, the 400-µm-thick slice should be resectioned into 50-µm-thick sections for further histological processing. For this, the slices are embedded in 10% gelatin (3,5,15), fixed for 1 hour with 4% paraformaldehyde, 0.1% glutaraldehyde, and 15% picric acid in PB, and resectioned at 50 µm with a Vibratome.

3. Sections should be treated with 1% sodium borohydride in PBS and washed extensively in PBS (see Protocol for the In Vivo Approach for details). For subsequent processing, PBS with 0.2% Triton X-100 (PBS+T) is used to dilute the immunoreagents and for washing. The sections are incubated with 5% normal donkey serum in PBS+T to reduce background staining, followed by a rabbit antivesicular acetylcholine transporter (VAChT) antibody (14) and Rhodamine-tagged donkey antirabbit IgG. The sections are mounted on gelatin-subbed glass slides,

air-dried in the dark, dehydrated in ascending alcohols, cleared with xylene, and covered with a coverslip with krystalon. Subsequently, they will be examined with a laser confocal microscope. In our case, the slices are examined under an LSM 510 Laser scanning microscope (Carl Zeiss, North York, ON, Canada) equipped with an argon and 2 helium–neon lasers.

4. A 2D and 3D reconstruction of the cell and processes can be obtained from serial optical sections obtained with the confocal microscope. The images obtained with the application of this method are comparable in quality to camera lucida drawings, but have the following advantages: (*i*) they can be acquired very quickly; and (*ii*) both the morphological properties of the cell and the general distribution of the transmitter-specific innervation can be analyzed. These 2D and 3D representations allow one to rapidly obtain an evaluation of changes in the morphology of the cell and areas of predominant cholinergic input in the different groups of animals (Figure 5). It can be followed by a more detailed analysis of the innervation of the cell, as described in some publications (see e.g., Reference 24), using stereology if required.

5. The serial optical sections can also be used to perform a 3D reconstruction of the neuron. This can be useful for rotation of the cell in space to obtain a view in different projection planes. Proper morphological identification of

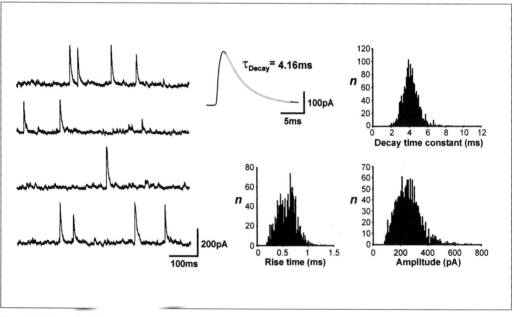

Figure 4. **Tight seal whole cell patch clamp recording from a layer V cortical pyramidal neuron in a 60-day-old rat.** The advantage of the whole cell technique is that it allows one to perform high resolution voltage clamp recordings and thus resolve miniature synaptic currents (i.e., action potential independent synaptic events recorded in the presence of the Na channel blocker tetrodotoxin). In this example, the membrane potential was held at 0 mV, the reversal potential for glutamate receptor-mediated events, to record exclusively miniature inhibitory postsynaptic currents (mIPSCs). Because the reversal potential for these events is approximately -65 mV, holding the membrane at 0 mV also helps amplify the events for optimal detection. The traces on the left are continuous to show on-going mIPSCs. Note the high signal-to-noise ratio obtained with this recording configuration, yielding clear detection of the synaptic events. The inset in the middle is an average of 300 consecutive mIPSCs to illustrate the kinetics of the events. The histograms show the results of the detailed analysis that can be performed on these events to characterize the rising and decaying kinetics as well as the amplitude distribution of the events.

certain cell types can require a view of the cell in several planes (4). Rotation can also serve to test whether an axonal bouton, which appears in contact with a dendrite in a certain plane of view, is indeed in true contact with the dendrite by viewing it in a perpendicular plane (Figure 6).

Processing of Slices for Electron Microscopy

If the slices are to be processed for EM, the first 2 initial steps to follow are the same as for confocal microscopy (see above). Triton X-100 should be omitted from all solutions. Other differences to consider are:

1. Endogenous peroxidase activity should be removed by incubating sections with 0.3% H_2O_2 in PBS for 15 minutes.
2. After inhibition of intrinsic peroxidase activity, the 50-µm-thick sections are infiltrated for 2 hours in 30% sucrose and 10% glycerol for cryoprotection.
3. The tissue is quickly frozen in liquid nitrogen-cooled isopentane and thawed at room temperature, as described above in the Protocols for the In Vivo Approach.
4. After washing 2 × 15 minutes in PBS, the signal from the intracellularly labeled neuron (byocitin or neurobiotin) is revealed by incubating the sections for 2 hours in an avidin–biotin complex (1:1000), diluted in PBS. Following 2 washes in PBS, a DAB reaction is performed with double intensification as described above.
5. After 2 washes in PBS, the sections can be temporarily covered by a coverslip with PBS for photographic purposes. Then, those containing the cell are processed for EM immunocytochemistry. Following incubation in the primary antibody (in this case a rabbit anti-VAChT serum) (14) for 48 hours at 4°C and washing, tissue should be placed for 2 hours in a biotinylated goat antirabbit IgG antibody, washed again, and finally incubated for 2 hours in an ABC complex.
6. The subsequent steps (DAB reaction without intensification, osmication, dehydration, and flat-embedding in Epon) should be performed as described above (see Protocol for EM in the Whole Animal Approach).
7. For analysis of the material at the light and electron microscopic levels, including the strategy to observe the same fields by light and electron microscopy, see the section above on Processing of Material from the Whole Animal Experiments.

RESULTS AND DISCUSSION

The protocols described in this chapter provide different types of information concerning the physiological properties of neurons. These range from identifying the types of inputs the neurons respond to (e.g., the types of natural stimuli that evoke responses in the recorded neuron and/or the pharmacological and biophysical properties of the synaptic events mediating specific inputs to these cells), to detailed analysis of membrane conductances specific to the type of cell studied (Figures 2 and 4). In the ultrastructural protocols, we propose the use of DAB-based reactions to reveal both the intracellular labeling and the immunocytochemical signal. While this approach may at first glance appear to have the drawback of potentially confusing the two signals, in most situations this does not happen. At the light microscopic level, it is easy to distinguish the homogenous black to grey-black labeling of the cell from the less homogeneous brown immunocytochemical labeling. At the EM level, such distinction is also possible because the homogenous intracellular deposits of intensified DAB contrast with the more variable intensity of the boutons labeled by the nonintensified DAB reaction. Even in cases in which the distinction would be difficult at the EM level, the observation of the same fields by light microscopy and EM allows the discrimination of the 2 signals. The real limitation of using DAB to reveal both signals is in situations when the 2 signals are colocalized. Alternative approaches to reveal the intracellular signal at the EM level have been proposed to avoid such a limitation (2).

In Vivo Approach

Over the years, we have used the whole animal approach with success in the cat, monkey, and rat. Recently, we modified the protocol to enhance the penetration of the immunoreagents. We originally used freeze-thawing of the entire tissue fragment containing the cell by direct immersion in liquid nitrogen followed by PB at room temperature; however, tissue fragmentation was a frequent problem. An alternative we proposed in a previous publication (10) was to use immersion in isopentane at -70°C (either cooled by dry ice or kept in a low temperature freezer). However, while isopentane at -70°C would completely prevent tissue cracking, the morphology would suffer as well. Therefore, the revised protocol involves the freezing and thawing of Vibratome sections instead of the originally proposed freezing and thawing of blocked tissue about 5 mm in thickness. The revised protocol (described in the Methods section above) provides excellent morphological preservation.

Two different antigenic signals can be easily detected using a pre-embedding method, and a third can be added subsequently on ultrathin sections by a post-embedding approach. We have exploited such possibilities in previous publications (20). In the past, we have used extensively internally radiolabeled monoclonal antibodies for the detection of one of the signals (19,30). However, such antibodies are not widely available, as they have to be radiolabeled during biosynthesis. Furthermore, they require the use of EM radioautography, an approach that requires long

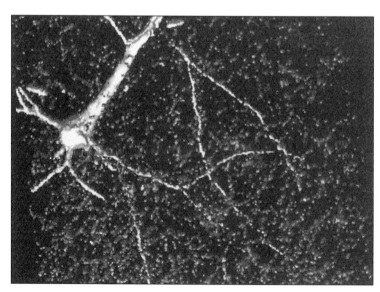

Figure 5. Confocal reconstruction of a layer V pyramidal neuron with surrounding VAChT-IR fibers illustrating the pattern of innervation of the cell. Note the abundance of immunoreactive boutons in the area of the basal dendrites. This 3D reconstruction was based on ten 1-µm-thick serial optical sections. The reconstruction was performed with the aid of the MCID-M4 image analysis system. Scale bar = 50 µm. (See color plate A7.)

photographic exposure times of 1 to 6 months. This is unfortunate, as such antibodies possess unique characteristics, such as superb penetration in the tissue and do not require any developing antibodies or complexes. The most logical approach today would be to use a DAB-based reaction for the detection of the first antigenic site and a pre-embedding immunogold approach for the detection of the second site. However, there is an important limitation: the penetration of the 1 to 1.4 nm gold-labeled secondary antibodies in the section (28) is much less satisfactory than that of the internally radiolabeled antibodies. This lack of penetration when applying the pre-embedding gold method is a major drawback when investigating the innervation of an intracellularly labeled neuron because the parts of the cell located deeply in the Vibratome section cannot be studied. It should be stressed that DAB-based protocols applying bispecific monoclonal antibodies or modified (4 step) peroxidase-antiperoxidase (PAP) approaches allow a complete (or virtually complete) labeling of the entire thickness of the section (31), and that even ABC approaches allow a reasonable penetration. The use of DAB-based pre-embedding immunocytochemistry for one antigenic site and post-embedding immunogold for the other site avoids most problems of penetration. However, many antigenic sites cannot be detected using post-embedding immunocytochemistry, unless osmication is omitted and special resin embedding is applied (26) at the cost of ultrastructural preservation. Therefore, an ideal approach does not exist. The combination of DAB-based immunocytochemistry with the application of internally radiolabeled monoclonal antibodies still represents the best alternative for someone with access to internally radiolabeled antibodies or with the technical capabilities to prepare them. For protocols used in the preparation of radiolabeled monoclonal antibodies see Reference 8.

One of the advantages of the use of the camera lucida reconstructions of the neurons from flat-embedded sections followed by serial semithin sections (that are re-embedded and re-cut for EM) is that we can observe the same fields by the light and electron microscope. An example of the establishment of such light–electron microscopic correlation is given in Figure 2.

In the in vivo protocol, we have used almost exclusively HRP-filled pipet. How-

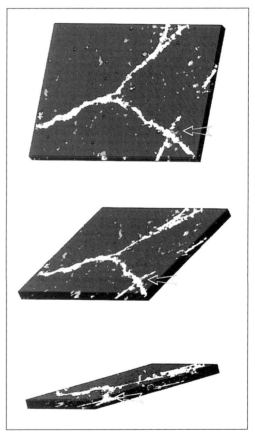

Figure 6. Confocal reconstruction of a dendrite from a layer V cortical pyramidal neuron illustrating the close association with surrounding VAChT-IR terminals. This 3D reconstruction was based on four 1-μm-thick serial optical sections. The reconstruction was performed using an image analysis system, as for Figure 5. The progressive 3D rotation from the top to the bottom of the panel illustrates how this approach can allow us to identify close association (but not synapses) of boutons with postsynaptic dendrites of the cell. (See color plate A8.)

ever, the injection of biocytin or neurobiotin has been used for the intracellular labeling of neurons (6,16,27,33). They have a high affinity for avidin, which allows greater flexibility for combination with other labeling techniques (e.g., fluorochromes, HRP-DAB reaction protocols). Also, biocytin and neurobiotin may be preferred to HRP when the study of the axon is important, as they give a better and more complete labeling of axons. This said, we still prefer HRP when ultrastructural studies are to be performed because of the simplicity of the approach (a single step reaction process to reveal it) and the absence of background labeling. Therefore, we focused on the use of HRP here.

Although in this chapter we exemplified the use of confocal microscopy in combination with the use of the slice approach, there is no reason not to use it in combination with the whole animal intracellular approach. In this case, the cell could be filled intracellularly with Lucifer yellow or biocytin. In the case of biocytin, the best way is to reveal it with streptavidin combined with a fluorochrome. The protocols described in the Methods for slices can be easily adapted to this end. However, one major drawback of confocal microscopy is that it is not possible to study whether a synapse is present, because of the limitations in resolution of the approach compared to EM. This, combined with the lower success rate of in vivo intracellular labeling make it almost a requirement to be able to combine confocal microscopy and EM. Fortunately, methods have been developed that allow the observation of the same fields by confocal microscopy and

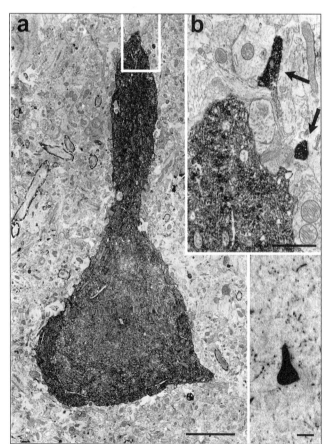

Figure 7. Electron micrographs of a biocytin-labeled layer V pyramidal neuron illustrating the concurrent detection of VAChT-IR fibers. (a) EM micrograph of a section of the cell body at low magnification. Scale bar = 5 µm. (b) EM micrograph taken from an adjacent serial section showing the area of the apical dendrite delineated in panel a at higher magnification. Scale bar = 1 µm. Note the VAChT-IR boutons (arrows) in close proximity to the cell. The inset shows the 4-µm-thick semithin Epon section, which was re-embedded to obtain the ultrathin sections from which the EM micrographs in panels a and b were obtained. Scale bar = 10 µm.

11. Electrophysiology/Morphology of CNS Neurons

EM, methods that can be easily adapted to intracellular labeling approaches (34). However, this combined confocal-EM approach is complex (see Chapter 15) and therefore the best compromise is still to use the combined light and electron microscopic approaches we describe in the Methods.

Slice Approach

The protocol that we described above is the result of an optimization by trial and error to obtain ideal conditions for both electrophysiological recording and morphological preservation (Figures 4–8). With our optimized approach, we have been able to obtain success rates of over 90%, a result that can be considered exceptional, particularly when comparing with the low success rates of the whole animal approach. An important factor determining the success rate in slices is to aim for cells that are deeper (>100 µm deep) in the tissue. With both the blind or visual patch clamp approaches, one tends to record from the first cell encountered at the sur-

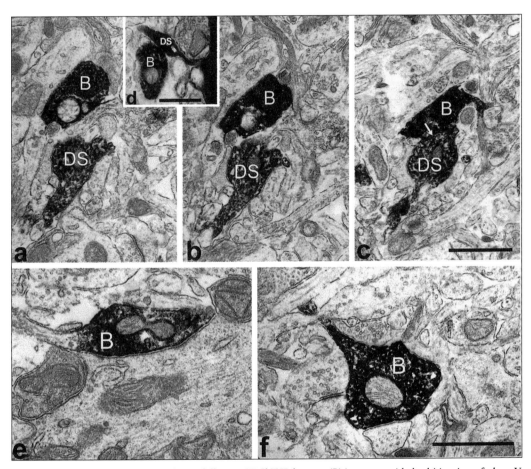

Figure 8. Electron micrographs in panels a to d illustrate VAChT-IR boutons (B) in contact with dendritic spines of a layer V pyramidal neuron (same cell as in Figure 6). The micrographs in panels a, b, and c were taken from serial sections illustrating a VAChT-IR bouton approaching a dendritic spine (DS) of the cell and forming a synapse in panel c (arrow). Scale bar = 1 µm. (d) Another example of a VAChT-IR bouton in contact with a dendritic spine of the cell which can be seen here in continuity with the dendritic shaft. Scale bar = 0.5 µm. Electron micrographs (e–f) illustrate VAChT-IR terminals (B) presynaptic to a cell body (in panel e) or a dendritic profile (in panel f). Note the quality of the ultrastructural preservation of this slice preparation after several hours of incubation and recording. Scale bar = 1 µm.

face of the slice. In this case, however, upon withdrawal of the electrode, the cell is often ripped out of the slice. Presumably, cells that are embedded deeper into the tissue are more firmly held in place. The use of thicker (400 µm thick) slices is also preferable because it provides better preservation of the dendritic tree of the cells. This is important for the quality of the physiology of neurons (e.g., maintain the integrity of synaptic events) and the electrophysiological recordings (13,32).

The morphological preservation that we achieve can be considered excellent for confocal microscopy (see Figures 5 and 6) and quite satisfactory for EM (see Figures 7 and 8). In Figure 5, we show a confocal reconstruction of the cholinergic innervation of a physiologically and morphologically characterized neuron, as obtained with a Zeiss confocal microscope and the 3D module of an image analysis system (MCID-M4; Imaging Research, St. Catharines, ON, Canada). In Figure 6, we illustrate how rotation of a 3D-reconstructed dendrite can help in the definition of proximity of cholinergic varicosities to the dendrite. However, such an approach cannot provide evidence of the occurrence of synapses. EM analysis is still required.

Figures 7 and 8 provide examples of the morphological preservation that can be attained routinely with the approach described in the methods. In Figure 7, observe the overall quality of the fixation of the cell, and in Figure 8, note the clear histological features of synapses established by cholinergic boutons.

ACKNOWLEDGMENTS

The studies in this chapter were supported by grants from the Canadian Medical Research Council (MRC) to A.R.S. and Y. De K., the National Institute for Aging to A.R.S., and the National Institute of Neurological Disorders and Stroke to Y. De K. We also acknowledge a Team Grant from the Quebec FCAR. Some experimental parts of these studies were performed in collaboration with Drs. J.L. Henry and A.C. Cuello. The authors are especially grateful to T.P. Wong for help with patch clamp experiments, to Marie Ballak for help with electron microscopy, and to Alan Forster for photographic help. We are also grateful to Drs. P. Somogyi and E.H. Buhl for valuable advice on how to improve the morphological preservation of slices. Y. De K. is a Scholar of the Canadian MRC.

REFERENCES

1. Alvarez, F.J., A.M. Kavookjian, and A.R. Light. 1993. Ultrastructural morphology, synaptic relationships, and CGRP immunoreactivity of physiologically identified C-fiber terminals in the monkey spinal cord. J. Comp. Neurol. *329*:472-490.
2. Branchereau, P., E.J. Van Bockstaele, J. Chan, and V.M. Pickel. 1995. Ultrastructural characterization of neurons recorded intracellularly in vivo and injected with lucifer yellow: advantages of immunogold-silver vs. immunoperoxidase labeling. Microsc. Res. Tech. *30*:427-436.
3. Buhl, E.H., Z.S. Han, Z. Lorinczi, V.V. Stezhka, S.V. Karnup, and P. Somogyi. 1994. Physiological properties of anatomically identified axo-axonic cells in the rat hippocampus. J. Neurophysiol. *71*:1289-1307.
4. Chéry, N. and Y. De Koninck. 2000. Visualisation of lamina I of the dorsal horn in live adult rat spinal cord slices. J. Neurosci. Methods *96*:133-142.
5. Cobb, S.R., K.Halasy, I. Vida, G. Nyiri, G. Tamas, E.H. Buhl, and P. Somogyi. 1997. Synaptic effects of identified interneurons innervating both interneurons and pyramidal cells in the rat hippocampus. Neuroscience *79*:629-648.
6. Côté, S., A. Ribeiro-da-Silva, and A.C. Cuello. 1993. Current protocols for light microscopy immunocytochemistry, p. 147-168. *In* A.C. Cuello, (Ed.), Immunohistochemistry II. John Wiley & Sons, Chichester.
7. Cowan, R.L., S.R. Sesack, E.J. Van Bockstaele, P. Branchereau, J. Chan, and V.M. Pickel. 1994. Analysis of synaptic inputs and targets of physiologically characterized neurons in rat frontal cortex: combined in vivo intracellular recording and immunolabeling. Synapse *17*:101-114.
8. Cuello, A.C. and A. Côté. 1993. Preparation and application of conventional and non-conventional monoclonal antibodies, p. 107-145. *In* Cuello, A.C. (Ed.), Immunohistochemistry II. John Wiley & Sons, Chichester.
9. De Koninck, Y., A. Ribeiro-da-Silva, J.L. Henry, and

A.C. Cuello. 1992. Spinal neurons exhibiting a specific nociceptive response receive abundant substance P-containing synaptic contacts. Proc. Natl. Acad. Sci. USA 89:5073-5077.

10. De Koninck, Y., A. Ribeiro-da-Silva, J.L. Henry, and A.C. Cuello. 1993. Ultrastructural immunocytochemistry combined with intracellular marking of physiologically identified neurons in vivo, p. 369-393. In Cuello, A.C. (Ed.), Immunohistochemistry II. John Wiley & Sons, Chichester.

11. Dodt, H.U., A. Frick, K. Kampe, and W. Zieglgänsberger. 1998. NMDA and AMPA receptors on neocortical neurons are differentially distributed. Eur. J. Neurosci. 10:3351-3357.

12. Dodt, H.U. and W. Zieglgänsberger. 1994. Infrared videomicroscopy: a new look at neuronal structure and function. Trends Neurosci. 17:453-458.

13. Edwards, F.A. and A. Konnerth. 1992. Patch-clamping cells in sliced tissue preparations. Methods Enzymol. 207:208-222.

14. Gilmor, M.L., N.R. Nash, A. Roghani, R.H. Edwards, H. Yi, S.M. Hersch, and A.I. Levey. 1996. Expression of the putative vesicular acetylcholine transporter in rat brain and localization in cholinergic synaptic vesicles. J. Neurosci. 16:2179-2190.

15. Han, Z.-S., E.H. Buhl, Z. Lorinczi, and P. Somogyi. 1993. A high degree of spatial selectivity in the axonal and dendritic domains of physiologically identified local-circuit neurons in the dentate gyrus of the rat hippocampus. Eur. J. Neurosci. 5:395-410.

16. Horikawa, K. and W.E. Armstrong. 1988. A versatile means of intracellular labeling: injection of biocytin and its detection with avidin conjugates. J. Neurosci. Methods 25:1-11.

17. Jochem, W.J., A.R. Light, and D. Smith. 1981. A high voltage electrometer for recording and iontophoresis with fine-tipped, high resistance microelectrodes. J. Neurosci. Methods 3:261-269.

18. Light, A.R. and H.H. Willcockson. 1999. Spinal laminae I-II neurons in rat recorded in vivo in whole cell, tight seal configuration: properties and opioid responses. J. Neurophysiol. 82:3316-3326.

19. Ma, W., A. Ribeiro-da-Silva, Y. De Koninck, V. Radhakrishnan, A.C. Cuello, and J.L. Henry. 1997. Substance P and enkephalin immunoreactivities in axonal boutons presynaptic to physiologically identified dorsal horn neurons. An ultrastructural multiple-labelling study in the cat. Neuroscience 77:793-811.

20. Ma, W., A. Ribeiro-da-Silva, Y. De Koninck, V. Radhakrishnan, J.L. Henry, and A.C. Cuello. 1996. Quantitative analysis of substance P immunoreactive boutons on physiologically characterized dorsal horn neurons in the cat lumbar spinal cord. J. Comp. Neurol. 376:45-64.

21. Maddox, P.H. and D. Jenkins. 1987. 3-Aminopropyltriethoxysilane (APES): a new advance in section adhesion. J. Clin. Pathol. 40:1256-1260.

22. Merighi, A. and J.M. Polak. 1993. Post-embedding immunogold staining, p. 229-264. In A.C. Cuello, (Ed.), Immunohistochemistry II. John Wiley & Sons, Chichester.

23. Moore, C.I. and S.B. Nelson. 1998. Spatio-temporal subthreshold receptive fields in the vibrissa representation of rat primary somatosensory cortex. J. Neurophysiol. 80:2882-2892.

24. Naim, M., R.C. Spike, C. Watt, S.A. Shehab, and A.J. Todd. 1997. Cells in laminae III and IV of the rat spinal cord that possess the neurokinin-1 receptor and have dorsally directed dendrites receive a major synaptic input from tachykinin-containing primary afferents. J. Neurosci. 17:5536-5548.

25. Nelson, S., L. Toth, B. Sheth, and M. Sur. 1994. Orientation selectivity of cortical neurons during intracellular blockade of inhibition. Science 265:774-777.

26. Nusser, Z., J.D. Roberts, A. Baude, J.G. Richards, W. Sieghart, and P. Somogyi. 1995. Immunocytochemical localization of the alpha 1 and beta 2/3 subunits of the GABAA receptor in relation to specific GABAergic synapses in the dentate gyrus. Eur. J. Neurosci. 7:630-646.

27. Paré, D., H.C. Pape, and J. Dong. 1995. Bursting and oscillating neurons of the cat basolateral amygdaloid complex in vivo: electrophysiological properties and morphological features. J. Neurophysiol. 74:1179-1191.

28. Pickel, V.M., J. Chan, and C. Aoki. 1993. Electron microscopic labeling of endogenous and/or transported antigens in rat brain using silver intensified one nanometer colloidal gold conjugated immunoglobulins, p. 265-280. In A.C. Cuello (Ed.), Immunohistochemistry. John Wiley & Sons, Chichester.

29. Pinault, D. 1996. A novel single-cell staining procedure performed in vivo under electrophysiological control: morpho-functional features of juxtacellularly labeled thalamic cells and other central neurons with biocytin or Neurobiotin. J. Neurosci. Methods 65:113-136.

30. Ribeiro-da-Silva, A., Y. De Koninck, A.C. Cuello, and J.L. Henry. 1992. Enkephalin-immunoreactive nociceptive neurons in the cat spinal cord. Neuroreport 3:25-28.

31. Ribeiro-da-Silva, A., J.V. Priestley, and A.C. Cuello. 1993. Pre-embedding ultrastructural immunocytochemistry, pp. 181-227. In A.C. Cuello, (Ed.), Immunohistochemistry II. John Wiley & Sons, Chichester.

32. Soltesz, I. and I. Mody. 1995. Ca(2+)-dependent plasticity of miniature inhibitory postsynaptic currents after amputation of dendrites in central neurons. J. Neurophysiol. 73:1763-1773.

33. Tasker, J.G., N.W. Hoffman, and F.E. Dudek. 1991. Comparison of three intracellular markers for combined electrophysiological, morphological and immunohistochemical analyses. J. Neurosci. Methods 38:129-143.

34. Todd, A.J. 1997. A method for combining confocal and electron microscopic examination of sections processed for double- or triple-labelling immunocytochemistry. J. Neurosci. Methods 73:149-157.

12 Tract Tracing Methods at the Light Microscopic Level

Marina Bentivoglio and Giuseppe Bertini
Department of Morphological and Biomedical Sciences, University of Verona, Verona, Italy, EU

OVERVIEW

Tract tracing techniques provide tools for the study of the termination or origin of central neural pathways or peripheral nerves. The understanding of the organization of neural circuits has represented one of the major goals in neuroscience since its birth. In the second half of the 19th century, the pioneers in tract tracing discovered that retrograde degeneration of neuronal cell bodies and anterograde degeneration of fibers could be used to trace pathways in the nervous system. Thus, "the earliest way to identify the neurons sending their axons to a given neural structure was to destroy the structure" (20). In addition, the Golgi method, which impregnates random subsets of neuronal cell bodies and processes in their entirety, played a crucial role not only in unraveling the basic structure of the nervous system, but also in pioneering investigations on its connectivity. Cajal's seminal studies on the wiring of the nervous system were in fact based on Golgi-impregnated material (5). It is very hard, however, to effectively impregnate the axons and to reconstruct their trajectory over long distances with the Golgi staining. This technique, which is also very capricious, is thus best suited for the study of local neural connections, i.e., at a limited distance from the cell body.

Before the introduction in the 1970s of methods taking advantage of the axonal transport of molecules (7,15,17), the degeneration techniques represented the main tools for tract tracing. Anterograde tracing exploits the Wallerian degeneration of entire axons or their distal portion that follows lesion of the parent cell bodies or disconnection from the axon proximal portion. These techniques became especially effective after the introduction of stains for the selective silver impregnation of degenerating axons (9,21). In the retrograde degeneration technique, the loss of neuronal perikarya consequent to lesions of the target and/or of the efferent axons can be detected with the Nissl cell staining. Degeneration methods still represent the only techniques available for the investigation of long distance neural connections in adult human post-mortem material. This can be of interest, for example, in the case of restricted lesions in autoptic human specimens.

The study of short (intrinsic, local) connections or of long connections (extrinsic, in which axons terminate at some distance away from the area that contains the parent cell bodies) requires distinct methodological approaches. Besides the Golgi staining, intrinsic connections can be investigated using technologically sophisticated techniques based on the intracellular (including intra-axonal) injection of dyes. Direct staining can potentially visualize entire pathways, and new perspectives in the direct labeling of connections have been opened in recent years by advances in molecular genetics, leading to the development of a new generation of axonal tracers, the so-called reporter-based tracers (6). For example, β-galactosidase, a reliable histochemical marker (the reporter molecule), is fused by genetic recombination to proteins that associate with cytoskeletal components, thus enabling specific targeting of axons, dendrites, or synapses (6). Due to space limitations, this promising experimental strategy will not be discussed in the present chapter.

The classical tract tracing techniques are suited for the study of extrinsic monosynaptic connections. To this day, experimental investigation of polysynaptic connections has proven harder than that of each separate set of connections. Here, we wish to simply mention that neurotropic viruses can be used as retrograde transsynaptic tracers. Briefly, viruses are transported to and replicate in the cell bodies through chains of interconnected neurons, and their presence can be detected with immunohistochemistry (16). The use of viruses as tracers requires special safety rules in the laboratory (26). In transsynaptic tracing with viruses, the virus is in general injected at the periphery (muscles or peripheral organs).

Chemically identified connections can also be investigated selectively using labeled (e.g., isotope-conjugated) neurotransmitters (8), but in this chapter, we will focus on the experimental, nonselective study of extrinsic monosynaptic connections in vivo.

In particular, we will outline the general procedure for tract tracing in the central nervous system, while it should be mentioned that tracer injections in peripheral tissues (muscles, skin, and other peripheral organs) may require special methodological approaches. For tracing in the peripheral and autonomic nervous systems, or for transganglionic tracing, it is therefore advisable to run pilot experiments in order to test the efficacy of the selected tracer(s) when their properties in the paradigm under study are not well established in the literature.

Lipophilic dyes, which diffuse along the neuron membranes (14), are particularly suited for in vitro tracing of neural connections during development, as well as for the study of connections in tissue slices and dissociated cells (13). A detailed description of these techniques is beyond the scope of the present discussion, but general principles for tracing in fixed tissue specimens will be briefly mentioned.

BACKGROUND

A large number of retrograde and anterograde tracers are available nowadays, and the list keeps growing, especially for those suited for particular experimental purposes such as developmental studies, tracing of connections in vitro, multiple tracing, tracing combined with immunohistochemistry, or in situ hybridization, tracing combined with intracellular dye filling.

The characteristics and properties of commonly used tracers are summarized in Tables 1 and 2.

Classes of Tracers

General Concepts

Retrograde tracers, i.e., molecules transported through axons to the parent cell bodies, are taken up by the axon terminals but can also be incorporated by the axon itself, as it occurs when the axon is damaged. The possibility of tracer uptake from intact fibers of passage, however, has been documented for several tracers and should always be taken into account and controlled experimentally if possible, especially when it could be crucial for the interpretation of the experimental findings.

Anterograde tracers are taken up by neuronal cell bodies and/or axons. The only anterograde tracers that are certainly not taken up by fibers of passage, are the isotope-labeled amino acids (7), since amino acids are used for protein synthesis, whose machinery is located in the neuronal cell bodies. Anterograde tracing with tritiated amino acids (most commonly a mixture of proline and leucine) using autoradiography was the first technique introduced for anterograde tracing based on axonal transport (7). Autoradiographic methods will not be discussed in the present chapter, but the basic protocol, still in use, is detailed in the original paper (7).

Retrograde Tracers

Two main classes of retrograde tracers are listed in Table 1: the horseradish peroxidase (HRP)-based tracers [of which the subunit B of cholera toxin (CTb)-HRP is an example], and fluorescent tracers: fluorescent microspheres (beads), fast blue (FB), diamidino yellow (DY), propidium iodide (PI), Evans blue (EB) and nuclear yellow (NY). For the combination of retrograde tracing with immunohistochemistry, wheat germ agglutinin (WGA) conjugated with inactivated HRP (WGA-apoHRP; Sigma, St. Louis, MO, USA) can also be effectively used (see below).

Anterograde Tracers

Several classes of anterograde tracers are indicated in Table 2: *Phaseolus vulgaris* leucoagglutinin (PHA-L), CTb, dextrans, and carbocyanine derivatives. As examples of fluorescent dextrans we indicated tetramethyl rhodamine-dextran amine, called Fluoro-Ruby (FR), and fluorescein-conjugated dextran, called Fluoro-Emerald (FE). In addition, a widely used tracer is represented by biotinylated dextran amine (BDA). Other biotin-based tracers are biocytin (η-biotinoyl-L-lysine), and neurobiotin [N-(2-aminoethyl) biotinamide hydrochloride]. Lipophilic carbocyanine dyes include DiI and DiA, fluorescing in different colors. DiA can be substituted by DiO, which, however, is less effective (26). As an alternative to DiI, Fast DiI is also available (26).

Criteria for Tracer Selection

The first criterion for the choice of the tracer lays, of course, in the experimental discussion itself, i.e., whether an anterograde or a retrograde tracer, or more than one tracer, are needed. Keep in mind that the axonal transport of most tracers is bidirectional, but all tracers exhibit a preference for either retrograde or anterograde transport.

Decide in advance whether the light microscopy observations will be complemented by ultrastructural studies, since tract tracing in electron microscopy has special requirements, both in terms of tracer selection and tissue fixation procedure (see Chapter 13).

For single labeling, a crucial parameter is the spread of the tracer at the injection sites (Tables 1 and 2). Animal size and length of the pathways under study should also be considered. Retrograde fluorescent tracers with limited diffusion at the injection sites, e.g., FB and DY, are very useful in rodents, but in order to fill a relatively large territory (for example in the monkey cortex) tracers resulting in more widespread diffusion, such as HRP and HRP conjugates, or NY, may be preferable. On the other hand, Fluorogold (FG), a tracer that combines sensi-

Table 1. Parameters for Retrograde Tracing

	Tracer	Beads	CTB-HRP	FG	FB	DY	PI	EB	NY
	Supplier	c,d	b,e	a	e	e	e	e	e
Compatibility	Histochemistry	yes	yes	yes (1)	-	-	-	-	-
	Immunoperoxidase	yes	yes	yes (1)	-	-	-	-	-
	Immunofluorescence	yes	-	yes	yes	yes	yes	yes	-
Injection	Procedure	P	P, pellet	P, I	P, I	P	P	P	P
	Diffusion at injection site	+/-	+++	+++	++	+	+++	+++	+++
Filling	Soma	++	+++	+++	+++	nuclear	++	+++	nuclear
	Dendrites	+/-	+++	++	++	-	+	+	-
Post Injection and Section Processing Parameters	Survival	2 d–mos	2–3 d	2 d–mos	4 d–mos	4–8 d	2–3 d	4–14 d	2–3 d
	Fixation	PAF	PAF + GA	PAF	PAF	PAF	PAF	PAF	PAF
	Stability over time	+++	+	+++	+++	+	++	+	-
	Detection	F	LM	F	F	F	F	F	F
	Color	red, green (2)	brown (3)	yellow	blue	yellow	red	red	yellow

Suppliers: a, Fluorochrome, Englewood, CO, USA; b, List Biological, Campbell, CA, USA; c, Lumafluor, New York, NY, USA; d, Molecular Probes, Eugene, OR, USA; e, Sigma; f, Vector Laboratories.

Abbreviations: BDA, biotin dextran amine; CTB, cholera toxin subunit B; CTB-HRP, CTB conjugated with horseradish peroxidase; DY, diamidino yellow; EB, Evans blue; F, fluorescence; FB, fast blue; FE, fluoro-emerald; FG, fluorogold; FR, fluoro-ruby; GA, glutaraldehyde; I, iontophoresis; LM, light microscopy; NY, nuclear yellow; P, pressure; PAF, paraformaldehyde; IF, immunofluorescence; PHA-L, *Phaseolus vulgaris* leucoagglutinin.

Notes: (1) fluorescent labeling may decrease; (2) other colors are available; (3) depends on the chromogen; (4) in IF, depends on the fluorochrome-conjugated secondary antibody.

12. Tract Tracing Methods at the Light Microscopic Level

Table 2. Parameters for Anterograde Tracing

	Tracer	FE/FR	CTB	BDA	Biocytin	Neurobiotin	PHA-L	DiI	DiA
	Supplier	d,e	b,e	d,e	d,e	f	e,f	d	d
Compatibility	Histochemistry	yes	yes	yes	yes	yes	yes	-	-
	Immunoperoxidase	-	-	yes	yes	yes	-	-	-
	Immunofluorescence	yes	yes	-	-	-	yes	yes	yes
Injection	Procedure	P	P	P, I, pellet	P, I, pellet	P, I, pellet	I	P, deposits	P, deposits
	Diffusion at injection site	++	++	++	+	++	+/-	++ (P), +/- (deposits)	++ (P), +/- (deposits)
Postinjection and Section	Survival	2–14 d	3–4 d	7–15 d	1–2 d	1–2 d (deposits)	2–7 wks	days (P), wks	days (P), wks
Processing Parameters	Fixation	PAF	PAF	PAF ± GA	PAF ± GA	PAF ± GA	PAF	PAF	PAF
	Stability over time	++	+++	+++	+++	+++	+++	+/-	+/-
	Sensitivity in dehydration	decrease	stable	stable	stable	stable	stable	disappear	disappear
	Detection	F	LM, IF	LM	LM	LM	LM, IF	F	F
	Color	green/red	brown (3,4)	brown (3)	brown (3)	brown (3)	brown (3,4)	red	green

Suppliers: a, Fluorochrome, Englewood, CO, USA; b, List Biological, Campbell, CA, USA; c, Lumafluor, New York, NY, USA; d, Molecular Probes, Eugene, OR, USA; e, Sigma; f, Vector Laboratories.

Abbreviations: BDA, biotin dextran amine; CTB, cholera toxin subunit B; CTB-HRP, CTB conjugated with horseradish peroxidase; DY, diamidino yellow; EB, Evans blue; F, fluorescence; FB, fast blue; FE, fluoro-emerald; FG, fluorogold; FR, fluoro-ruby; GA, glutaraldehyde; I, iontophoresis; LM, light microscopy; NY, nuclear yellow; P, pressure; PAF, paraformaldehyde; IF, immunofluorescence; PHA-L, *Phaseolus vulgaris* leucoagglutinin.

Notes: (1) fluorescent labeling may decrease; (2) other colors are available; (3) depends on the chromogen; (4) in IF, depends on the fluorochrome-conjugated secondary antibody.

tivity of labeling and good filling of the injected sites, is not effective in the monkey brain, possibly due to its resemblance with autofluorescence (Bentivoglio, unpublished observations).

For multiple labeling, a crucial parameter for the selection of tracers is represented by the compatibility (both in terms of tissue fixation procedure and label visualization) between markers (see below).

Visualization of Tracers

Neuronal labeling obtained with fluorescent tracers can be visualized directly at the microscope, using the light excitation wavelength appropriate for the fluorochrome. For most other tracers, however, labeling must be revealed indirectly. The techniques suitable for revealing tracers at the light microscopic level can be summarized as follows:

1. If the tracer is an enzyme or is conjugated to an enzyme, its presence in labeled cells can be revealed by specific histochemical procedures. This is the case for HRP, the prototype of retrograde tracers, which can be delivered either free or conjugated. The molecules most commonly conjugated with HRP to obtain sensitive tracers are the lectin WGA and CTb (choleragenoid). Free HRP and HRP conjugates (WGA-HRP, CTb-HRP) are revealed through a histochemical reaction in which a chromogen is oxidized by HRP resulting in colored precipitates in the cell. Therefore, not only the tracer sensitivity but also the staining procedure is a crucial parameter for successful labeling.
2. The tracer can be conjugated with a nonenzymatic probe that allows its visualization. This can be a fluorochrome, but the most effective ones are isotopes, gold particles, or biotin. The conjugation of a tracer with an isotope requires the use of autoradiography to visualize the labeling. Gold conjugates are detectable in light microscopy after silver intensification, for which kits are commercially available. Biotin-based tracers are revealed through histochemistry, exploiting their high affinity for avidin–peroxidase complexes.
3. The tracer can be revealed by immunohistochemical techniques, using antibodies raised against the tracer molecule. This approach allows the flexibility of immunohistochemistry (based on immunoperoxidase or immunofluorescence protocols), but of course it requires that antibodies against the tracer are available. The most widely used tracer revealed by immunohistochemistry is PHA-L (10).

PROTOCOLS

In outlining tract tracing protocols, we will follow the steps of a general procedure (outlined in the flow chart of Figure 1) providing clues for retrograde and anterograde tracing strategies.

Materials and Reagents

Tracers

- A list of tracers and their suppliers is indicated in Tables 1 and 2.

Tracer Preparation

To dissolve the tracer:
- Distilled water or saline or phosphate-buffered saline (PBS), 0.01 M, pH 7.2 to 7.4.
- Dimethylsulfoxide (DMSO; Sigma).

To prepare gold-conjugated tracers:
- Colloidal gold solution, 6 to 12 nm particles (Electron Microscopy Sci-

ence, Fort Washington, PA, USA).
- Polyethylene glycol (Sigma).

Injection

- Anesthetics (we commonly use sodium pentobarbital, 50 mg/kg i.p.).
- Hamilton microsyringes (1, 5, and 10 µL) or glass micropipets prepared with a pipet puller.
- Stereotaxic equipment (David Kopf Instruments, Tujunga, CA, USA).
- Iontophoresis apparatus (Stoelting, Wood Dale, IL, USA).

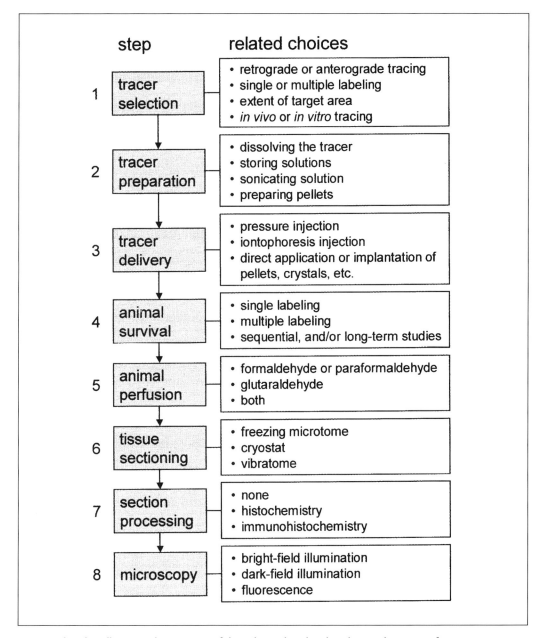

Figure 1. Flow chart illustrating the main steps of classical procedures based on the axonal transport of tracers.

Perfusion

- Anesthetics (see above).
- PBS.
- Fixative: 1% to 4% paraformaldehyde (Fluka, Neu-Ulm, Switzerland) or 0.5% to 2.5% glutaraldehyde (Fluka) in 0.1 M phosphate buffer, pH 7.2 to 7.4.

Tissue Sectioning

- Sucrose (10% to 30%) in PBS.
- Freezing microtome or cryostat (e.g., Leitz, Stuttgart, Germany).
- Vibratome (e.g., Ted Pella, Redding, CA, USA).

Section Processing

For HRP histochemistry:
- 3-3′ diaminobenzidine (DAB; Sigma).
- Nickel ammonium sulphate (Sigma).
- 3,3′,5,5′ tetramethyl benzidine (TMB; Sigma).
- Sodium nitroferricyanide (Sigma).
- Ammonium heptamolybdate or ammonium paratungstate (Sigma).
- Tris-HCl buffer.
- Hydrogen peroxide (Fluka).

For immunohistochemistry:
- Primary antibodies against the tracer. For example, anti-FG polyclonal antibodies raised in rabbit (Sigma); anti-PHA-L polyclonal antibodies raised in goat (Vector Laboratories, Burlingame, CA, USA).
- Secondary antibodies (Vector Laboratories), selected according to the immunohistochemical protocol. For immunoperoxidase: biotinylated (e.g., goat antirabbit, rabbit antigoat, etc.) IgGs. For immunofluorescence: fluorescein isothiocyanate (FITC)- or tetramethyl rhodamine (TRITC)-conjugated IgGs.
- Avidin–biotin peroxidase complex (ABC kit, Vectastain; Vector Laboratories).

For biotin-based tracers:
- ABC kit (see above).

For silver intensification (to reveal gold-conjugated tracers):
- IntenSe™-M kit (Amersham Pharmacia Biotech, Piscataway, NJ, USA)

For mounting and covering with a coverslip:
- Ethanol (at different concentrations) and xylene.
- Glass slides and coverslips.
- DPX or Entellan (Merck, Darmstadt, Germany), glycerin.

Microscopy

- Light microscope, possibly equipped with dark-field condenser.
- Fluorescence microscope and appropriate filter systems. In our laboratory, we use a Leitz Ploemopack epifluorescence illumination system (mercury lamp source: 100 W HBO), equipped with the following filter blocks: A [band-pass (BP) filter: 340–380 nm]; D (BP: 355–425 nm); N2 (BP: 530–560 nm).

Procedure

Preparation of the Tracer(s)

1. Dissolve the tracer in distilled water or PBS unless different solvents are specifically required. For example, biocytin (5%) is usually dissolved in Tris buffer, pH 8.0, and carbocyanine derivatives can be dissolved in ethanol (up to 5%), or 3% DMSO. Some tracers (e.g., DY, biocytin) are difficult to dissolve and remain in suspension. Therefore, pay attention to potential clogging of the

syringe or pipet during tracer delivery. Alternatively, pellet preparation may be desired (see below).

Notes:

a. For the use of HRP, keep in mind that the enzymatic activity decays over time (check the expiration date of the chemical).

b. For the preparation of gold-conjugated tracers, the protocol of Basbaum and Menétrey (2) can be followed (but see also Bentivoglio and Chen, Reference 3). This implies the use of a colloidal gold solution, to which polyethylene glycol is added, mixed with the tracer solution, and centrifuged.

c. Store the tracer solutions at 4°C (the solutions of most tracers are stable and can be stored for months at 4°C) or store it frozen at -20°C (this is necessary for HRP and its conjugates).

Tracer Delivery

When the procedure implies tracer injection:

2. Anesthetize and fix the animal on the stereotaxic apparatus.
3. Drill holes on the skull at the stereotaxic coordinates.
4. Fix the injection device to the stereotaxic holder.
5. Sonicate the tracer solutions just prior to filling the syringe or pipet.
6. Inject the tracer.

Notes:

a. The tracers can be injected by direct mechanical or air-pressure through Hamilton microsyringes or glass micropipets. Take care to avoid plugging the syringe or pipet tip with the tracer solution or suspension. Clean carefully the syringe or pipet tip before penetrating the tissue in order to avoid spurious tracer deposit along the injection track. This is especially important when the target is located deeply in the brain and must be reached stereotaxically.

b. While injecting tracer solutions or suspensions, divide the total volume into smaller (e.g., 0.05–0.1 µL) deposits in multiple tracks or at different heights in the same track (if possible), since this procedure helps the tracer penetration in the tissue and its uptake. To avoid leakage along the needle track while withdrawing the syringe or pipet, wait for 1 to 3 minutes at the end of each deposit.

c. Some tracers can also be applied by iontophoresis, a technique particularly indicated for small injection sites, restricted to the chosen target. For each tracer, iontophoretic delivery parameters are well defined. For example, injection of PHA-L (which, incidentally, can only be administered iontophoretically) requires a positive 4 to 7 µA current applied to a 2.5% solution in PBS, pH 8.0.

d. When localized delivery of a highly concentrated tracer is necessary, solid pellets of some tracers can be prepared and inserted in the target (25,26) (Tables 1 and 2). Finally, HRP can be delivered in several different ways (HRP gelfoam, polyacrylamide gels containing HRP, or implantation of HRP crystals) (25).

e. As mentioned above, developmental studies have special requirements that will not be addressed in detail here. It is worth mentioning, however, that for tracing axon bundles in embryos, a pin bearing a dextran amines crystal can be directly applied to cut axons (11).

f. For the deposit of carbocyanines (DiI, DiA; see Table 2), which are especially suited for anterograde tracing in paraformaldehyde-fixed tissue, insert the crystals directly into the tissue with a pin or a small forceps.

Postinjection Survival

7. After injection, suture the surgical wound, and let the animal recover before returning it to the animal care facilities. Tracer injections are usually well tolerated and do not endanger the animal's survival. The anesthesia is often responsible for postinjection complications.

8. Allow the animal to survive for the time appropriate for an effective labeling of each tracer.

Notes:

a. Most tracers (especially the retrograde ones) are transported through axons by fast transport. However, the survival required for effective labeling is highly variable (Tables 1 and 2). In general, HRP is degraded by lysosome enzymes in the labeled neurons within a few days, and the postinjection survival should, therefore, be limited. Biocytin also requires short postinjection survival (24–48 h). For most of the other tracers, metabolic degradation in the labeled cells does not seem to be crucial. In particular, PHA-L requires a long postinjection survival (10–20 days), and fluorescent tracers are very stable over time and for some of them (e.g., FB) a relatively long survival (a few days to 2 or more weeks, depending on the length of the pathway) is necessary to obtain brilliant labeling. However, changes in the size of the injection site with time are to be expected. In addition, some fluorescent tracers, and in particular DY and NY, may leak over time from labeled neurons into the adjacent glial cells, and this can result in spurious labeling. To avoid this drawback, relatively short survival should be adopted after injections of DY or NY (the latter, especially). This implies that for multiple retrograde labeling, or in protocols of combined retrograde and anterograde labeling, the injections may have to be performed in 2 different surgical sessions. In these cases, HRP or HRP conjugates, NY or DY should be injected in the second session, a few days before the animal's sacrifice.

b. Sequential injections of dyes may be desired for special experimental paradigms in which the rearrangement of circuits is studied over time (for example, during development or in the study of postlesional axonal regeneration). In these cases, a long-lasting tracer is injected first, and a second tracer is injected in the same or a different target after days, weeks, or even months.

c. Axonal transport of tracers is much slower in cold-blooded animals than in mammals, and cold-blooded animals should be kept in a warm environment during the survival period.

Perfusion

9. Anesthetize the animal with a lethal dose and proceed to transcardial perfusion (through the left ventricle or the aorta). Flush out the blood first with saline or PBS, and then perfuse with the fixative solution.

Notes:

a. Adequate pressure of the solutions is necessary to ensure thorough perfusion of the parenchyma. Heparin may be added to the perfusion solution to prevent blood clotting. Carefully washing out the blood is especially important in the use of all tracers that imply oxidation of a chromogen with hydrogen peroxide (e.g., HRP and its conjugates), since red blood cells possess endogenous catalases that precipitate the chromogen. This may not represent a problem for the data analysis (erythrocytes can be easily distinguished from neurons), but it results in low quality photography.

b. Formaldehyde (4%, i.e., 10% formalin) or paraformaldehyde (most commonly 4%), usually in phosphate-buffered solutions, are suited for the fixation of all tracers for light microscopy (Tables 1 and 2), but may at least in part inactivate the HRP enzymatic activity. Glutaraldehyde fixation is optimal for HRP labeling and may be required when the experiments also imply an electron microscopic investigation (see Chapter 13). Perfusion with mixed aldehydes (e.g., 1%–2% paraformaldehyde and 1%–2% glutaraldehyde) can provide a good compromise in some cases. Glutaraldehyde fixation results, however, in high background fluorescence and should be avoided when using fluorescent dyes or tracers to be visualized with immunohistofluorescence. On the other hand, glutaraldehyde fixation does not affect gold-conjugated tracers, for which, therefore, fixation is not a critical step. Strong formaldehyde fixation is optimal for a brilliant fluorescence (but may also increase autofluorescence).

c. For in vitro tracing, the animal is perfused prior to tracer delivery, and the tissue samples required for the experiment are dissected out and stored in formaldehyde or paraformaldehyde, prior to and during tracer diffusion.

Cutting Sections

10. Cut frozen or Vibratome sections from the areas of interest.

Notes:

a. The use of frozen sections (cut with a freezing microtome or a cryostat) requires prior cryoprotective treatment of the tissue blocks in sucrose solutions (10%–30% in phosphate buffer) until they sink. Vibratome sectioning can be performed even immediately after perfusion and is required when the experiment also implies further processing for electron microscopy.

b. In general, sections should be 20 to 50 µm thick (30 µm is optimal in most cases), but advance planning of section thickness and section sampling (the interval between processed sections) is critical when quantitative data analysis has to be performed.

Mounting Sections

11. Mount sections onto glass slides subbed with chrome alum or gelatin. Immediate mounting is not crucial for sections processed for HRP histochemistry, but for fluorescent tracers it is advisable to mount the sections on slides soon after cutting.
12. Let sections air-dry.
13. Dehydrate and mount.

Notes:

a. Dehydration should, in general, be very quick, as most tracers are sensitive to prolonged treatment in ethanol. Sections can also simply be dipped in xylene just prior to covering with a coverslip. Permount (Electron Microscopy Sciences, Fort Washington, PA, USA), Entellan (Merck, Darmstadt, Germany), or DePeX (Merck) are mounting media suited for most tracing techniques. Glycerin is also used as mounting medium.

Section Processing for Tracer Visualization

14. Sections derived from material injected with fluorescent dyes can be observed directly at the microscope (see below). When nonfluorescent tracers are employed, immunoperoxidase or immunofluorescence protocols can be used.

HRP-Based Tracers

For HRP histochemistry, the standard DAB recipe (0.05% DAB, 0.003% H_2O_2 in TB) produces a brown reaction product. DAB is carcinogenic (requires caution in handling and should be inactivated after use) and is not a sensitive chromogen for HRP histochemistry, but is optimal for the definition of the boundaries and extent of the HRP injection sites. DAB-processed sections can be lightly counterstained with cresyl violet. TMB (19), on the other hand, is a very sensitive chromogen but is less stable than DAB over time and is sensitive to dehydration; TMB-processed sections can be lightly counterstained with neutral red. Several different recipes for HRP histochemistry are available (22,25,27). Nickel-intensified DAB reaction (0.05% DAB, 0.2% nickel ammonium sulfate, and 0.003% H_2O_2 in Tris-HCl buffer) results in black reaction products and also enhances the signal. Modified TMB stabilization (22,27) makes the reaction products more stable, allows counterstaining with cresyl violet, and is suited for further EM processing.

Biotin-Based Tracers

For biotin-based tracers, the detection of labeling is achieved through the avidin–biotin complex (ABC)-peroxidase reactions, using DAB (or metal-intensified DAB) as chromogen.

Microscopic Study

- Use bright-field illumination for the study of free or conjugated HRP, biotin-based tracers, or immunostained labels. The labeling visualized by some chromogens (e.g., DAB) is also well visible under dark-field illumination (which increases the signal-to-noise ratio), but TMB, for example, is hard to identify in dark-field.

- A fluorescence microscope is obviously needed for the study of the labeling of fluorescent or fluorochrome-conjugated dyes and immunofluorescent labels. Before starting the study, make sure that the fluorescence microscope is equipped with the proper filter systems (excitation with the appropriate light wavelength is critical for fluorescence). If the fluorescent label is not visible at relatively low power (e.g., using a 10× objective), always check the field at higher power before discarding the experiment (the label or conditions of observation may not be sufficiently sensitive at low power).

In multiple labeling, when the protocol does not allow the simultaneous visualization of 2 fluorescent tracers in the same light excitation wavelength, the same field must be observed using different filter systems. Fluorescence and dark-field or even bright-field illumination can also be combined during the observation. Observe first under fluorescence, then turn on the microscope's main light bulb and look for the best compromise for observing both signals at the same time.

- If labeling is analyzed with a digital camera and software for image analysis, make sure that the camera is sufficiently sensitive to record all the signal you see under the microscope.

Tips for Tracing with Fluorescent Dyes

Fluorescent dyes or fluorochrome-conjugated tracers are very appealing due to the simplicity of their use, which requires fluorescence microscopy but does not imply complicated protocols of tissue processing. Fluorescence, however, is not stable over time; the quality of the preparations is better preserved if the slides are stored in dark boxes at 4°C. Since fading during the observation may occur, it is always advisable to prepare more than one series of sections, examining only one series to check the experimental findings. If the results are good, the extra series can be used for charting the label and/or for photography. Although fluorescent counterstains can be used (1,23), it is a good policy to prepare adjacent Nissl-stained serial sections, especially if the cytoarchitectonic study (e.g., of the boundaries of a given structure) presents delicate problems of interpretation.

Autofluorescence, exhibiting granular features, may occur in the cytoplasm of neuronal cell bodies and increases with the animal's age. This may interfere with the detection of fluorescent labeling, especially when using orange-red or yellow fluorescent tracers. However, autofluorescence can be distinguished from the specific label since, at variance with the latter, it remains visible through all the different filter sets.

It should also be mentioned that fluorescent labeling can, in some instances, be transformed into a stable DAB reaction product through photoconversion (4,18, 24). Photoconversion could be due to oxidation of DAB by photoexcited molecules, but the physicochemical mechanism of the reaction is still unknown. In the procedure, the section is covered with a drop of freshly made DAB solution (1.5 mg/mL Tris buffer 0.1 M, pH 8.2, without any hydrogen peroxide) and exposed to fluorescence illumination. It should, however, be kept in mind that: *(i)* only fluorescent labels in the field visualized by the microscope objective (e.g., 16×, 25×) are photoconverted; *(ii)* in our experience, photoconversion of the labels resulting from fluorescent retrograde axonal tracers is not sensitive, whereas DiI photoconversion is effective; and *(iii)* photoconversion can be used to proceed towards the electron microscopic study of some of the labels.

Criteria for Multiple Labeling

Multiple labeling implies the visualization of two or more markers in the same tissue. The markers can be used to label different subsets of the same neuronal population or fiber tract or different neuronal populations or fibers simultaneously. Multiple markers are needed to investigate simultaneously more than one set of connections (double retrograde, double anterograde, and retrograde combined with anterograde) and/or the occurrence of branched connections reaching more than one target through axon collaterals. Multiple markers can also be represented by tracer(s) combined with other molecules, which characterize the connection(s). For example, for the study of chemically characterized circuits, tracing is combined with immunohistochemistry or in situ hybridization analysis of endogenous molecules (such as neurotransmitters or their synthetic enzymes, neuromodulator and neuroactive molecules, and receptors).

For the study of functional activation of circuits, tracing is commonly combined with the immunohistochemical revelation of Fos, the protein encoded by the immediate early gene *c-fos*, which is induced in response to a variety of stimuli, such as sensory stimulation or drug administration (12).

Whatever the experimental requirements, in multiple labeling, the recipes of the fixatives and revelation protocols of each of the markers should be compatible.

The optimal goal for multiple labeling is the simultaneous visualization of the labels in the same sections. This can be achieved following three main criteria:

1. Different colors of the markers. This is the most commonly adopted approach and is especially suited for fluorescence microscopy, since fluorochromes fluorescing in different colors can be easily differentiated with specific filter sets. Multiple colors can be used for the visualization of different populations of labeled cell bodies or different fiber bundles (Figure 2E), as well as for the

simultaneous visualization of cell bodies and fibers in anterograde–retrograde combined techniques or to study exogenous tracers and endogenous molecules (for example, the organization of fibers terminating on cell bodies characterized by their immunoreactivity to a given antigen). These approaches are becoming increasingly used, since they are especially suited for studies at high resolution in confocal microscopy.

2. Different intraneuronal features of distribution of the markers, i.e., cytoplasmic labeling versus nuclear labeling. Although this strategy may seem difficult to apply, it may actually result in very distinct signals of different labels (Figure 2B), allowing the detection of both markers even at low power microscopic observation. This is also facilitated if the two markers bear different colors. Such strategy can be used for double retrograde labeling, combining a tracer that labels preferentially the neuronal nucleus, such as DY, with a tracer that labels preferentially the neuronal cytoplasm, such as FB (Figure 2B).

3. Different features of cytoplasmic labeling of different markers, e.g., granular labeling versus homogenously diffuse staining or granules of different sizes; the latter is, however, the approach of choice in electron microscopy, whereas particles of different sizes are difficult to distinguish in light microscopy. Again, the visualization of different features of labeling in the cytoplasm is enhanced by different colors of the labels. For example, this strategy is very sensitive when visualizing the retrograde labeling of gold-conjugated tracers such as CTb-gold and WGA-apoHRP-gold (black granules in the cytoplasm resulting from silver enhancement) combined with immunoreactivity revealed by DAB reaction products (brown diffuse immunostaining) (Figure 2C).

RESULTS AND DISCUSSION

The procedures described in the Protocols sections, which have to be adapted to each experimental design, result in effective retrograde labeling of neuronal cell bodies (Figure 2, A, B, C, and D) or anterograde labeling of axons (Figure 2, E and F) and their preterminal and terminal arborizations in vivo. It is peculiar, in this respect, that the HRP techniques, which belong to the first generation of tract tracing methods, are still among the most widely used tracing strategies (25).

Tracing in vitro with lipophilic carbocyanine derivatives (13,14) is especially effective in developing (embryonic or early postnatal) fixed tissues. In the evaluation of the findings, the possibility of transmembrane dye diffusion should be considered (26). Carbocyanines can also be used in vivo, but they do not present advantages over other tracers.

It should be also be taken into account that the bidirectional transport of tracers may result in problems of interpretation in the case of reciprocal connections (i.e., when both anterograde and retrograde labeling are found in the area under study). For example, in tracing corticocortical connections, both retrogradely labeled cell bodies and anterogradely labeled terminals can be found in the areas interconnected via association or commissural fibers. After tracer injection in a cortical field, thalamocortical neurons and terminals of corticothalamic fibers can also be found in the interconnected thalamic area. Bidirectional labeling is especially marked in HRP tracing. In general, in these cases, caution should be taken in the interpretation of the neuropil labeling, which could be represented by cross-sectioned dendrites, fibers, and terminal elements.

By combining the different features of labeling of retrograde tracers (Figure 2, B and D), the simultaneous visualization of two tracers in same neuronal cell body can be achieved with high sensitivity. This

12. Tract Tracing Methods at the Light Microscopic Level

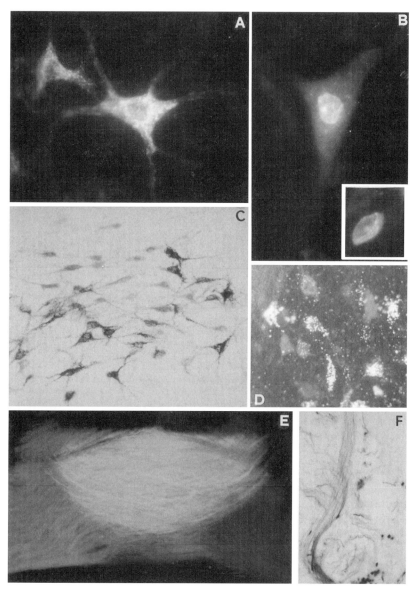

Figure 2. Photomicrographs that illustrate different retrograde (A–D), anterograde (E,F) tracers, as well double retrograde (B,D), double anterograde (E) labeling, and retrograde labeling combined with an endogenous protein immunoreactivity. Panels A, B, and E show fluorescent dyes, panels C and F are in bright-field illumination, and panel D illustrates a combination of fluorescence and dark-field illumination. Panels A through D and F were derived from experiments in rats and panel E in the frog. (A) FB retrogradely labeled reticulospinal neurons; note the labeling of the cytoplasm and proximal dendrites. (B) Neuron of the brainstem reticular core double labeled after FB injection in the spinal cord and DY injection in the thalamus; FB labeling is revealed by blue fluorescence in the cytoplasm and DY labeling by gold fluorescence in the nucleus; the inset illustrates a single DY-labeled neuron. (C) Neurons of the substantia nigra retrogradely labeled by WGA-apoHRP-gold injection in the striatum (black granular labeling of the cytoplasm revealed by silver intensification) and combined with calbindin immunohistochemistry (brown diffuse cell immunostaining resulting from the use of DAB as chromogen). (D) Thalamic neurons double labeled after FB injection in the cortex and WGA-apoHRP-gold injection in the thalamic reticular nucleus; FB labeling is revealed by blue fluorescence and the gold-conjugated tracer by the granular labeling (resulting from silver intensification) that assumes a gold color in dark-field illumination (courtesy of Dr. Roberto Spreafico, Neurological Institute Besta, Milan, Italy). (E) Double anterograde labeling of the optic chiasm with fluorescent dextran amines: Texas Red® conjugated dextran amine (red fluorescent) in the left optic tract and fluorescein-conjugated dextran amine (green fluorescent) on the right side; the superimposition of the two colors (double exposure photograph) results in the yellowish labeling at the fiber crossing (courtesy of Dr. Bernd Fritzsch, Creighton University, Omaha, NE, USA). (F) Biotin–dextran amine fiber labeling. (**See color plate A9.**)

reveals that the double labeled cell projects to both the areas injected with each tracer. The possibility to use not only two fluorescent tracers (Figure 2B) but also the combination of fluorescence and dark-field observation (Figure 2D) allows flexibility in the choice of the experimental strategy.

The combined approaches that exploit nuclear staining versus cytoplasmic labeling are also well suited for the characterization of neurons whose functional activation to a given stimulus is revealed by induction of Fos immunopositivity. This strategy is increasingly used when the experiments require the definition of circuits in which Fos, which is a nuclear protein, is elicited as functional marker.

The combination of tract tracing with immunohistochemistry allows one to characterize the molecular (e.g., chemical) identity of retrogradely labeled neurons (Figure 2C).

By the use of anterograde tracers of different colors, different pathways can be traced simultaneously; in fluorescence routine photography, they can be illustrated by double exposure photographs (Figure 2E).

Confocal microscopy opened novel perspectives for the use and analysis of fluorescent tracers.

Overall, the strategies for tract tracing are very versatile, offer a wide range of tools and procedures, most of which are relatively simple, and are at a relatively low cost. For a successful application of the techniques, the experiment requirements and planning are crucial for the choice of the tracing protocol. Basic training in the histological processing of the nervous tissue is undoubtedly of help in applying the different protocols, and a good tissue preservation and section quality are very important for adequate analysis and illustration of the findings.

ACKNOWLEDGMENTS

Experiments referred to in this chapter and the preparation of this manuscript were supported by MIUR grants.

REFERENCES

1. Alvarez-Buylla, A., C.Y. Ling, and J.R. Kirn. 1990. Cresyl violet: a red fluorescent Nissl stain. J. Neurosci. Methods 33:129-133.
2. Basbaum, A.I. and D. Menétrey. 1987. Wheat germ agglutinin-apoHRP gold: a new retrograde tracer for light- and electron-microscopic single- and double-label studies. J. Comp. Neurol. 261:306-318.
3. Bentivoglio, M. and S. Chen. 1993. Retrograde neuronal tracing combined with immunocytochemistry, p. 301-328. In A.C. Cuello (Ed.), IBRO Handbook Series: Methods in Neurosciences, Vol. 14, Immunohistochemistry II. John Wiley & Sons, Chichester.
4. Bentivoglio, M. and H.-S. Su. 1990. Photoconversion of fluorescent tracers. Neurosci. Lett. 113:127-133.
5. Cajal, S.R. 1911. Histologie du Système Nerveux de l'Homme et des Vertébrés. Maloine, Paris.
6. Callahan, C.A., S. Yoshikawa, and J.B. Thomas. 1998. Tracing axons. Curr. Opin. Neurobiol. 8:582-586.
7. Cowan, W.M., D.I. Gottlieb, A.E. Hendrickson, J.L. Price, and T.A. Woolsey. 1972. The autoradiographic demonstration of axonal connections in the central nervous system. Brain Res. 37:21-51.
8. Cuénod, M., P. Bagnoli, A. Beaudet, A. Rustioni, L. Wiklund, and P. Streit. 1982. Transmitter-specific retrograde labeling of neurons, p. 17-44. In V. Chan-Palay and S. Palay (Eds.), Cytochemical Methods in Neuroanatomy. A.R. Liss, New York.
9. Fink, R.P. and L. Heimer. 1967. Two methods for selective silver impregnation of degenerating axons and their synaptic endings in the central nervous system. Brain Res. 4:369-374.
10. Gerfen, C.R., P.E. Sawchenko, and J. Carlsen. 1989. The PHA-L anterograde axonal tracing method, p. 19-47. In L. Heimer and L. Zaborsky (Eds.), Neuroanatomical Tract-Tracing Methods. Plenum Press, New York.
11. Glover, J.C. 1995. Retrograde and anterograde axonal tracing with fluorescent dextran-amines in the embryonic nervous system. Neurosci. Prot. 30:1-13.
12. Herrera, D.G. and H.A. Robertson. 1996. Activation of c-fos in the brain. Prog. Neurobiol. 50:83-107.
13. Honig, M.G. and R.I. Hume. 1986. Fluorescent carbocyanine dyes allow living neurons of identified origin to be studied in long term cultures. J. Cell Biol. 103:171-187.
14. Honig, M.G. and R.I. Hume. 1989. DiI and DiO: versatile fluorescent dyes for neuronal labelling and pathway tracing. Trends Neurosci. 12:333-341.
15. Kristensson, K. and Y. Olsson. 1971. Retrograde axonal transport of protein. Brain Res. 29:363-365.
16. Kuypers, H.G. and G. Ugolini. 1990. Viruses as transneuronal tracers. Trends Neurosci. 13:71-75.
17. LaVail, J.H. and M.M. LaVail. 1972. Retrograde axonal transport in the central nervous system. Science 176:1416-1417.

18. **Lübke, J.** 1993. Photoconversion of different fluorescent substances for light and electron microscopy. Neurosci. Prot. 5:1-13.
19. **Mesulam, M.M.** 1978. Tetramethyl benzidine for horseradish peroxidase neurohistochemistry: a noncarcinogenic blue reaction product with superior sensitivity for visualizing neural afferents and efferents. J. Histochem. Cytochem. 26:106-117.
20. **Nauta, W.J.H. and M. Feirtag.** 1986. Fundamental Neuroanatomy. Freeman & Co., New York.
21. **Nauta, W.J.H. and P.A. Gygax.** 1951. Silver impregnation of degenerating axon terminals in the central nervous system: (1) technic (2) chemical notes. Stain Technol. 26:3-9.
22. **Olucha, F., G.F. Martinez, and G.C. Lopez.** 1985. A new stabilizing agent for the tetramethyl benzidine (TMB) reaction product in the histochemical detection of horseradish peroxidase (HRP). J. Neurosci. Methods 13:131-138.
23. **Schmued, L.C.** 1994. Anterograde and retrograde neuroanatomical tract tracing with fluorescent compounds. Neurosci. Prot. 5:1-15.
24. **Schmued, L.C. and L.F. Snavely.** 1993. Photoconversion and electron microscopic localization of the fluorescent axon tracer fluoro-ruby (rhodamine-dextranamine). J. Histochem. Cytochem. 41:777-782.
25. **van der Want, J.J., J. Klooster, B.N. Cardozo, H. de Weerd, and R.S. Liem.** 1997. Tract-tracing in the nervous system of vertebrates using horseradish peroxidase and its conjugates: tracers, chromogens and stabilization for light and electron microscopy. Brain Res. Brain Res. Protocol 1:269-279.
26. **Vercelli, A., M. Repici, D. Garbossa, and A. Grimaldi.** 2000. Recent techniques for tracing pathways in the central nervous system of developing and adult mammals. Brain Res. Bull. 51:11-28.
27. **Weinberg, R.J. and S.L. van Eyck.** 1991. A tetramethylbenzidine/tungstate reaction for horseradish peroxidase histochemistry. J. Histochem. Cytochem. 39:1143-1148.

13 | Tract Tracing Methods at the Ultrastructural Level

Isaura Tavares, Armando Almeida, and Deolinda Lima
Institute of Histology and Embryology, Faculty of Medicine and IBMC, University of Oporto, Porto, Portugal, EU

OVERVIEW

One of the major goals of research in neurobiology is the characterization of neuronal circuits, which are the basis of the functioning of the entire nervous system. Tract tracing methods are among the best approaches to elucidate the connections between neurons located in distinct areas of the nervous system since they allow an easy recognition of large neuronal populations and their structural characterization. Although the bulk of information is achieved at light microscope (LM) level, only electron microscopy allows one to demonstrate of the occurrence of biological interaction between neurons and characterize the synaptic circuitry involved. In the choice of tracers for ultrastructural analysis of neuronal connections, neurobiologists need to consider specific characteristics of the tracer, in addition to those addressed for tracing at LM (see Chapter 12). High sensitivity is an essential requisite for tracers in ultrastructural studies, since synaptic contacts are mainly established with small diameter neuronal processes. Further important demands are that the tracer resists to histological processing for ultrastructural analysis and that the tracing method can further be combined with other neurobiological techniques aimed to achieve a more complete characterization of the neuronal circuit. Under these perspectives, the most suitable tracers to perform retrograde, anterograde, or bidirectional tracing at the electron microscope (EM) level, are discussed. The elected techniques are presented in a sole protocol that allows a successful combination of retrograde, anterograde, and neurochemical analysis at LM and EM levels. Each technique can, however, be used separately or designed to achieve different goals, as pointed out at the various steps of the protocol.

BACKGROUND

Tracing studies at the EM are mainly aimed at the identification and characterization of the synaptic specialization within the circuit studied, which is essential for the recognition of a neuronal circuit and gives some insights onto its excitatory or inhibitory nature. Inspection of labeled neurons in the LM prior to their examination in the EM (correlated light and electron microscopy analysis) is the most valuable procedure to fully characterize neuronal circuits (11). It avoids the time-consuming blind search for labeled structures at the EM and allows to relate the ultrastructural observations with thorough LM characterization of the same neuronal elements, as to their structure and location.

In ultrastructural tract tracing studies, neurobiologists have to deal with problems specifically related to tissue preparation. The two main questions are the establishment of procedures that enable a satisfactory penetration of reagents without compromising structural preservation and visualization of multiple tracers. The requirements for maintenance of ultrastructural integrity are difficult to conciliate with detection of most tracers. The classical inclusion of glutaraldehyde in the aldehyde perfusion mixture reduces the activity of enzymes revealed by histochemical reactions, masks antigenic sites, causes unspecific staining by fixing antibodies to free aldehyde groups, and hinders penetration of reagents into the tissue (10,63). It may be useful to restrict the presence of glutaraldehyde to the first minutes of vascular perfusion (e.g., a 2% glutaraldehyde and 2% paraformaldehyde solution) so that the bulk of the perfusion as well as the postfixation use only paraformaldehyde (e.g., a 4% solution). This procedure has proven to render a fair compromise between ultrastructural preservation and tracer visualization (72). Whenever this reduction in glutaraldehyde concentration is unable to produce such an aim, it may be helpful to treat thick sections, before the onset of tracer detection, with sodium borohydride (e.g., a 0.1%–1% solution for 10–30 min), which neutralizes Schiffs bases (18,32). Although osmication by immersion in osmium tetroxide (OsO_4) provides additional fixation by linkage to components unreacted by aldehydes and counterstains the tissue, it occasionally removes tracers from neurons (41) and alters the ultrastructural appearance of some chromagens (69,82).

In ultrastructural tract tracing studies of nervous areas where axons course predominantly in one direction (e.g., spinal cord), it may be useful to cut sections perpendicularly to that plane in order to maximize penetration of reagents. This problem is particularly overwhelming for tracers detected immunocytochemically, since the detergents, which are commonly used at LM tracing to allow penetration of immunoreagents, are incompatible with ultrastructural preservation (23,38,39,63,79). Triton® X-100, one of the most effective detergents, should be present only in solutions containing primary antibodies in concentrations not exceeding 0.1% (3,72,91). In some instances, it turns useful to replace the detergent by ethanol treatment before the onset of tracer detection (38).

As to the question of visualization of multiple tracers, it may be recalled that the majority of ultrastructural studies of neuronal connections use peroxidase or immunoperoxidase methods with diaminobenzidine (DAB) as chromagen, which results in an amorphous electrondense precipitate. The search for substances that can be distinguished from DAB led to the use of particulate markers, such as colloidal-gold or ferritin (15,77) and crystalline markers like tetramethylbenzidine (TMB) (40,82,85). Other peroxidase substrates with suitable ultrastructural appearances are available such as Vector® VIP (92) (Vector Laboratories, Burlingame, CA, USA) and benzidine dihydrochloride (BDHC) (69,79). However, Vector VIP can obscure ultrastructural details (92), whereas BDHC is highly toxic and produces inconsistent results (40,69).

Retrograde Tracers

There is a multitude of retrograde tracers available for the ultrastructural study of neuronal connections. The tracers can be grouped into four classes: *(i)* horseradish peroxidase (HRP)-based tracers [HRP and wheat germ agglutinin (WGA)-HRP]; *(ii)* colloidal gold-labeled tracers (WGA-gold, WGA-HRP-gold, and WGA-apo-HRP-gold); *(iii)* fluorescent tracers (mainly Fluorogold; Fluorochrome Inc., Englewood, NJ, USA) and *(iv)* cholera toxin subunit b (CTb)-based tracers (unconjugated CTb, CTb-HRP, and CTb-gold). The choice of the tracer depends on the purpose of the study, but it should take into account the requisites addressed above.

CTb-tracers, mainly in its unconjugated or gold-linked forms, have been elected for most neuroanatomical studies. Due to a particularly low rate of diffusion, CTb produces confined injections sites, a feature of major importance in tracing studies, especially those dealing with small nuclei (24,36,71). The mechanisms of neuronal uptake are well established and rely on binding to the GM1 ganglioside receptor (30,57). It is disputable whether CTb-based tracers are transported by passing fibers. This has been discarded based on the lack of labeling following pressure (36,71) or iontophoretic (47) injections in several axonal tracts. However, a few neurons were labeled after injections of CTb-tracers in other neuronal pathways (17,42). It is possible that molecular features such as the GM1 distribution along different portions of the neuronal membrane account for the observed variability of CTb uptake by passing fibers.

Several questions need to be considered before electing the form of CTb-tracer to use. Due to proteolytic degradation of HRP by lysosome enzymes, CTb-HRP can be detected only for a few days upon injection. On the contrary, the long-term retention of unconjugated CTb and CTb-gold by neurons renders flexibility to the experiments and allows time for other surgical manipulations to be carried out, namely injections of other tracers. CTb-gold is particularly suitable for such multiple labeling experiments due to the distinct ultrastructural aspect of gold particles and the possibility of combining the silver intensification procedures used for detecting CTb-gold with methods for tracer or neurotransmitter identification based on peroxidase histochemistry. An additional advantage of CTb-gold for ultrastructural studies is that glutaraldehyde fixation does not interfere with its visualization while it affects immunodetection of unconjugated CTb (24,72) and reduces the enzymatic activity of CTb-HRP. In spite of being insensitive to glutaraldehyde concentration, CTb-gold is affected by postfixation with OsO_4, which removes gold particles, unless gold toning is performed (41,42).

The main problem of CTb-gold for ultrastructural tracing studies is that, as with colloidal gold labeled tracers (7,53,66), distal dendrites are seldom filled. Unconjugated CTb, on the contrary, fills the dendritic tree up to small distal dendrites (3,14,24,72), the neuronal structures which receive the majority of synaptic input (11). This prompts neurobiologists to develop methods to overcome the drawbacks discussed above (see Protocol), thus allowing the successful use of unconjugated CTb for EM studies.

Correlated confocal and electron microscopy emerges as a new tool in the ultrastructural study of retrogradely labeled neurons (see also Chapter 15). Optical sections of neuronal profiles to be observed at the EM can be used for three-dimensional reconstruction of the neurons, which allows for their visualization in various perspectives, a procedure of particular value for regions where the similar orientation of neuronal processes imposes, as discussed above, the use of a single plane of sectioning. Another advantage is the easier match between optical and ultrathin sections, as compared to the thick sections of conventional correlated light and electron microscopy (21,70,76). Briefly, the technique relies on confocal observation of neurons retrogradely labeled by a tracer identified by a fluorochrome, and the subsequent replacement of this marker by an electrondense product suitable for ultrastructural observation. The conversion procedure can be performed by photoconversion (6,45,46) or replacement of immunofluorescence reagents by immunoperoxidase products (55, 56,60,76), a process which can be achieved due to the higher affinity of the latter antiserum to antigenic complexes as compared to immunofluorescent substances.

Fluorescent tracers may be useful in correlated confocal and ultrastructural analysis, since they are directly observed without any sort of histochemical treatment. Among the numerous fluorescent tracers available (9), Fluorogold stands out by its ability to produce extensive filling of the dendritic arbor either upon photoconversion (6) or immunoperoxidase detection of the tracer (16,22,81). It must, however, be noted that the tracer massively diffuses from the injection site and is taken up by passing fibers even after iontophoretic delivery (20,59,65). Unconjugated CTb, detected by the sequential immunofluorescence–immunoperoxidase procedure described above, appears to be the method of choice for correlated confocal and ultrastructural analysis (Figure 1).

Another recent achievement in retrograde tracing at the ultrastructural level is the use of "suicide" tracers. Some lectins, like ricin and volkensin, bind to oligosaccharides in axonal membranes and are internalized and transported retrogradely to the cell body, where they inactivate protein synthesis, leading to neuronal death (86,88). Degenerating profiles and reduction of the density of mitochondria and synapses is then easily detected at the EM (13,37,54,64). Other toxins, like saporin, which are devoid of binding properties, can be used for the same purpose upon conjugation with substances that bind to neuronal surface molecules. In this case, binding to neurochemical markers specific for some neuronal populations allows the selective destruction of particular afferents of the injected area (86,87). Recently, the use of saporin conjugated with CTb produced a very lethal "killer-tracer" that was successfully used at LM studies (43). Suicide tracing allows one to correlate behavioral analysis with specific lesion of neuronal pathways (48,50), thus opening interesting possibilities of functional evaluation of neuronal circuits characterized at the ultrastructural level.

Anterograde Tracers

Currently, anterograde tracing of neuronal connections at the ultrastructural level is mainly performed by the use of two types of tracers: *(i) Phaseolus vulgaris* leucoagglutinin (PHA-L); and *(ii)* biotin-based tracers [biotinylated dextran amine (BDA), biocytin, and neurobiotin]. In a manner similar to retrograde tracing, anterograde tracers should fill several requirements to be used in ultrastructural studies. High sensitivity is not a decisive issue in the choice of tracer as all appear to fill axons in a Golgi-like fashion, enabling the examination of fine morphological features (12,26,29,31,34,58,68,83). However, contrary to the long immunocytochemical procedure needed to detect PHA-L, detection of biotin-based tracers involves a simple and fast avidin–biotin complex (ABC)-peroxidase reaction, favoring the use of the latter tracer class. In fact, ultrastructural preservation is excellent, because higher glutaraldehyde concentrations are allowed during the fixation, and the time between vascular perfusion and osmication is shorter (1,19,33,61,75,84,89). Moreover, multiple labeling studies can be performed more easily since the risk of cross-reaction between immunoreagents is decreased. BDA is preferable to biocytin and neurobiotin since it is not transported transneuronally (28) and is suitable for tracing studies of long pathways (25,35,67,68,78).

The disadvantages of biotin-based tracers over PHA-L need, however, to be envisaged. Contrary to PHA-L, biotin-based tracers are taken up by fibers of passage in variable degrees and can be transported in the retrograde direction (12,26–28,35,67,68,90). In the case of BDA, the latter drawback can be overcome by decreasing the concentration of the tracer and diameter of the micropipets used for iontophoresis (61).

Bidirectional Transport of Tracers

Areas of the central nervous system are often reciprocally connected, which prompts a careful evaluation of the methodology available to study reciprocal circuits. The bidirectional transport of a single tracer proved to be preferable to the use of retrograde and anterograde transports of two different tracers from the same area (52). Tracers commonly have distinct rates of diffusion and transport, which impairs the precise evaluation of the site of origin of labeled structures and makes difficult to find a postsurgery period suitable for the detection of both anterogradely and retrogradely labeled structures. Furthermore, the uptake and migration of one tracer can be influenced by the other (22). Detecting one single tracer instead of two improves ultrastructural preservation, since it may reduce the time of incubation in reagents and consequently decrease the lap between vascular fixation and osmication. Finally, the combination with other neurobiology methods becomes easier, since the same reagents are used in the detection of the anterograde and retrograde tracing, which decreases the possibilities of cross-reactivity.

BDA and biocytin have been used for the study of reciprocal connections, but the transport of those tracers in the retrograde direction is quite discrete (44,49). Unconjugated CTb is becoming the tracer of election for such studies (3,4,73,74) since it produces an extensive filling of the dendritic tree and the anterograde transport can be considerably improved by introducing simple changes in the method of detection (see Protocol), namely: *(i)* increase of the concentration of primary antibodies; *(ii)* extension of the incubation time in the immunoreagents; *(iii)* use of graded series of ethanol to increase penetration of reagents; and *(iv)* incubation with sodium borohydride to counteract the deleterious effects of glutaraldehyde (5,32,38,76). In spite of the punctate appearance of anterograde labeling due to the lack of intense staining of axonal strands, anterogradely labeled terminals can be clearly recognized in correlated light and ultrastructural observation (Figure 1), and the method is exquisite for studies aimed at the morphological characterization of the neurons involved in reciprocal circuits, particularly if the correlated confocal and ultrastructural method described before is employed.

Further important advantages of unconjugated CTb to detect reciprocal connections is that it appears not to fill retrogradely axonal collaterals of labeled neurons, in which case axonal arborizations of retrogradely labeled neurons could be taken as anterograde labeling. Moreover, it is not transported transynaptically, which discards the possibility that labeled boutons are generated by local circuit neurons targeted by the afferent fibers (3,5,47,80).

PROTOCOL

We describe a method in which the excellent features of unconjugated CTb for retrograde tracing and BDA for anterograde tracing at the ultrastructural level were explored to investigate the occurrence of a dysynaptic pathway in the rat and combined with immunocytochemical processing to characterize neurochemically one component of the circuit. The general protocol is also valid for the single use of each tracer, as well as for bidirectional tracing with CTb, provided that the changes, referred to in the Notes, are introduced.

Materials and Reagents

- Alkaline phosphatase streptavidin (Vector Laboratories, Burlingame, CA, USA).
- Ammonium paratungstate (Fluka Chemicals, Buchs, Switzerland).
- Ammonium chloride (Sigma, St. Louis, MO, USA).

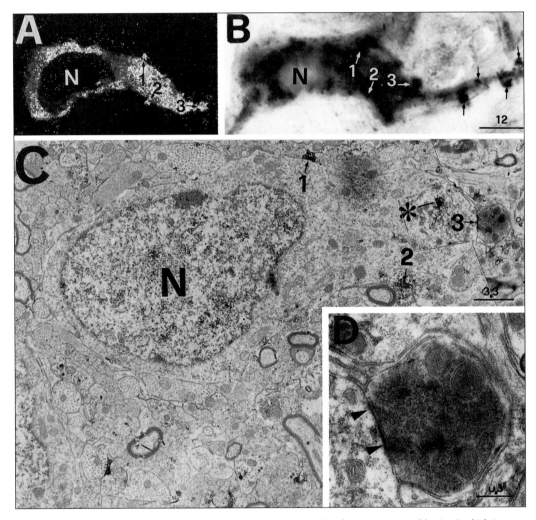

Figure 1. Confocal (A), light (B), and ultrastructural (C and D) photographs of a neuron in spinal lamina I, which is retrogradely labeled by CTb from the VLM and receives appositions from CTb-labeled fibers from the same injection site (bidirectional tracing with CTb). The nucleus (N) is used as a landmark. In panel A, a single confocal section, in which CTb was revealed by immunofluorescence using a streptavinin-Alexa 594 antibody (Molecular Probes), shows the neuron and boutons 1 to 3 in apposition with the perikarya. The same neuron is depicted in panel B in a thick section following immunoperoxidase reaction with the ABC method. Four additional boutons, marked with arrows, are seen in apposition with a dendrite. In panel C, an ultrathin section of the neuron was taken from a plane close to that depicted in the confocal image. The asterisk indicates CTb labeling at the perikaryon. In panel D, bouton 3 is shown at higher magnification. It establishes an asymmetrical synapse with the perikaryon (arrowheads). Scale bars are expressed in micrometers.

13. Tract Tracing Methods at the Ultrastructural Level

- Anti-CTb antiserum raised in goat (List Biological Laboratories, Campbell, CA, USA).
- Anti-dopamine-β-hydroxylase (DBH) antiserum raised in rabbit (Affinity, Excester, UK).
- BDA (mw 10 kDa)(Molecular Probes, Eugene, OR, USA).
- Biotinylated antirabbit antiserum raised in swine (Dakopatts, Glostrup, Denmark).
- Cobalt chloride (Sigma).
- CTb low salt (List Biological Laboratories).
- Current source device (e.g., model CS-3 from Transkinetics Systems, Canton, MA, USA).
- DAB (Sigma).
- Donkey antigoat antiserum (Vector Laboratories).
- Epon (Fluka Chemicals).
- Ethanol (Merck, Darmstadt, Germany).
- Fast Blue (Sigma).
- Glass capillaries (e.g., 1.5 x 0.75 mm)(Clark Electromedical Instruments, Reading, England, UK).
- Glass microelectrode puller (e.g., model PE-21 from Narishige Scientific Instruments, Tokyo, Japan).
- Glucose-oxidase (Sigma).
- Hydrogen peroxide (Merck).
- Naphtol AS-MX-phosphate (Sigma).
- Normal swine serum (Sigma).
- OsO_4 (Merck).
- Peroxidase-antiperoxidase (PAP) antiserum raised in goat (Dakopatts).
- Phosphate buffer (PB; 0.1 M, pH 7.2–7.4).
- PB (0.1 M, pH 6.2).
- Propylenoxide (Merck).
- Phosphate-buffered saline (PBS; 0.1 M, pH 7.2).
- Silver wire (e.g., 200 μm x 7.6 m) (A-M Systems, Carlsborg, WA, USA).
- Stereotaxic frame (e.g., Model 900 from David Kopf, Tujunga, CA, USA).
- Tris-HCl buffer (0.05 M, pH 7.6).
- Tris-HCl buffer (0.1 M, pH 8.2).
- TMB (Sigma).
- Vibrating microtome (e.g., model VT 1000S from Leica Microsystems, Wetzlar, Germany).

Procedure

1. Under halothane anesthesia (4% for induction and 1%–1.5% for maintenance) and stereotaxic control, perform iontophoretic injections of BDA using micropipets with 20 to 30 μm inner diameter tips and positive 7 μA continuous current for a time to be settled up according to the aimed injection site (in general 10–30 min).

2. Seven to twelve days later, reanesthetize animals with halothane and inject iontophoretically 1% low salt CTb using positive continuous current. Settle up the current intensity (2–5 μA), time of deliverance (3–20 min) and micropipet inner diameter tip (15–30 μm) for the desired injection site.

3. Four to seven days later, reanesthetize animals with chloral hydrate (0.35 g/kg b.w. i.p.) and perfuse them through the ascending aorta with PBS. After the fluid exiting from the right atrium is clear, change to a fixative mixture composed by 2% glutaraldehyde and 2% paraformaldehyde in 0.1 M PB, pH 7.2 to 7.4 (100 mL), followed by 4% paraformaldehyde in the same buffer (1000 mL). A mixing chamber ("Y" shape glass tube) is useful to achieve a gradual change from the vascular rinse to the fixative. The rate of perfusion should be estimated in order to allow fast delivery of the vascular rinse and

first fixative (20–30 mL/min) and slower release of the final fixative (10–15 mL/min). Dissect areas of interest and postfix in 4% paraformaldehyde in 0.1 M PB, pH 7.2 to 7.4, for 2 hours, followed by overnight washing in 8% sucrose, in the same buffer, both at 4°C.

Note: For single anterograde tracing with BDA, use a different fixative both on the perfusion and postfixation with higher glutaraldehyde concentration (e.g., 2.5% glutaraldehyde and 4% paraformaldehyde).

4. Section nervous tissue pieces with a vibrating microtome at 75 to 100 µm through the injection sites and 40 to 60 µm trough the areas containing the retrograde and anterograde labeling.

5. Wash in 0.1 M PB, pH 7.2 to 7.4, for 15 minutes. In order to keep the possibility of further ameliorate labeling by introducing changes in the histo- or immunocytochemical reactions, it may be convenient to store some sections at -20°C in an antifreeze mixture composed of 160 mL of 0.05 M PB, pH 7.2, 120 mL ethylene glycol, and 120 mL glycerol. Before reacting these sections, wash them thoroughly with 0.1 M PB, pH 7.2 to 7.4, 3 x 10 minutes.

Note: For bidirectional tracing, incubate sections for 30 minutes with 50% ethanol in PB, pH 7.4, wash as in step 5, incubate with 1% sodium borohydride (Sigma) for 10 minutes and wash as in step 5, 3 times.

6. Begin the histochemical reaction for BDA by incubating sections in ABC diluted in PBS at 1:200 for 2 hours.

7. Wash in PBS 3 x 15 minutes.

8. Wash in 0.05 M Tris-HCl buffer, pH 7.6, 2 x 10 minutes.

9. Incubate in a solution composed by 10 mg DAB, 4 µL H_2O_2 in 20 mL 0.05 M Tris-HCl, pH 7.6. Control reaction speed under light microscope observation. Stop the reaction by washing as in step 8. Wash in PBS 2 x 10 minutes.

10. Begin the immunocytochemical reaction for CTb by incubating in goat anti-CTb antiserum at 1:40 000 to 1:60 000 in PBS during 48 hours at 4°C. Repeat washes as in step 7.

Note: For single retrograde or bidirectional tracing with CTb, retitrate anti-CTb antiserum to, respectively, 1:10 000 or 1:7500 to 1:12 000 and, in the latter case, prolong incubation for 72 hours in a PBS solution containing 0.1% sodium azide.

11. Incubate in donkey antigoat diluted at 1:200 in PBS for 1 hour. Repeat washes as in step 7.

12. Incubate in goat PAP diluted at 1:100 in PBS for 1 hour. Repeat washes as in step 7.

Note: For bidirectional tracing with CTb, incubation in steps 11 and 12 should last for 2 hours.

13. Wash in PB, pH 6.2, 2 x 10 minutes.

14. Incubate for 20 minutes in a solution composed by 500 µL of 1% ammonium paratungstate, 125 µL of 0.2% TMB, 100 µL of 0.4% NH_4Cl, 100 µL of 20% D-glucose, and 10 mL 0.1 M PB, pH 6.2 .

Note: For single retrograde or bidirectional tracing with CTb, it is preferable to perform the peroxidase reaction as in step 9.

15. Replace the former mixture by a new solution to which 10 µL of glucose oxidase was added. Control the reaction speed under light microscope observation. Stop the reaction by washing as in step 13. Vigorous stirring and extension of washing may be necessary to reduce the background.

16. Incubate for 10 minutes in a solution composed by 10 mg of DAB, 100 µL of 0.4% NH_4Cl, 100 µL of 20% D-glucose, 200 µL of 1% $CoCl_2$, and 10 µL glucose oxidase in 10 mL PB, pH 6.2. Wash as in step 13 and in PBS 2 x 10 minutes.

17. Incubate in 10% normal swine serum for 1 hour.
18. Begin the immunocytochemical reaction for DBH by incubating in rabbit anti-DBH antiserum at 1:2000 in PBS containing 1% normal swine serum overnight. Repeat washes as in step 7.
19. Incubate in biotinylated swine antirabbit antiserum at 1:200 for 1 hour. Repeat washes as in step 7.
20. Incubate in alkaline phosphatase streptavidin at 1:150 for 1 hour. Repeat washes as in step 7. Wash in 0.1 M Tris-HCl, pH 8.2, 2 times for 10 minutes.
21. Incubate in a filtered solution containing 200 µL 10% naphtol, 20 µL 1 M levamisole, 10 mg Fast Blue in 9.8 mL 0.1 M Tris-HCl, pH 8.2. Control the reaction speed under microscope observation. Stop the reaction by washing in 0.1 M Tris-HCl, pH 8.2, 2 x 5 minutes. Wash in PBS 2 x 5 minutes.
22. Mount between slides in a glycerol-based medium (3:1 in 0.1 M PB, pH 7.2–7.4) and search for neurons containing CTb and DBH and apposed by BDA-fibers. Carefully draw and photograph selected areas and "landmarks" (e.g., blood vessels). Remove the coverslips and wash slides in 0.1 M PB, pH 7.2 to 7.4, for 10 to 30 minutes.
23. Postfix in 1% OsO_4 0.1 M PB, pH 7.2 to 7.4, for 1 hour at 4°C. Dehydrate through graded series of ethanol in water: 70% for 5 minutes, 80% for 5 minutes, 90% for 10 minutes, 95% for 10 minutes, and 100% 2 x 15 minutes. Embed with 1:1 epon/propylenoxide for 30 minutes, followed by 3:1 overnight. Include in Epon for 30 minutes at 60°C.
24. With the aid of tissue and toothpick, remove excess resin and flat-embed between dymethylclorosilane-coated slides (2). Use clips to hold slides together and allow polymerization to take place at 60°C overnight.
25. Separate the slides and select the areas identified and photographed in step 19. Fill up flat-bottom embedding capsules with Epon and invert over selected section. Allow to polymerize at 60°C for at least 24 hours. Alternatively, detach the section from the slide and glue it into a blank tube of resin. Trim the area of interest and collect ultrathin sections on single-slot electron microscope grids. Counterstain the material with lead citrate (62) for 1 to 3 minutes and observe the sections in the EM.

RESULTS AND DISCUSSION

By applying the protocol described above following stereotaxic injections of BDA in the caudal ventrolateral medulla (VLM) and unconjugated CTb in the spinal dorsal horn, it was possible to conclude that axonal boutons from fibers originated in the VLM (DAB chromagen; brown) apposed to neurons of the pontine A_5 noradrenergic group (immunoreactive to DBH-Fast Blue as chromagen; blue), which project to the spinal dorsal horn (TMB chromagen; black granules) (Figure 2A).

Although processing for ultrastructural analysis removes Fast Blue from the tissue (Figure 2B), previous drawing and photographing of CTb plus DBH double-labeled neurons allows the recognition of these cells. At low power EM magnifications, these neurons can be easily identified by the characteristic aspect of TMB crystals (Figure 2C), which are often associated with the microtubuli and rough endoplasmic reticulum. Anterogradely labeled boutons are filled with amorphous DAB material, which is deposited between synaptic vesicles and around mitochondria (Figure 2D).

Another chromagen, which can be used in combination with BDA histochemistry and CTb immunodetection, is pyronin, which uses 1-naphtol as substrate for peroxidase reaction and gives a pink color to perikarya and dendrites (data not shown and Reference 51). Due to their high sensitivity and clear-cut difference from both DAB and TMB reaction products, Fast Blue and pyronin are markers of choice for correlated light and electron microscopy studies aimed at identifying the origin of the input received by neurochemically characterized neurons that project to a certain area. Fast Blue presents the additional advantage of using a nonperoxidase detection system, which greatly reduces the risk of "cross-talk" with DAB and TMB reactions, although it does not avoid that adequate control experiments are performed. Whatever the chromagen chosen, it should be kept in mind that both Fast Blue and pyronin reaction products are washed away during dehydration, implying that only neurons double labeled with another tracer (e.g., CTb with TMB as chromagen) can be followed up at the EM.

Another caution is that the Fast Blue reaction should be the last to be performed in the multiple labeling procedures, since the chromagen may alter its appearance or even be removed after prolonged incubations (8). The higher sensitivity of TMB over other chromagens could produce cross-reactivity between peroxidase reactions, but in our experiments TMB crystals were not present over BDA-labeled boutons. If this problem arises, it can be over-

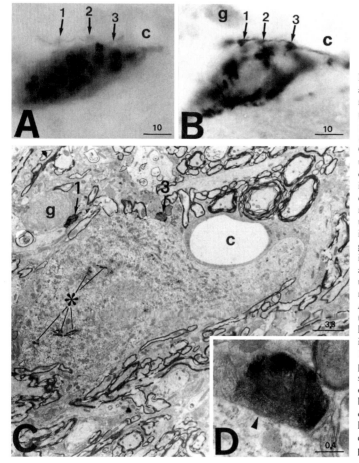

Figure 2. Correlated light (A and B) and electron (C and D) analysis of a neuron in the pontine A$_5$ noradrenergic cell group. A capillary (c) and a glial cell (g) are used as landmarks. The neuron is retrogradely labeled with CTb (chromagen TMB; black granules in panels A and B and electron dense crystals marked with an asterisk in panel C) from the cervical spinal cord dorsal horn. Its noradrenergic nature is demonstrated in panel A by the diffuse blue staining due to immunoreactivity for DBH (chromagen Fast Blue removed in panel B during processing for ultrastructural analysis). The neuron receives appositions from axonal varicosities (chromogen DAB: brown staining in panel A, black product in panel B, and electrondense mark in panels C and D) labeled with BDA from the VLM (1–3 in the thick sections in panels A and B; 1 and 3 in the ultrathin section in panel C). In panel D, bouton 3 is shown at a higher magnification. An obliquely cut synapse is established between the bouton and the perikaryon (arrowhead) of the retrogradely labeled neuron. Scale bars are expressed in micrometers. (See color plate A10.)

come by inverting the sequence of peroxidase reactions. Another important aspect to be aware of in the design of the protocol is that the chromagen TMB should be used only for immunoreactions with high specificity and low background. Retitrating the primary antibodies in the immunoreaction, as discussed by Llewellyn-Smith et. al. (40), can help in reducing background, but the replacement of hydrogen peroxide by glucose oxidase, as proposed by the authors, did not appear to produce the desired effect in our hands.

ACKNOWLEDGMENTS

We are very grateful to Dr. Marianne Wikström, University of Göteborg, Sweden, who kindly offered us antibodies against CTb when they were not yet commercially available. Isaura Tavares acknowledges the advises of Professors J.P. Bolam (MRC Anatomical Neuropharmacology Unit, Oxford, England, UK) and Trevor Batten (Department of Cardiovascular Studies, University of Leeds, England, UK) regarding the use of TMB and Fast Blue, respectively.

REFERENCES

1. Aicher, S.A., D.J. Reis, R. Nicolae, and T.A. Milner. 1995. Monosynaptic projections from the medullary gigantocellular reticular formation to sympathetic preganglionic neurons in the thoracic spinal cord. J. Comp. Neurol. *363*:563-580.
2. Aldes, L.D. and T.B. Boone. 1984. A combined flat-embedding, HRP histochemical method for correlative light and electron microscopic study of single neurons. J. Neurosci. Res. *11*:27-34.
3. Almeida, A., I. Tavares, D. Lima, and A. Coimbra. 1993. Descending projections from the medullary dorsal reticular nucleus make synaptic contacts with spinal cord lamina I cells projecting to that nucleus: an electron microscopic tracer study in the rat. Neuroscience *55*:1093-1106.
4. Almeida, A., I. Tavares, D. Lima, and A. Coimbra. 1998. Reciprocal connections between the spinal dorsal horn and the dorsal part of the medullary dorsal reticular nucleus. Society Neurosci. Abstr. *24*:393.
5. Angelucci, A., F. Clascá, and M. Sur. 1996. Anterograde axonal tracing with the subunit B of cholera toxin: a highly sensitive immunohistochemical protocol for revealing fine axonal morphology in adult and neonatal brains. J. Neurosci. Methods *65*:101-112.
6. Balercia, G., S. Chen, and M. Bentivoglio. 1992. Electron microscopic analysis of fluorescent neuronal labeling after photoconversion. J. Neurosci. Methods *45*:87-98.
7. Basbaum, A.I. and D. Menétrey. 1987. Wheat germ agglutinin-apoHRP gold: a new retrograde tracer for light- and electron microscopic single-and double-label studies. J. Comp. Neurol. *261*:306-318.
8. Batten, T.F.C., K. Appenteng, and S. Saha. 1988. Visualization of CGRP and ChAT-like immunoreactivity in identified trigeminal neurones by combined peroxidase and alkaline phosphatase enzymatic reactions. Brain Res. *447*:314-324.
9. Bentivoglio, M. and S. Chen. 1993. Retrograde neuronal tracing combined with immunocytochemistry, p. 301-328. *In* A.C. Cuello (Ed.), Immunocytochemistry II. John Wiley & Sons, New York.
10. Bolam, J.P. 1992. Preparation of central nervous system tissue for light and electron microscopy, p. 1-29. *In* J.P. Bolam (Ed.), Experimental Neuroanatomy. A Practical Approach. The Practical Approach Series, D. Rickwood and B.D. Hames (Eds.). Oxford University Press, Oxford.
11. Bolam, J.P. and C.A. Ingham. 1990. Combined morphological and histochemical techniques for the study of neuronal microcircuits, p. 125-198. *In* A. Björklund, T. Hökfelt, F.G. Wouterlood, and A.N. van den Pol (Eds.), Handbook of Chemical Neuroanatomy. Elsevier Science Publishers, Amsterdam.
12. Brandt, H.M. and A.V. Apkarian. 1992. Biotin-dextran: a sensitive anterograde tracer for neuroanatomic studies in rat and monkey. J. Neurosci. Methods *45*:35-40.
13. Brauer, K., G. Seeger, W. Härtig, S. Rossner, R. Poethke, J. Kacza, R. Schliebs, G. Brückner, and V. Bigl. 1998. Electron microscopic evidence for a cholinergic innervation of GABAergic parvalbumin-immunoreactive neurons in the rat medial septum. J. Neurosci. Res. *54*:248-253.
14. Bruce, K. and I. Grofova. 1992. Notes on a light and electron microscopic double-labeling method combining anterograde tracing with *Phaseolus vulgaris* leucoagglutinin and retrograde tracing with cholera toxin subunit B. J. Neurosci. Methods *45*:23-33.
15. Chan, J., C. Aoki, and, V.M. Pickel. 1990. Optimization of differential immunogold-silver and peroxidase labeling with maintenance of ultrastructure in brain sections before plastic embedding. J. Neurosci. Methods *33*:113-127.
16. Chang, H.T., H. Kuo, J.A. Whittaker, and N.G. Cooper. 1990. Light and electron microscopic analysis of projection neurons retrogradely labeled with Fluoro-Gold: notes on the application of antibodies to Fluoro-Gold. J. Neurosci. Methods *35*:31-37.
17. Chen, S. and G. Aston-Jones. 1995. Evidence that cholera toxin B subunit (CTb) can be avidly taken up and transported by fibers of passage. Brain Res. *674*:107-111.

18. Clancy, B. and L.J. Cauller. 1998. Reduction of background autofluorescence in brain sections following immersion in sodium borohydride. J. Neurosci. Methods 83:97-102.
19. Cucchiaro, J.B. and D.J. Uhlrich. 1990. *Phaseolus vulgaris* leucoagglutinin (PHA-L): a neuroanatomical tracer for electron microscopic analysis of synaptic circuitry in the cat's dorsal lateral geniculate nucleus. J. Electron Microsc. Tech. 15:352-368.
20. Dado, R.J., R. Burstein, K.D. Cliffer, and G.J. Giesler, Jr. 1990. Evidence that Fluoro-Gold can be transported avidly through fibers of passage. Brain Res. 533:329-333.
21. Deitch, J.S., K.L. Smith, J.W. Swann, and J.N. Turner. 1991. Ultrastructural investigation of neurons identified and localized using confocal scanning laser microscope. J. Electron Microsc. Tech. 18:82.
22. Deller, T., C. Leranth, and M. Frotscher. 1994. Reciprocal connections of lateral septal neurons and neurons in the lateral hypothalamus in the rat: a combined *phaseolus vulgaris*-leucoagglutinin and Fluoro-Gold immunocytochemical study. Neurosci. Lett. 168:119-122.
23. Eldred, W.D., C. Zucker, H.J. Karten, and S. Yazulla. 1983. Comparison of fixation and penetration enhancement techniques for use in ultrastructural immunocytochemistry. J. Histochem. Cytochem. 31:285-292.
24. Ericson, H. and A. Blomqvist. 1988. Tracing of neuronal connections with cholera toxin subunit B: light and electron microscopic immunohistochemistry. J. Neurosci. Methods 24:225-235.
25. Erisir, A., S.C. Van Horn, and S.M. Sherman. 1997. Relative numbers of cortical and brainstem inputs to the lateral geniculate nucleus. Proc. Natl. Acad. Sci. USA 94:1517-1520.
26. Gerfen, C.R. and P.E. Sawchenko. 1984. An anterograde neuroanatomical tracing method that shows the detailed morphology of neurons, their axons and terminals: immunohistochemical localization of an axonally transported plant lectin, *Phaseolus vulgaris* leucoagglutinin (PHA-L). Brain Res. 290:219-238.
27. Gerfen, C.R., P.E. Sawchenko, and J. Carlsen. 1989. The PHA-L anterograde axonal tracing method, p. 19-47. *In* L. Heimer and L. Záborszky (Eds.), Neuroanatomical Tract Tracing Methods 2. Plenum Press, New York.
28. Huang, Q., D. Zhou, and M. DiFiglia. 1992. Neurobiotin, a useful neuroanatomical tracer for in vivo anterograde, retrograde and transneuronal tract tracing and for in vitro labeling of neurons. J. Neurosci. Methods 41:31-43.
29. Izzo, P.N. 1991. A note on the use of biocytin in anterograde tracing studies in the central nervous system: application at both light and electron microscopic level. J. Neurosci. Methods 36:155-166.
30. Joseph, K.C., S.U. Kim, A. Stieber, and N.K. Gonatas. 1978. Endocytosis of cholera toxin into GERL. Proc. Natl. Acad. Sci. USA 75:2815-2819.
31. King, M.A., P.M. Louis, B.E. Hunter, and D.W. Walker. 1989. Biocytin: a versatile anterograde neuroanatomical tract-tracing alternative. Brain Res. 497:361-367.
32. Kosaka, T., I. Nagatsu, J.-Y. Wu, and K. Hama. 1986. Use of high concentrations of glutaraldehyde for immunocytochemistry of transmitter-synthesizing enzymes in the central nervous system. Neuroscience 18:975-990.
33. Lan, C.T., W.C. Wu, E.A. Ling, and C.Y. Chai. 1997. Evidence of a direct projection from the cardiovascular-reactive dorsal medulla to the intermediolateral cell column of the spinal cord in cats as revealed by light and electron microscopy. Neuroscience 77:521-533.
34. Lanciego, J.L. and F.G. Wouterlood. 1993. Dual anterograde axonal tracing with *Phaseolus vulgaris*-leucoagglutinin (PHA-L) and biotinylated dextran amine (BDA). Neurosci. Prot. 94-050:06-01-13.
35. Lapper, S.R. and J.P. Bolam. 1991. The anterograde and retrograde transport of neurobiotin in the central nervous system of the rat: comparison with biocytin. J. Neurosci. Methods 39:163-174.
36. Lima, D., J.A. Mendes-Ribeiro, and A. Coimbra. 1991. The spino-latero-reticular system of the rat: projections from the superficial dorsal horn and structural characterization of marginal neurons involved. Neuroscience 45:137-152.
37. Ling, E.A., C.Y. Wen, J.Y. Shieh, T.Y. Yick, and W.C. Wong. 1991. Ultrastructural changes of the nodose ganglion cells following an intraneural injection of Ricinus communis agglutinin-60 into the vagus nerve in hamsters. J. Anat. 179:23-32.
38. Llewellyn-Smith, I.J., and J.B. Minson. 1992. Complete penetration of antibodies into vibratome sections after glutaraldehyde fixation and ethanol treatment: light and electron microscopy for neuropeptides. J. Histochem. Cytochem. 40:1741-1749.
39. Llewellyn-Smith, I.J., J.B. Minson, and P.M. Pilowsky. 1992. Retrograde tracing with cholera toxin B-gold or with immunocytochemically detected cholera toxin B in central nervous system, p. 180-201. *In* P.M. Conn (Ed.), Methods in Neuroscience, Vol. 8. Academic Press, London.
40. Llewellyn-Smith, I.J., J.B. Minson, and P.M. Pilowsky. 1993a. The tungstate-stabilized tetramethylbenzidine reaction for light and electron microscopic immunocytochemistry and for revealing biocytin-filled neurons. J. Neurosci. Methods 46:27-40.
41. Llewellyn-Smith, I.J., P.M. Pilowsky, and J.B. Minson. 1993b. Retrograde tracers for light and electron microscopy, p. 31-59. *In* J.P. Bolam (Ed.), Experimental Neuroanatomy. A Practical Approach. The Practical Approach Series, D. Rickwood, and B.D. Hames (Eds.). Oxford University Press, Oxford.
42. Llewellyn-Smith, I.J., J.B. Minson, A.P. Wright, and A.J. Hodgson. 1990. Cholera toxin B-Gold, a retrograde tracer that can be used in light and electron microscopic immunocytochemical studies. J. Comp. Neurol. 294:179-191.
43. Llewellyn-Smith, I.J., C.L. Martin, L.F. Arnolda, and J.B. Minson. 1999. Retrogradely transported CTB-saporin kills sympathetic preganglionic neurons. Neuroreport 10:307-312.
44. Lozsádi, D.A. 1995. Organization of connections between the thalamic reticular and the anterior thalamic nuclei in the rat. J. Comp. Neurol. 358:233-246.
45. Lübke, J. 1993a. Photoconversion of diaminobenzi-

46. Lübke, J. 1993b. Photoconversion of different fluorescent substances for light and electron microscopy. Neurosci. Prot. *93-050*:06-01-13.
47. Luppi, P.-H., P. Fort, and M. Jouvet. 1990. Iontophoretic application of unconjugated cholera toxin B subunit (CTb) combined with immunohistochemistry of neurochemical substances: a method for transmitter identification of retrogradely labelled neurons. Brain Res. *534*:209-224.
48. Manthy, P.W., S.D. Rogers, P. Honoré, B.J. Allen, J.R. Ghillardi, J. Li, R.S. Daughters, D.A. Lappi, R.G. Wiley, and D.A. Simone. 1997. Inhibition of hyperalgesia by ablation of spinal lamina I neurons expressing substance P receptor. Science *278*:275-279.
49. Marini, G., L. Pianca, and G. Tredici. 1996. Thalamocortical projection from the parafascicular nucleus to layer V pyramidal cells in frontal and cingulate areas of the rat. Neurosci. Lett. *203*:81-84.
50. Martin, W.J., N.K. Gupta, C.M. Loo, D.S. Rohde, and A.I. Basbaum. 1999. Differential effects of neurotoxic destruction of descending noradrenergic pathways on acute and persistent nociceptive processing. Pain *80*:57-65.
51. Mauro, A., I. Germano, G. Giaccone, M.T. Giordana, and D. Schiffer. 1985. 1-naphtol basic dye (1-NBD). An alternative to diaminobenzidine (DAB) in immunoperoxidase techniques. Histochemistry *83*:97-102.
52. May, P.J., W. Sun, and W.C. Hall. 1997. Reciprocal connections between the zona incerta and the pretectum and superior colliculus of the cat. Neuroscience *77*:1091-1114.
53. Ménétrey, D. and C.L. Lee. 1985. Retrograde tracing of neural pathways with a protein-gold complex. II: electron microscopic demonstration of projections and collaterals. Histochemistry *83*:525-530.
54. Milner, T.A., J.R. Hammel, T.T. Ghorbani, R.G. Wiley, and J.P. Pierce. 1999. Septal cholinergic deafferentation of the dentate gyrus results in a loss of a subset of neuropeptide Y somata and an increase in synaptic area on remaining neuropeptide Y dendrites. Brain Res. *831*:322-336.
55. Naim, M.M., R.S. Spike, C. Watt, S.A.S. Shehab, and A.J. Todd. 1997. Cells in laminae III and IV of the rat spinal cord that possess the neurokinin-1 receptor and have dorsally directed dendrites receive a major synaptic input from tachykinin-containing primary afferents. J. Neurosci. *17*:5536-5548.
56. Naim, M.M., S.A.S. Shehab, and A.J. Todd. 1998. Cells in laminae III and IV of the rat spinal cord which possess the neurokinin-1 receptor receive monosynaptic input from myelinated primary afferents. Eur. J. Neurosci. *10*:3012-3019.
57. Okada, E., T. Maeda, and T. Watanabe. 1982. Immunohistochemical study on cholera toxin binding sites by monoclonal anti-cholera toxin antibody in neuronal culture. Brain Res. *242*:233-241.
58. Paré, D. and Y. Smith. 1996. Thalamic collaterals of corticostriatal axons—their terminal field and synaptic targets in cats. J. Comp. Neurol. *372*:551-567.
59. Pieribone, V.A. and G. Aston-Jones. 1988. The iontophoretic application of Fluoro-Gold for the study of afferents to deep brain nuclei. Brain Res. *475*:259-271.
60. Pólgar, E., S.A.S. Shehab, C. Watt, and A.J. Todd. 1999. GABAergic neurons that contain neuropeptide Y selectively target cells with the neurokinin-1 receptor in laminae III and IV of the rat spinal cord. J. Neurosci. *19*:2637-2646.
61. Reiner, A., C.L. Veenman, and M.G. Honig. 1993. Anterograde tracing using biotynilated dextran amine. Neurosci. Prot. *93-050*:14-01-14.
62. Reynolds, E.S. 1963. The use of lead citrate at high pH as an electron opaque stain in electron microscopy. J. Cell Biol. *17*:208-212.
63. Ribeiro-da-Silva, A., J.V. Priestley, and C. Cuello. 1993. Pre-embedding ultrastructural immunocytochemistry, p. 181-227. *In* A.C. Cuello (Ed.), Immunohistochemistry II. John Wiley & Sons, New York.
64. Roberts, R.C., C. Strain-Saloum, and R. Wiley. 1995. Effects of suicide transport lesions of the striatopallidal or striatonigral pathways on striatal ultrastructure. Brain Res. *701*:227-237.
65. Schmued, L.C. and J.H. Fallon. 1986. Fluoro-Gold: a new fluorescent retrograde axonal tracer with numerous unique properties. Brain Res. *377*:147-154.
66. Seeley, P.J. and P.M. Field. 1988. Use of colloidal gold complexes of wheat germ agglutinin as label for neural cells. Brain Res. *449*:177-181.
67. Sidibé, M. and Y. Smith. 1996. Differential innervation of striatofugal neurones projecting to the internal or external segments of the globus pallidus by thalamic afferents in the squirrel monkey. J. Comp. Neurol. *365*:445-465.
68. Smith, Y. 1992. Anterograde tracing with PHA-L and biocytin at the electron microscopic level, p. 61-79. *In* J.P. Bolam (Ed.), Experimental Neuroanatomy. A Practical Approach. The Practical Approach Series, D. Rickwood, and B.D. Hames (Eds.). Oxford University Press, Oxford.
69. Smith, Y. 1993. Benzidine dihydrochloride: a substrate for immunoelectron microscopy. Neurosci. Prot. *93-050*:03-01-07.
70. Sun, X.L., L.P. Tolbert, and J.G. Hildebrand. 1995. Using laser scanning confocal microscopy as a guide for electron microscopic study: a simple method for correlation of light and electron microscopy. J. Histochem. Cytochem. *43*:329-335.
71. Tavares, I. and D. Lima. 1994. Descending projections from the caudal medulla oblongata to the superficial or deep dorsal horn of the rat spinal cord. Exp. Brain Res. *99*:455-463.
72. Tavares, I., D. Lima, and A. Coimbra. 1996. The ventrolateral medulla of the rat is connected with the spinal cord dorsal horn by an indirect descending pathway relayed in the A5 noradrenergic cell group. J. Comp. Neurol. *374*:84-95.
73. Tavares, I., D. Lima, and A. Coimbra. 1997. The pontine A5 noradrenergic cells which project to the spinal cord dorsal horn are reciprocally connected with the caudal ventrolateral medulla in the rat. Eur. J. Neurosci. *9*:2452-2461.
74. Tavares, I., A. Almeida, F. Esteves, D. Lima, and A. Coimbra. 1998. The caudal ventrolateral medullary reticular formation is reciprocally connected with the

spinal cord. Society Neurosci. Abstr. *24*:1132.
75. Teune, T.M., J. Van der Burg, C.I. De Zeeuw, J. Voogd, and T.J. Ruigrok. 1998. Single Purkinje cell can innervate multiple classes of projection neurons in the cerebellar nuclei of the rat: a light and ultrastructural triple-tracer study in the rat. J. Comp. Neurol. *392*:164-178.
76. Todd, A.J. 1997. A method for combining confocal and electron microscopic examination of sections processed for double- or triple-labelling immunocytochemistry. J. Neurosci. Methods *73*:149-157.
77. Todd, A.J., C. Watt, R.C. Spike, and W. Sieghart. 1996. Colocalization of GABA, glycine, and their receptors at synapses in the rat spinal cord. J. Neurosci. *16*:974-982.
78. Totterdell, S. and G.E. Meredith. 1997. Topographical organization of projections from the entorhinal cortex to the striatum of the rat. Neuroscience *78*:715-729.
79. Totterdell, S., C.A. Ingham, and J.P. Bolam. 1992. Immunocytochemistry I: pre-embedding staining, p. 103-128. *In* J.P. Bolam (Ed.), Experimental Neuroanatomy. A Practical Approach. The Practical Approach Series, D. Rickwood, and B.D. Hames (Eds.). Oxford University Press, Oxford.
80. Trojanowsky, J.Q. and M.L. Schmidt. 1984. Interneuronal transfer of axonally transported proteins: studies with HRP and HRP conjugates of wheat germ agglutinin, cholera toxin and the B subunit of cholera toxin. Brain Res. *311*:366-369.
81. Van Blockstaele, E.J., A.M. Wright, D.M. Cestari, and V.M. Pickel. 1994. Immunolabeling of retrogradely transported Fluoro-Gold: sensitivity and application to ultrastructural analysis of transmitter-specific mesolimbic circuitry. J. Neurosci. Methods *55*:65-78.
82. Van der Want, J.J.L., J. Klooster, B. Nunes Cardozo, H. de Weerd, and R.S.B. Liem. 1997. Tract tracing in the nervous system of vertebrates using horseradish peroxidase and its conjugates: tracers, chromogens and stabilization for light and electron microscopy. Brain Res. Protocols *1*:269-279.
83. Veenman, C.L., A. Reiner, and M.G. Honig. 1992. Biotynilated dextran amine as an anterograde tracer for single- and double-labeling studies. J. Neurosci. Methods *41*:239-254.
84. Weedman, D.L. and D.K. Ryugo. 1996. Projections from auditory cortex to the cochlear nucleus in rats: synapses on granule cell dendrites. J. Comp. Neurol. *371*:311-324.
85. Weinberg, R.J. and S.L. van Eyck. 1991. A tetramethylbenzidine/tungstate reaction for horseradish peroxidase histochemistry. J. Histochem. Cytochem. *39*:1143-1148.
86. Wiley, R.G. 1992. Neural lesioning with ribosome-inactivating proteins: suicide transport and immunolesioning. Trends Neurosci. *15*:285-290.
87. Wiley, R.G. and D.A. Lappi. 1993. Preparation of anti-neuronal immunotoxins for selective neural immunolesioning. Neurosci. Prot. *93-020*:02-01-12.
88. Wiley, R.G., W.W. Blessing, and D.J. Reis. 1982. Suicide transport: destruction of neurons by retrograde transport of ricin, abrin and modeccin. Science *216*:889-890.
89. Wouterlood, F.G. and B. Jorritsma-Byham. 1993. The anterograde neuroanatomical tracer biotinylated dextran-amine: comparison with the tracer *Phaseolus vulgaris*-leucoagglutinin in preparations for electron microscopy. J. Neurosci. Methods *48*:75-87.
90. Wouterlood, F.G., P.H. Goede, and H.J. Groenewegen. 1990. The in situ detectability of the neuroanatomical tracer *Phaseolus vulgaris*-leucoagglutinin (PHA-L). J. Chem. Neuroanat. *3*:11-18.
91. Wouterlood, F.G., Y.M.H.F. Sauren, and A. Pattiselanno. 1988. Compromises between penetration of antisera and preservation of ultrastructure in pre-embedding electron microscopic immunocytochemistry. J. Chem. Neuroanat. *1*:65-80.
92. Zhou, M. and I. Grofova. 1995. The use of peroxidase substrate Vector® VIP in electron microscopic single and double antigen localization. J. Neurosci. Methods *62*:149-158.

14 In Vivo Analysis of Cell Proliferation and Apoptosis in the CNS

Laura Lossi[1,2], Silvia Mioletti[1], Patrizia Aimar[1,2], Renato Bruno[1], and Adalberto Merighi[1,2,3]

[1]Department of Veterinary Morphophysiology, [2]Neuroscience Research Group, and [3]Rita Levi Montalcini Center for Brain Repair, Università degli Studi di Torino, Torino, Italy, EU

OVERVIEW

The balance between cell proliferation and death is fundamental in several morphogenetic processes and ultimately determines the mass, shape, and function of the various tissues and organs that form the animal body. Apoptosis is a gene-regulated process of programmed cell death (PCD) that plays fundamental roles in several normal and pathological conditions (56,126). This form of "cell suicide" is most often detected during embryonic development, but is also found in normal cell and tissue turnover (26,77,81,133). Although the nervous tissue is traditionally regarded as being fundamentally constituted by postmitotic nonproliferating cells, analysis of cell proliferation and apoptosis in vivo has recently gained an increasing importance mainly considering that: *(i)* proliferative and/or apoptotic events have been extensively characterized not only during embryonic development but also in several areas of the postnatal and adult brain (9,10, 46,63–65,79,97,101); *(ii)* trophic factor deprivation often results in apoptotic cell death of target neurons (25,82,131); and *(iii)* links have been hypothesized between apoptosis and signal transduction (31,60).

We describe here a series of techniques that are currently employed in our laboratory for the detection of proliferating and apoptotic cells in the central nervous system (CNS) directly on tissue sections both at light and electron microscopic level. We will also briefly consider some biochemical assays that may be of relevance for a more in-depth characterization of the apoptotic process in neural cells.

BACKGROUND

Identification of Proliferating Cells In Vivo

Direct observation of mitotic cells (i.e., the estimation of the mitotic index) in tissue sections is clearly insufficient for a correct estimation of the proliferating index (percentage of dividing cells) due to the (relatively) short duration of the M phase of the cell cycle and asynchronism of mitoses. Routinely used methods rely on the possibility of: *(i)* labeling proliferating cells during the S phase of their cycle by incorporation of exogenously administered nucleotide analogues into the newly synthesized DNA. These exogenous labels are subsequently visualized by appropriate procedures (see below); and *(ii)* directly visualize in situ certain molecules that are expressed during different phases of the cell cycle once the cell is committed to division and are therefore regarded as specific markers of cell proliferation.

The former methods are based on the use of radiolabeled nucleotides (usually tritiated tymidine) or nucleotide analogues such as the 5-bromodeoxyuridine (BrdU) or the 5-iododeoxyuridine (IdU) that are specifically incorporated into the DNA and are subsequently visualized by autoradiography or immunocytochemistry. These methods are highly specific and allow for a precise identification of the proliferating cells and a correct estimation of the proliferating index. Historically, the isotopic methods were developed first and have been widely employed for analysis of proliferative events into the developing and immature CNS (see for example References 2–9). Nonradioactive BrdU labeling is now generally employed as the technique of choice because of its simplicity and lack of safety restrictions linked to the handling of radioactive material (22,36,69,99,100). The isotopic and nonisotopic methods are perfectly overlapping in terms of specificity and, after an initial suspicion that the immunocytochemical labeling was less sensitive, can be safely considered as equivalent also under this aspect. Several anti-BrdU monoclonal antibodies are now commercially available that can be successfully employed to label in situ proliferating cells following the systemic administration of the tracer, usually following the intraperitoneal route. In addition, some antibodies allow for the specific detection of BrdU or IdU and can be employed for dual labelings in time window experiments (see also Results and Discussion), which have opened the possibility to perform cell cycle kinetic studies (17,24,57,68,69,93,130).

An interesting feature of all these methods is that once the marker has been incorporated into the parent cell DNA, it is then equally divided between the two daughter cells. Therefore, the label can be traced over several cell generations, according to the cell cycle kinetic parameters and the length of animal survival. By calculating the time necessary for the halving of the tracer into the newly generated cells, it is thus possible to obtain reliable information on the length of the different phases of the cell cycle. To this purpose, quantitative estimation of silver grains after autoradiography has proved to be very useful and highly reliable (16).

The immunocytochemical methods relying on the visualization of specific markers for proliferating cells have encountered some favor in the recent past, mainly for studies on human material or retrospective animal samples in which the experimental administration of the isotopic and/or nonisotopic nucleotides is not practicable for obvious reasons. A number of molecules which are associated with different phases of the cell cycle have been isolated so far. Specific antibodies raised against at least some of them, such as for example the Ki67 nuclear antigen (50,89,91) and the proliferating cell nuclear antigen (PCNA) (103,121), allow for detection on tissue sections of frozen and/or wax-embedded material. Other molecules, which have been proposed as cell proliferation markers, are more indirectly linked to the cell cycle events and, in general terms, offer an estimate of certain metabolic activities that may be increased during the proliferative status. A list of the most commonly used markers of cell proliferation is reported in Table 1. There are some obvious advantages and disadvantages linked to the use of these endogenous markers as a tool for identifying proliferating cells in compari-

14. In Vivo Analysis of Cell Proliferation and Apoptosis in the CNS

son with the methods relying on the labeling of newly synthesized DNA. Among the advantages one has first to consider the technical simplicity and, second, that experimental labeling in vivo is not required. The disadvantages include some uncertainty regarding the real specificity of these markers considering the possibility that at least some of them may not be expressed only in cells which are in the process of their division and/or committed to division. This caution may be substantiated by the observations indicating that: *(i)* some of these molecules persist for a more or less extended period of time in postmitotic elements; and *(ii)* fixation and tissue processing may significantly alter the pattern of staining (24,90). This aspect is particularly relevant under certain experimental conditions in which one has the need to carefully label a proliferating cell population in a restricted temporal window. Further details are given in Table 1.

Identification of Apoptotic Cells In Vivo

Apoptotic cells can be directly visualized in tissue sections by several means. Available methods include the use of fluorescent nuclear, cytoplasmic and membrane stains, Annexin V binding assay, ultrastructural analysis, single-stranded DNA monoclonal antibodies, immunostaining for molecular markers of apoptosis, and molecular biology techniques. Because no single parameter defines apoptosis in all systems, reliable detection (and quantification) of apoptotic cells may require a combination of strategies.

Fluorescent Nuclear Stains

Several fluorescent nuclear stains including propidium iodide (PI), 4′,6-diamidino-2-phenylindole (DAPI), a number of Hoechst dyes, SYTOX® Green, several SYTO® stains, and YO-PRO™-1 iodide (the last three from Molecular Probes, Eugene, OR, USA), and ethidium homodimer-2 (EthD-2) have been proposed as tools for the identification of apoptotic cells in the fluorescence–confocal microscope and/or (less relevant to the purpose of this discussion) by flow and laser scanning cytometry. Basically, these dyes bind to the cell DNA and can be divided into 2 main categories according to their capability of penetrating (permeant dyes) or not penetrating (impermeant dyes) the intact plasma membrane of living cells.

The cell impermeant PI (59) and EthD-2 can be used to stain dead cells. Hoechst 33342 (Bisbenzimide) is a cell permeant DNA binding dye that shows intense fluorescence when bound to the condensed chromatin in apoptotic cells (100). The permeability of this stain to cell membranes theoretically allows for the staining of cells at very early stages of the apoptotic program when plasma membranes are still intact. YO-PRO-1 iodide is a nucleic acid stain that crosses the plasma membranes of apoptotic cells brightly staining dying cells with a green fluorescence. YO-PRO-1 stain enters apoptotic cells at a stage when other dyes such as PI are still excluded (33).

SYTOX Green is virtually nonfluorescent when free in solution, but becomes brightly fluorescent when binding to nucleic acids with excitation and emission peaks similar to fluorescein.

SYTO 13 was used in conjunction with PI to study apoptosis in cerebellar granule cells (11).

Fluorescent Cytoplasmic and Membrane Stains

As for the nuclear stains, many of the cytoplasmic and membrane dyes have been developed for use in flow cytometric analysis. However, the possibility to image slice preparations of living tissue by laser scanning confocal microscopy (see Chapter 15)

Table 1. Markers of Cell Proliferation in Use (or of Potential Interest) for CNS Analysis

Marker	Main Characteristics and Biological Activity	References	Comments
PCNA/Cyclin (mAb PC10)	36 kDa polypeptide which acts as an auxiliary protein of DNA polymerase δ and is expressed at the transition of the phases G_1/S of the cell cycle.	18,24,27,72, 103,128	• PCNA persists for a certain period of time in daughter cells after division has occurred. • There is the possibility that under certain conditions of fixation the molecule is detected in cells which are not committed to division (phase G_0).
Ki-67 nuclear antigen (MAbs Ki-67, KiS5, and mib-1)	Nonhistone protein containing 10 ProGluSerThr (PEST) motifs which are associated with high turnover proteins and plays a pivotal role in maintaining cell proliferation. It is expressed in the G_1, S, G_2 and M phases of the cell cycle, but not in G_0.	38,50,89,92, 95,114,117, 118	• Ki-67 is likely the most reliable marker of cell proliferation. • The Ki-67 antibody (mib-1) recognizes an epitope detected, besides to humans, in several mammals (rat, rabbit, calf, lamb, dog). In other species, staining is weak (mouse) or totally absent (swine, cat, chicken, pigeon).
DNA polymerase α	Key enzyme in DNA synthesis whose levels of activity gradually decrease during development.	84	
Topoisomerase IIα (MAb kiS4)	DNA-modifying enzyme. The α isozyme has a role in cell proliferation and is expressed in the S, G_2, and M phases.	115,122,134	• Transcription of the topoisomerase II αmRNA closely correlates with expression of BrdU in the developing rat brain. βmRNA has a generalized distribution in the developing brain.
MAb kiS2	The antibody recognizes a 100 kDa protein (p100) highly expressed in cells at the G_1/S transition and persisting through G_2 and M phase.	87	
repp-86	Proliferation-specific protein (p86) expressed exclusively in the S, G_2, and M phases of the cell cycle.	20	
Telomerase	Cellular reverse transcriptase that helps to provide genomic stability by maintaining the integrity of the chromosome ends, the telomeres.	54,116,125, 129	• Telomerase is expressed in highly proliferative normal, immortal, and tumor cells.
Casein-kinase 2α	Growth-related serine/threonine protein kinase.	70	• Expression of the molecule is mainly related to cell growth.
Transferrin receptor	Glycoprotein participating in transferrin uptake.	105	• The transferrin receptor is expressed in all cells (except for erytrocytes) but it is particularly abundant in proliferating cells.
Statin	Nuclear protein specifically expressed in quiescent (norcycling) G_0-phase cells.	113	
MAb Th-10a	The antibody is specific for a 95 kDa nuclear protein appearing in the S phase of the cell cycle in normal mouse thymocytes.	71	
Argyrophilic nucleolar organizer region proteins (AgNORs)	Proteins that accumulate in highly proliferating cells. Expression is low for G_1 phase and high for $S-G_2$ phase.	98,112,139	• AgNORs are estimated in situ by quantification of the level of silver staining using morphometry and image analysis.

makes them of potential interest also for the detection of apoptotic cells in situ. These stains can be used to visualize apoptotic cells, basically by taking advantage of certain metabolic alterations which more or less specifically take place during PCD.

Reduction of the intracellular pH often occurs as an early event in apoptosis and may be preceding DNA fragmentation (66). The fluorescent carboxy dye SNARF®-1 (Molecular Probes) is a dual pH emission indicator that shifts from deep red (about 640 nm) in basic conditions to yellow-orange (about 580 nm) under acidic pH. The cell permeant acetoxymethyl (AM) esther can be added directly to the incubation medium, and the dye becomes trapped into the cells after hydrolyzation of the esther groups within the cytosol. Shifting of the emission spectrum allows for monitoring changes in the intracellular pH.

A similar principle is at the basis of the use of the calcium indicators fluo-3, fura-2, and indo-1-AM (see Chapter 15) to study changes of intracellular Ca^{2+} concentration ($[Ca^{2+}]_i$) in living cells. Changes in $[Ca^{2+}]_i$ are associated with numerous functional events within the cell, among which is apoptosis (13,75,85).

Increased oxidative activity is another early event in apoptosis. Numerous dihydro, colorless, nonfluorescent leuco-dye derivatives of fluorescein and rhodamine are readily oxidized back to the fluorescent parent dye and can thus be used as fluorogenic probes for detecting apoptosis (34,44,53).

Other fluorescent probes can be used to monitor apoptosis-induced mitochondrial membrane depolarization (Mito Tracker® Red CMX Ros; Molecular Probes) or depletion of reduced glutathione in apoptotic cells (monochlorobimane [mBCl] and monobromobimane [mBBr]; see References 12 and 41).

Finally, the dye Merocyanine 450 undergoes an increase in fluorescence with the loss of cell membrane asymmetry in correlation to the translocation of membrane phosphatidylserine (PS), which occurs in apoptotic cells (see the section Annexin V Binding Assay and Reference 32).

All the above nuclear, cytoplasmic, and membrane fluorescent stains are useful to distinguish live cells, dead cells, and apoptotic cells by flow cytometry. However, in general terms, their properties make them not satisfactory if not unsuitable for in situ detection, if one needs to stain fixed cells.

Annexin V Binding Assay

Annexin V is a 35.8 kDa protein that possesses a strong anticoagulant activity and belongs to a family of proteins with high calcium-dependent activities for aminophospholipids, among which is PS (32). This latter is a major component of plasma membranes and is usually almost completely segregated to the cytoplasmic face of the membrane. Early in apoptosis, PS is externalized as a means of recognition and uptake of apoptotic cells by phagocytes (39). The usefulness of fluorochrome-Annexin V conjugates (Annexin V-fluorescein isothiocyanate [FITC] and Annexin V-Cy3™) as tools for the visualization of apoptotic cells has been demonstrated by several authors, mainly in cultured cells (136), but they can be potentially utilized also for the in vivo analysis of apoptotic cells in unfixed tissues as discussed above for the fluorescent stains.

Ultrastructural Analysis

Apoptosis is characterized by a series of stereotyped ultrastructural changes in cell morphology, and thus, electron microscopy represents an ideal tool for an unequivocal identification of apoptotic cells. Dying cells may adopt one of at least three different morphological types which have been referred to as apoptotic, autophagic, and nonlysosomal vesiculate (30). Changes in cell ultrastructure during apoptosis affect both nuclear and cytoplasmic morphology. Nuclear changes include chromatin condensation, blebbing of nuclear membranes, and fragmentation of the nucleus into highly electron-dense apoptotic bodies. Cytoplasmic changes consist of condensation and fragmentation of the cell body into membrane-bound vesicles, which contain ribosomes, morphologically intact mitochondria, and nuclear material. To the purpose of the present discussion, it must be stressed that all of these changes take place in rather late stages of the apoptotic program (see the section Molecular Biology Techniques and Results and Discussion), and that apoptotic cells are rapidly cleared from tissue by phagocytes. This explains why ultrastructural analysis is often insufficient to the detection of apoptosis in tissue sections unless a very large population of cells are undergoing PCD. Therefore, although ultrastructural analysis is perhaps the most specific tool for the identification of apoptotic cells in situ, it suffers from several pitfalls mainly related to the dramatically low degree of sensitivity of this procedure.

Single-Stranded DNA Monoclonal Antibodies

Specific monoclonal antibodies to stain single-stranded DNA can be used to localize in situ cells undergoing PCD. Single-stranded DNA is produced in apoptotic cells as a consequence of an increased sensitivity of the nucleic acid to thermal denaturation (for review see Reference 45). The critical step in this procedure is the heating of tissues in conditions that prevent thermal denaturation of DNA in situ in non-apoptotic cells but induce DNA denaturation in apoptotic nuclei. Under optimal experimental conditions, staining of single-stranded DNA with specific monoclonal antibodies nicely correlates with the apoptotic morphology of positive cells. Moreover, at least in certain cell types, it detects early apoptotic events already at a stage prior to chromatin condensation and in the absence of intranucleosomal DNA fragmentation and allows for the distinction of apoptosis and necrosis.

Immunostaining for Molecular Markers of Apoptosis

A few intracellular proteins normally expressed inside cells have been shown to be translocated onto the cell surface during PCD, thereby allowing apoptotic cells to be distinguished from normal cells, as previously described for PS. A number of antibodies have been raised against the various molecules that appear to be induced and/or activated during apoptosis, including tissue glutaminase (42,80), the CD44 cell surface adhesion molecule (73), the activated type 1 insulin-like growth factor receptor (86), and a mitochondrial membrane antigen that is specifically recognized by the 7A6 monoclonal antibody, which was developed by immunizing mice with apoptotic Jurkat cells (135). Finally, a series of antibodies and in situ hybridization probes for a number of positive or negative effector proteins of apoptosis such as bcl-2 (19,58) or caspase 3 (see the section Activation of Apoptotic Caspases and References 35 and 102) have been made available.

As it is often the case, the search for the "Holy Grail" marker of apoptosis is still ongoing, and the results obtained with this approach need to be interpreted with caution, since in most cases the specificity of at least some of the markers mentioned above needs further confirmation.

Molecular Biology Techniques

DNA fragmentation in small size oligomers (127) is a biochemical hallmark of apoptosis producing what has been referred to as nucleosomal DNA "ladders". This feature can be appreciated after electrophoretic separation of DNA in agarose gels and is a consequence of the action of DNA nucleases on the chromatin to produce double-stranded DNA fragments with a size reflecting oligomers of the nucleosomes (see the section Biochemical Assays for Apoptosis).

An important advance in the detection of dying cells in situ by light and fluorescence microscopy was the development of a series of molecular biology protocols which can be generally referred to as in situ end-labeling (ISEL) techniques. Each of these ISEL techniques utilizes a different DNA modifying enzyme to attach labeled nucleotides to the free ends of fragmented DNA. In fact, cleavage of the DNA during the apoptotic process may yield double-stranded mono- and/or oligonucleosomes of low molecular weight, as well as "nicks" (single strand breaks) in high molecular weight DNA. These DNA strand breaks can be detected by enzymatic labeling of the free 3′ OH termini with modified nucleotides, usually dUTP linked to a fluorochrome, biotin, or digoxigenin (DIG).

The ISEL techniques include the in situ nick translation with DNA polymerase I (51,55) or unmodified T7 polymerase (123), the end labeling with terminal deoxynucleotidyl transferase (TdT)(49, 51), and the ligation of DIG-labeled double-stranded DNA fragments with T4 DNA ligase (37).

DNA polymerase I catalyzes the template-dependent addition of nucleotides when one strand of a double-stranded DNA molecule is nicked. Theoretically, by such an approach, not only the apoptotic DNA is detected, but also the random fragmentation of DNA by multiple endonucleases occurring in cellular necrosis and as a consequence of nonproper tissue handling.

Unmodified T7 DNA polymerase is a highly processive DNA template-dependent 5′-3′ polymerase and a 3′-5′ exonuclease for 3′ OH acceptor groups at recessed, blunt, or overhanging ends. To detect apoptotic cells, the enzyme is utilized, first, to generate 5′ overhangs from any 3′ OH acceptor groups at DNA strand breaks, and second, to end-fill the overhangs and incorporate biotinylated dATP (123).

TdT is an unusual type of DNA polymerase found only in lymphocyte precursors at early stages of their differentiation (28,108). It catalyzes the template-independent addition of deoxyribonucleoside triphosphates to the 3′ OH ends of double- or single-stranded DNA. The nick end labeling with TdT (TUNEL = TdT-mediated dUTP nick end labeling) is considered to be more sensitive, faster, and specific than the template-dependent nick translation (108). This method for labeling apoptotic cells in situ has been successfully employed also at the electron microscope level. For ultrastructural examination, TdT was used to enzymatically link 11-digoxigeninated-dUTP to fragmented DNA in apoptotic cells in conjunction with anti-DIG gold-labeled antibodies (67,76).

T4 DNA ligase was used to ligate in situ double-stranded DNA fragments labeled in vitro with DIG (or Texas Red®) to DNA double-strand breaks with single base 3′ overhangs as well as blunt ends in apoptotic nuclei (37,63). This approach has been shown to be very specific for apoptotic DNA, avoiding the false positive results that can derive by DNA damage from necrosis, in vitro autolysis, peroxide toxicity, and heating.

Biochemical Assays for Apoptosis

Biochemical assays have been mainly developed for assessment of apoptosis in cultured cell systems and/or tissue extracts. Although these methods are not directly related to the main topic of this chapter, we will briefly review the most widely employed protocols, considering that, in many instances, they are of value to confirm the true apoptotic nature of dead cells after in situ detection (see Results and Discussion).

Detection of DNA Fragmentation

As previously mentioned, agarose gel electrophoresis of DNA extracted from cells or tissues represents an extremely specific method to ascertain the presence of apoptosis. The formation of nucleosomal DNA ladders during PCD is a consistent finding in those experimental systems where a significant population of cells can simultaneously be induced to die through apoptosis (for review see Reference 29). However, nucleosomal DNA ladders have not been generally or routinely observed in several adult tissues where apoptosis is ongoing as a part of the normal cell turnover, or during normal tissue development, and were not consistently detected in a number of experimental cell systems (14,96). Indeed, the technique has low sensitivity, and millions of cultured cells or milligrams of tissue are required to extract sufficient amounts of DNA for analysis. In addition, the quantitative estimation of the extent of intranucleosomal DNA fragmentation in a stained gel relies on optical densitometry and is not so technically straight. DNA nucleosomal ladders might not have been detected in normal tissues because apoptosis was asynchronous and/or it occurred in a comparatively small population relative to the majority of surviving cells (for discussion see Reference 29).

To at least partly overcome these problems, some alternatives have been proposed such as the direct gel-based DNA fragmentation assays using radioisotopes or biotinylated nucleotides (96,123), the ligation-mediated polymerase chain reaction (PCR) fragmentation assay (29), and an enzyme-linked immunosorbent assay (ELISA) TdT detection method for cells prelabeled with BrdU (53). Commercially available kits also include a cell death detection method based on the assessment of intranucleosomal DNA fragmentation by an antihistone monoclonal antibody using an ELISA procedure (23,43,119).

Activation of Apoptotic Caspases

The term caspases (caspase = cysteine aspartate protease) is used to indicate a family of cysteine proteases cleaving an aspartic acid residue which are thought to mediate very early stages of apoptosis (1,132). All these proteins are synthesized as pro-enzymes and are activated following a cleavage at aspartate residues that could be themselves the site of attack for other members of the family. Caspases are likely the most important effector molecules so far described for triggering the cellular apoptotic machinery. At present, the mammalian caspase family consists of ten members (1), the most intensively studied of which is caspase-3 (previously referred to as CPP32, Yama, or Apopain; see References 40 and 74). Caspase-3 is usually detected in the cell cytosol as an inactive precursor that is activated when cells undergo apoptosis (94,120). Several assays have been developed so far for the quantitative estimation of caspase-3 activity. The most widely employed substrate for detection is the tetrapeptide DEVD, which has been conjugated with colorimetric or fluorogenic substrates (137).

Cleavage of Poly-ADP-Ribose Polymerase

Poly-ADP-ribose polymerase (PARP) is a 113 kDa protein that binds specifically at DNA strand breaks and appears to be involved in DNA repair and genome surveillance and integrity at the onset of apoptosis. PARP is a substrate for certain caspases, including caspase-3 and -7, which, as mentioned above, are activated during early activation of the apoptotic pathway. The active form of caspase-3 is capable to cleave PARP at its consensus site, the DEVD tetrapeptide. Caspases cleave PARP to fragments of approximately 89 and 24 kDa (74). Detection of the 89 kDa PARP fragment with an anti-PARP antibody in western blotting represents a specific tool for detection of apoptosis.

PROTOCOLS

Protocol for Labeling Proliferating Cells In Vivo

Materials and Reagents

- BrdU (Sigma, St. Louis, MO, USA).
- IdU (Sigma).

Procedure

1. Dissolve BrdU or IdU in distilled water or physiological saline at a final concentration of 100 mg/mL. IdU is not readily soluble at physiological pH. Therefore predilution in a small amount of alkalinized distilled water (by addition of some drops of 4 N NaOH) is necessary.
2. Inject intraperitoneally the appropriate amount of the drug (0.1 mg/g body weight) and allow animals to survive for the required time.

Protocol for the Ultrastructural Visualization of BrdU/IdU-Labeled Cells

Materials and Reagents

- Syringe filters; pore size 0.22 μm (Whatman, Maidstone, England, UK).
- 200 to 300 Mesh nickel grids (Electron Microscopy Science, Fort Washington, PA, USA).
- Sodium metaperiodate (Sigma).
- Bovine serum albumin, Fraction V (BSA; Sigma).
- Triton® X-100.
- Normal goat serum (NGS; Sigma).
- Bispecific mouse monoclonal antibody to BrdU/IdU (Caltag Laboratories, Burlingame, CA, USA).
- Monospecific mouse monoclonal antibody to BrdU (Amersham Pharmacia Biotech, Buckinghamshire, England, UK).
- Goat antimouse IgG gold conjugates (20 or 30 nm) (British BioCell International, Cardiff, Wales, UK).

Procedure

1. Collect ultrathin sections onto uncoated nickel grids (see also Protocols in Chapter 10).

   Incubate sections for 5 minutes with sodium metaperiodate to remove osmium tetroxide.
3. Wash with 0.5 M TBS (Tris-buffered saline; pH 7.2–7.6) containing 1% Triton X-100 to facilitate contact between the hydrophobic surface of the section and immunoreagents.
4. Incubate grids 1 hour at room temperature in NGS to reduce aspecific background.
5. Transfer over drops of the bispecific mouse monoclonal antibody to BrdU/IdU diluted 1:10 in TBS/BSA and incubate overnight at room temperature.
6. Wash sections with 3 washes of TBS/BSA.
7. Incubate with 20 nm antimouse IgG gold conjugate diluted 1:15 in TBS/BSA for 1 hour at 37°C.
8. Rinse in TBS/BSA (3 times, for 10 minutes each).

9. Postfix in 2.5% glutaraldehyde for 10 minutes and wash in double-distilled water.
10. Leave grids to dry at room temperature protected from dust.
11. The same procedure is then applied on the other side of the grid using the undiluted monospecific mouse monoclonal antibody anti-BrdU and the anti-mouse IgG gold conjugate of 30 nm.

Protocol for Labeling Apoptotic Cells In Situ at the Light Microscopy Level

Materials and Reagents

- Proteinase K (Sigma).
- Digoxigenin-11′-2′-3′-dideoxy-uridine-5′ thriphosphate (DIG-11-ddUTP) (Roche Molecular Biochemicals, Mannheim, Germany).
- TdT (Roche Molecular Biochemicals).
- AntiDIG-alkaline phosphatase Fab fragments (Roche Molecular Biochemicals).
- Nitro blue tetrazolium (NBT; Sigma).
- 5-Bromo-3-inolylphosphate *p*-toluidine salt (BCIP; Sigma).
- BSA.

Procedure

1. Strip proteins from tissue sections by incubation in proteinase K (20 µg/mL) for 15 minutes.
2. Extensively wash in double-distilled water.
3. Rinse in terminal transferase buffer (TdT buffer: 130 mM Trizma base, pH 6.5, 140 mM sodium cacodylate, 1 mM cobalte chloride).
4. Incubate sections in TdT buffer containing terminal transferase (0.05 U/µL) and DIG-11-ddUTP (2 nM/mL) at 37°C for 120 minutes in a humid chamber.
5. Stop the reaction by transferring sections in terminal buffer (TB: 300 mM sodium chloride, 30 mM sodium citrate) for 15 minutes.
6. Rinse in double-distilled water and incubate in 2.5% BSA for 10 minutes. Wash in phosphate-buffered saline (PBS).
7. Incubate overnight in the anti-DIG-alkaline phosphatase Fab fragments diluted 1:5000 in PBS.
8. Reveal the alkaline phosphatase reaction using the NBT/BCIP procedure under continuous microscope monitoring.
9. Terminate the reaction by incubation in STM buffer (Tris 100 mM, NaCP 100 mM, MgCP 1 mM).
10. Slides are counterstained (optional), dehydrated, and mounted.

Protocol for Labeling Apoptotic Cells In Situ at the Electron Microscopy Level

Material and Reagents

- DIG-11-ddUTP.
- TdT.
- Anti-DIG gold-labeled conjugate with 10 nm particles (British BioCell International).
- BSA.

Procedure

1. Grids with epoxy sections are washed with TdT buffer (see previous protocol).
2. Incubate grids in TdT buffer containing terminal transferase (0.05 U/µL) and DIG-11-ddUTP (2 nM/mL) at 37°C for 120 minutes in a humid chamber.
3. Stop the reaction by transferring sections in terminal buffer (TB: 300 mM

14. In Vivo Analysis of Cell Proliferation and Apoptosis in the CNS

sodium chloride, 30 mM sodium citrate).

4. Rinse grids (3 times for 10 minutes each) with 0.5 M TBS, pH 7.2 to 7.6, containing 1% BSA (TBS/BSA).
5. Incubate in the same buffer with the anti-DIG gold-labeled conjugate at 37°C for 1 hour.
6. Wash sections again in TBS/BSA (3 times for 10 minutes each).
7. Postfix (10 minutes) in 2.5% glutaraldehyde in sodium cacodylate buffer.
8. Wash extensively in double-distilled water and leave to dry at room temperature.
9. Counterstain with Reynold's lead citrate and uranyl acetate.

Preparation of Cerebellum Extracts

Materials and Reagents

- Triton X-100.
- EDTA (Sigma).
- Phenenylmethylsulfonyl fluoride (PMSF; Sigma).
- Aprotinin (Sigma).
- Leupeptin (Sigma).
- Homogenizer (e.g., Polytron; Glen Mills, Clifton, NJ, USA).
- Ultracentrifuge (e.g., Beckman Instruments, Fullerton, CA, USA).

Procedure

1. After animal euthanasia, remove the cerebellum, strip the meninges, and wash 2 times with cold NaCl 0.9%.
2. Suspend the cerebellum in 2.5 volumes of a cold general lysis buffer (20 mM Tris, pH 8.0, 10 mM NaCl, 0.5% Triton X-100, 5 mM EDTA, 3 mM $MgCl_2$) containing protease inhibitors (1 mM PMSF, 1 mg/mL leupeptin, and 5 mg/mL aprotinin).
3. Homogenize gently.
4. Centrifuge the homogenate at $10^5 \times g$ for 1 hour at 4°C.
5. Remove the supernatant and suspend the pellet in homogenization buffer (1:1, vol/vol).

Immunochemical Detection of PARP

Materials and Reagents

- Modified Lowry protein assay reagent (Pierce Chemical, Rockford, IL, USA).
- Coomassie protein assay reagent (Pierce Chemical).
- Acrylamide (Sigma).
- N,N'-Methylene-bis-acrylamide (Sigma).
- TEMED (Sigma).
- Ammonium persulfate (Sigma).
- Glycine (Sigma).
- Lauryl sulfate, sodium salt (SDS; Sigma).
- Sample buffer, Laemmli (Sigma).
- Methanol (Sigma).
- Coomassie® Brilliant Blue dye (Sigma).
- Nitrocellulose membrane, electophoresis grade, 0.45 mm pore size (Sigma).
- Polyvinylidene fluoride (PVDF) membrane, electrophoresis grade (Sigma).
- Tween® 20.
- BSA.
- Rabbit polyclonal antibody to PARP (Santa Cruz Biotechnology, Santa Cruz, CA, USA).
- Peroxidase-labeled antirabbit antibody (Amersham Pharmacia Biotech).
- Color markers for sodium dodecyl sulfate polyacrylamide gel electrophoresis (SDS-PAGE) and protein transfer (Sigma).

- Enhanced chemiluminescent detection system (ECL Western blotting system; Amersham Pharmacia Biotech).
- Hyperfilm ECL™ (Amersham Pharmacia Biotech).
- Mini protean slab gel apparatus for electrophoresis and transblot unit (available from Bio-Rad, Hercules, CA, USA).

Procedure

1. Measure the protein concentration of resuspended pellet with modified lowry protein assay reagent or Coomassie Protein Assay Reagent.
2. Prepare a 9% polyacrilamide separating gel with a 4% polyacrilamide stacking gel.
3. Prepare a sample containing the color markers and samples containing 60 mg of pellet proteins in sample buffer; place samples in a boiling water bath for 3 to 5 minutes.
4. Load all samples onto the gel lanes and electrophorese the proteins at 180 V for 1 hour in a Tris-glycine running buffer (25 mM Tris, 192 mM glycine, 0.1% SDS, pH 8.3 ± 0.2).
5. Transfer proteins onto a nitrocellulose or a PVDF membrane using a Tris-glycine buffer (25 mM Tris, 192 mM glycine, 20% methanol) and the transblot unit at 100 V for 1 hour at 4°C.
6. To ensure consistency of loading, stain the gel for 1 hour in a solution consisting of 50% (vol/vol) methanol, 10% (vol/vol) acetic acid, 40% (vol/vol) water, and 0.1% Comassie Brilliant blue dye.
7. Following transfer of the proteins, block nonspecific sites of PVDF or nitrocellulose membrane with TBST buffer (20 mM Tris-HCl, pH 7.5, 150 mM NaCl, 0.2% Tween 20) containing 5% BSA overnight at 4°C.
8. Dilute the rabbit polyclonal antibody to PARP in blocking solution (1:500) and incubate the membrane for 3 hours.
9. After washing 3 times in TBST (1 × 15 minutes, 2 × 5 minutes) incubate the membrane with peroxidase-labeled antirabbit secondary antibody in blocking buffer (1:10 000) for 1 hour at room temperature.
10. Wash in TBST and detect proteins by ECL western blotting system with an exposure time of 1 hour.

RESULTS AND DISCUSSION

Identification of Proliferating Cells In Vivo

Light Microscopy

Figure 1 shows the localization of proliferating cells in the postnatal cerebellum at the light microscope level after immunocytochemical visualization of exogenously administered BrdU (or IdU) at different survival time. The immunocytochemical localization of the Ki67 nuclear antigen is shown in Figure 2, A and B. As expected, labeling was mainly observed at the level of the external granular layer (EGL) and, at a much lesser extent, within the molecular layer (ML) and the internal granular layer (IGL) of the postnatal cerebellar cortex. This pattern is consistent with the well known migratory pathway followed by granule cell precursors, which having been originated by the proliferation of progenitor cells within the EGL, migrate past to the ML and the Purkinje cell layer to reach their final destination, the IGL (9,83,88). The presence of proliferating cells within the ML and the IGL may also be linked to migration of the GABAergic cerebellar interneurons which are generated by interstitial proliferation within the white matter and, subsequently, migrate to reach the ML and the IGL following a direction opposite to the course of newly generated granule cells (138). The simultaneous existence of proliferative and migratory events during neurogenesis is rather common and somehow complicates the analysis of cell proliferation when studying central neurons in vivo (22,48,61,64). The crucial aspect of the problem is related to length of persistence of the label within postmitotic cells. As mentioned in the Background section, labeling of proliferating cells with tymidine analogues such as BrdU/IdU results in a permanent label of the DNA that, although progressively diluted in the subsequent cell generations, is retained by the cell until the quantity of the marker falls below the limit of detectability of the immunocytochemical techniques or the cell dies and is cleared up by tissue phagocytes (47). After BrdU injection and 1 hour survival time, the picture obtained represents a flash photograph of the cells which were in the S phase of their cycle (and therefore committed to division) at the moment in which the label became available to the tissue (17,36,99). This explains why the results obtained with

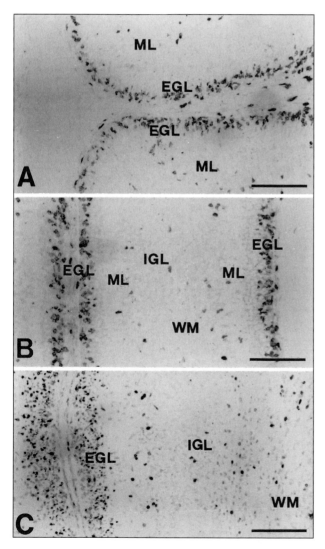

Figure 1. Light microscopic visualization of proliferating cells in the postnatal rabbit cerebellum by the in vivo BrdU labeling procedure described in the Protocols section. After 1 hour survival, labeled nuclei are detected mainly in the EGL both in P0 (A) and P5 (B) animals. Scattered positive nuclei are also apparent in the ML and IGL. Panel C shows the pattern of labeling in a P5 animal which received an injection of the marker immediately after birth. Note labeled nuclei in the inner (premigratory) part of the EGL and positive nuclei in the ML, which likely correspond to newly generated granule cells during the route of their migration to the IGL. Scattered positive cells are also detected in the WM. Abbreviations: BrdU, 5-bromodeoxyuridine; EGL, external granular layer; IGL, internal granular layer; ML, molecular layer; P0, postnatal day 0; P5, postnatal day 5; WM, white matter. Scale bars = 100 µm.

BrdU (IdU) labeling in several tissues and organs, including the nervous tissue, and those achieved after administration of tritiated tymidine are, in practical terms, equivalent to each other. However, BrdU (IdU) injection has the drawback of a higher variability between individuals in counts of labeled cells and has the additional problem of presenting some difficulties in counting cells that are scattered along a spatial gradient. To partly overcome this problem, multiple injections (21) or pellets of BrdU (IdU) may be implanted subcutaneously in experimental animals (47). However, if animals are left to survive for longer periods of time, the interpretation of the results becomes more difficult considering the possibility that labeled cells undergo subsequent divisions and that migratory phenomena also occur (22,79). In these conditions, the calculation of the proliferation index (percentage of proliferating cells) is not so straight. The difficulty in the evaluation of the dilution of the label in repeat-

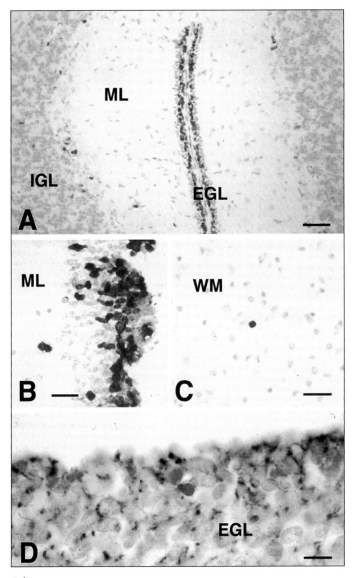

Figure 2. Visualization of proliferating (A and B) and apoptotic (C and D) cells in the postnatal cerebellum. Proliferating cells have been detected by immunocytochemical labeling for the ki-67 nuclear antigen. (A) Rabbit cerebellum at P5. Note that the pattern of staining is remarkably similar to that shown in Figure 1B. (B) Human cerebellum at P30. Dual color immunocytochemistry for ki-67 (peroxidase, brown) and bcl-2 protein (alkaline phosphatase, blue), shows numerous proliferating cells with nuclear ki-67 positivity in the outer (proliferative) part of the EGL expressing in their cytoplasm the bcl-2 protein, a negative modulator of apoptosis. (C) An apoptotic cell in the cerebellar white matter of a P5 rabbit. TUNEL method. (D) Combined visualization of apoptotic cells (TUNEL alkaline phosphatase—new fuchsin = red) and vimentin (a marker of undifferentiated neural precursors) immunoreactivity (nickel intensified peroxidase = black) in the EGL. Abbreviations: EGL, external granular layer; IGL, internal granular layer; ML, molecular layer; P5, postnatal day 5; P30, postnatal day 30; WM, white matter. Scale bars: panel A = 100 µm; panels B and C = 25 µm; panel D = 10 µm. (See color plate A11.)

edly dividing cells represents one of the most important drawbacks of the BrdU (IdU) labeling methods in comparison with the isotopic procedures relying on tritiated tymidine autoradiography. In the latter, in fact, halving of the label is apparent from the reduction of the number of silver grains that can be directly counted over the subsequent generations of labeled cells (2).

Another very important issue linked to the use of nonisotopic methods to label of proliferating cells is represented by the need to open double-stranded DNA in tissues and make incorporated BrdU (IdU) available to binding by specific antibodies. To this purpose, it is necessary to pretreat sections with proteinase K or other proteases to digest DNA-linked proteins and to incubate slides with HCl at different pH and molarity (47,99). All these procedures always have a more or less detrimental effect on tissue morphology and may also interfere with the possibility to successfully combine the visualization of BrdU (IdU)-labeled cells with the immunocytochemical staining for specific cell–tissue antigens (78,99). The availability of an anti-BrdU monoconal antibody, which is commercialized in a nuclease-containing buffer (Amersham Pharmacia Biotech), represents an useful tool to partly overcome the above problems.

The use of immunocytochemical markers of cell proliferation represents an alternative tool to the visualization of radioactive or nonradioactive exogenously introduced DNA labels.

The most appealing feature of this approach is represented by the fact that it does not involve the experimental administration of the label, and that, in general terms, it is technically less demanding. However, there are several concerns related to the use of this approach mainly linked to the specificity of the label and its persistence within the cells once they have stopped proliferating and are no longer committed to cell division (15,52,90,121).

Electron Microscopy

Theoretically, all the methods described so far for the in situ light microscopic detection of proliferating cells can be employed at the ultrastructural level. In practical terms, however, satisfactory results have been obtained by ultrastructural autoradiography (9) or post-embedding immunogold staining of incorporated BrdU (IdU) (57,104). Figure 3 shows the immunocytochemical detection of exogenously administered BrdU-IdU in the postnatal cerebellar cortex. Labeling is scattered into the nucleoplasm of progenitor cells and granule cell precursors within the EGL and appears to be associated with both eu- and eterochromatin. This pattern of localization is consistent with the data obtained after ultrastructural autoradiography. It seems to be of relevance to stress here that the post-embedding immunogold approach allows for the ultrastructural visualization of proliferating cells after conventional fixation (glutaraldehyde plus osmium tetroxide) and embedding. In our opinion, this represents a major advantage of the immunogold approach with respect to ultrastructural autoradiography, which is by far technically more demanding. Another significant advantage lies in the possibility to simultaneously visualize 2 labels (i.e., BrdU and IdU) in the same ultrathin section. This opens the way to perform ultrastructural cell kinetic studies in situ and represents a major implementation with respect to the isotopic procedures (62).

Identification of Apoptotic Cells In Vivo

Light Microscopy

Figure 2, C and D shows the detection of apoptotic cells in the postnatal cerebellar cortex by the TUNEL method for fragmented DNA. By this or similar approaches, several laboratories including ours have demonstrated the existence of an impressively high number of apoptotic neurons and glial cells in the postnatal cerebellum of altricial animals (59,63,106,123,124). As mentioned in the Background section of this chapter, several different alternatives are available to label apoptotic cells in situ. Nevertheless, the major concern, which is linked to all these techniques, regards their true specificity for apoptotic cells. In general terms, none of the existing methods can be considered one hundred percent specific, basically as a consequence that most of the steps involved in tissue preparation are likely to (or at least theoretically can) produce DNA damage that might produce false positive results. Indeed, the TUNEL method, which nowadays represents the most widely used procedure for the in situ detection of apoptotic cells, suffers from a number of potential pitfalls due to the possibility that a positive reaction is observed in the presence of necrotic or autolytic tissue, peroxide damage, or excessive heating of normal tissue (37). To overcome these problems, it is advisable to use more than one single approach to confirm the existence of true apoptotic phenomena, at least when they are found in unexpected locations or when one can not exclude that the experimental material might have been damaged by nonoptimal preparative procedures. For example, we have recently used both the TUNEL and the T4 DNA ligase methods to confirm the presence of apoptosis in the human postnatal cerebellum (63).

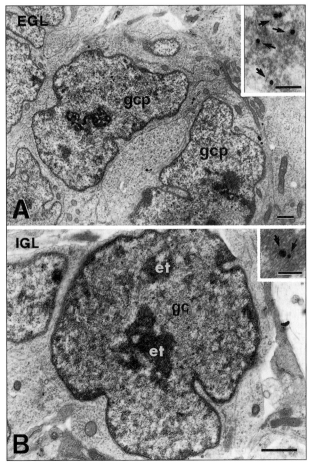

Figure 3. Ultrastructural visualization of proliferating cells in the P5 rabbit cerebellum after BrdU (A) or BrdU + IdU (B) administration. After 1 hour survival and BrdU injection (A), gold particles indicative of labeling are mainly apparent over the nucleus of the granule cell precursors within the EGL (insert). After sequential BrdU (1 hour survival) + IdU (12 hour survival) administration (B), double-labeled granule cells are present mainly in the IGL (insert). BrdU, 10 nm gold particles; IdU, 30 nm gold particles. Abbreviations: EGL, external granular layer; IGL, internal granular layer; et, eterochromatin; gc, granule cell; gcp, granule cell precursor; P5, postnatal day 5; Scale bars = 1 μm. Inserts: panel A = 0.1 μm; panel B = 0.25 μm.

As an alternative or in addition to the use of more than one protocol in situ, it is useful to confirm the existence of apoptosis by DNA extraction and electrophoresis or by western blotting and demonstration of specific cleavage of effector proteins (Figure 4). To do so it is important to stress the need for specific antibodies that consistently recognize the cleaved active form of the protein(s), since the precursor molecule may often have an ubiquitous distribution and thus be expressed also in nonapoptotic cells.

Electron Microscopy

The possibility to label apoptotic cells at the ultrastructural level relies on the availability of nucleic acids at the surface of epoxy sections. Under these conditions, the DNA can be specifically labeled with exogenous transferases and modified nucleotides (107–110). The use of nonisotopic nucleotide analogues has considerably improved the potential applications of the ISEL methods for ultrastructural analysis and has opened the way to successfully combine such an approach with the gold labeling procedures described in Chapter 10 for the immunocytochemical detection of cell and tissue antigens at the electron microscope level.

Although one has to consider that apoptotic cells show peculiar ultrastructural features (Figure 5) and can therefore be easily distinguished in the electron microscope in the absence of any labeling (30), the development of in situ methods allowed for the detection of cells with fragmented DNA, i.e., positive TUNEL labeling, in the absence of ultrastructural signs of apoptosis (see below). Molecular biology techniques at the ultrastructural level are 2-step procedures which are often referred to as transferase–immunogold techniques. When using this approach, an enzymatic (transferase) reaction is performed first, which allows for labeling of the nucleic acid (DNA) fragments; then an immunogold labeling procedure is used to reveal the incorporated nucleotide analogues (107–109). The procedure described in the Protocols section of this chapter is a modified TUNEL reaction which is based on the use of TdT to end-label DNA fragments with digoxigeninated dd-UTP. This latter is finally visualized in the electron microscope by using a gold-conjugated anti-DIG antibody (Figure 6). Similar protocols have been employed in other laboratories with comparable results. As mentioned above, a rather surprising observation following ultrastructural labeling of apoptotic cells in situ, was that some cells showed positive nuclei in the absence of the characteristic features of chromatin condensation, blebbing, and fragmentation into apoptotic bodies, i.e., characteristic ultrastructural features of apoptosis. Although these results should be interpreted with some caution in relation to the pitfalls of the TUNEL method, a series of recent observations indicate that fragmentation of DNA is a very early event during apoptosis and that it

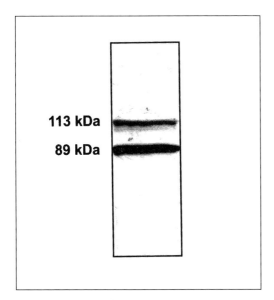

Figure 4. Western blot detection of activated PARP (89 kDa fragment) in P5 rabbit cerebellar extracts. Abbreviations: PARP, poly-ADP-ribose polymerase; P5, postnatal day 5.

occurs in situ prior to any ultrastructural modification of the cell morphology. It seems also of relevance to remark here that the ultrastructural visualization of apoptotic cells by the TUNEL procedure can be carried out on sections of osmicated epoxy-embedded material and is thus compatible with several immunogold staining protocols. We have recently used such a combined approach for the simultaneous visualization of proliferating and apoptotic cells in the postnatal rabbit cerebellum (62).

Combination with Immunocytochemical Labeling

Besides the need to specifically detect apoptosis in situ, one is often faced with the problem of identifying the nature of apoptotic cells in a given experimental system. Due to the fact that: *(i)* cells may undergo apoptosis very soon after they are generated; and *(ii)* they are rapidly removed from tissues by phagocytes, the detection of specific antigens in apoptotic cells is a quite demanding task. Moreover, the following possibility are likely to occur: *(i)* apoptotic cells are labeled at a stage of immaturity in which any specific antigen is not expressed yet; or *(ii)* they are already so heavily damaged that they do not retain specific antigens to give some clues on their origin or nature. The relevance of all the points raised so far becomes even more evi-

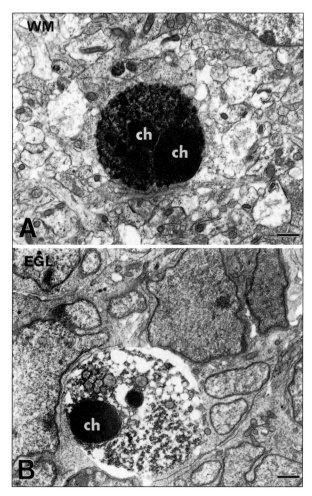

Figure 5. Apoptotic cells in the P5 rabbit cerebellum. The dying cells in panels A and B show chromatin masses (ch) typical of apoptosis and a variable degree of nuclear and cytoplasmic condensation. Abbreviations: ch, chromatin; EGL, external granular layer; P5, postnatal day 5; WM, white matter. Scale bars = 1 μm.

dent considering the relatively low degree of sensitivity, particularly at the ultrastructural level, of the immunocytochemical methods. The problem has assumed a major relevance following the demonstration that apoptosis in not a phenomenon restricted to neurons, but it also affects glial cells. Some difficulties are at least partly bypassed after ultrastructural examination, considering that different cell types have generally very distinctive cytological features, even at very early stages of differentiation. Nevertheless, ultrastructural studies are not always practical nor can they be employed in all experimental conditions. Some authors have tried to overcome the problem of identifying the nature of apoptotic cells in situ by combining immunocytochemistry and fluorescent dye nuclear staining for apoptosis. However, the specificity of these stains is often questionable as discussed in a previous section. Despite all difficulties, successful attempts have been made to combine in situ labeling of apoptotic cells with immunocytochemical labeling for markers of neuronal and/or glial differentiation (47,63,99,111). As mentioned, other technical drawbacks added further unpredictability to this combined approach, such as, for example, the need of tissue digestion for a proper exposure of the DNA to its modifying enzymes with

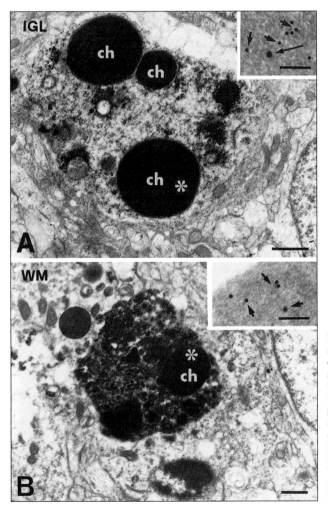

Figure 6. Ultrastructural detection of DNA fragmentation in apoptotic cells of the P5 rabbit cerebellum. Condensed chromatin masses (ch) after incorporation of digoxigeninated-dd-UTP following the TUNEL method described in the Protocols section are heavily labeled with colloidal gold particles (10 nm) (inserts, short arrows). The apoptotic cell in panel A also shows incorporation of BrdU (30 nm gold) (insert, long arrow) after 24 hour survival. Abbreviations: ch, chromatin; IGL, internal granular layer; P5, postnatal day 5; WM, white matter. The asterisks indicate the areas shown at higher magnifications in the inserts. Scale bars = 1 μm. Inserts = 0.1 μm.

obvious detrimental consequences on antigenicity. Nonetheless, an effective combination of the detection of apoptotic cells with immunocytochemistry would represent a valuable tool to answer a number of still unsolved questions in several neural systems and deserves further exploitation.

ACKNOWLEDGMENTS

Much of the work described here was supported by grants from the University of Torino, the Italian Ministero dell'Istruzione dell' Università e della Ricerca (MIUR—Cofin 2000, Cofin 2001), Ministero della Sanità (Progretto Finalizzato Alzheimer), and Consiglio Nazionale delle Ricerche (CNR).

REFERENCES

1. Alnemri, E.S., J.N. Livingston, D.W. Nicholson, G. Salvesen, N.A. Thornberry, W.W. Wong, and J. Yuan. 1996. Human ICE/CED-3 protease nomenclature. Cell 87:171-181.
2. Altman, J. 1966. Autoradiographic and histological studies of postnatal neurogenesis. A longitudinal investigation of the kinetics, migration and transformation of cells incorporating tritiated thymidine in infant rat, with special reference to postnatal neurogenesis in some brain regions. J. Comp. Neurol. 128:431-474.
3. Altman, J. 1969. Autoradiographic and histological studies of postnatal neurogenesis. Cell proliferation and migration in the anterior forebrain, with special reference to persisting neurogenesis in the olfactory bulb. J. Comp. Neurol. 137:433-458.
4. Altman, J. 1972a. Postnatal development of the cerebellar cortex in the rat. I. The external germinal layer and the transitional molecular layer. J. Comp. Neurol. 145:353-398.
5. Altman, J. 1972b. Postnatal development of the cerebellar cortex in the rat. II. Phases in the maturation of the Purkinje cells and of the molecular layer. J. Comp. Neurol. 145:399-464.
6. Altman, J. 1982. Morphological development of the rat cerebellum and some of its mechanism, p. 8-46. In S.L. Palay and V. Chan-Palay (Eds.), The Cerebellum—New Vistas. Springer-Verlag, New York.
7. Altman, J. 1992. Programmed cell death: the paths to suicide. Trends Neurol. Sci. 15:278-280.
8. Altman, J. and S.A. Bayer. 1978. Prenatal development of the cerebellar system in the rat. I. Cytogenesis and histogenesis of the deep nuclei and the cortex of the cerebellum. J. Comp. Neurol. 179:23-48.
9. Altman, J. and S.A. Bayer. 1997. Development of the Cerebellar System in Relation to Its Evolution, Structure and Functions. CRC Press, Boca Raton.
10. Alvarez-Buylla, A. 1997. Neurogenesis in the adult brain: prospects for brain repair, p. 86-100. In F.H. Gage and Y. Christen (Eds.), Isolation, Characterization and Utilization of CNS Stem Cells. Springer-Verlag, Berlin.
11. Ankarcrona, M., J.M. Dypbukt, E. Bonfoco, B. Zhivotovsky, S. Orrenius, S.A. Lipton, and P. Nicotera. 1995. Glutamate-induced neuronal death: a succession of necrosis or apoptosis depending on mitochondrial function. Neuron 15:961-973.
12. Banki, K., E. Hutter, E. Colombo, N.J. Gonchoroff, and A. Perl. 1996. Glutathione levels and sensitivity to apoptosis are regulated by changes in transaldolase expression. J. Biol. Chem. 271:32994-33001.
13. Barbiero, G., F. Duranti, G. Bonelli, J.S. Amenta, and F.M. Baccino. 1997. Intracellular ionic variations in the apoptotic death of L cells by inhibitors of cell cycle progression. Exp. Cell Res. 217:410-418.
14. Barres, B.A., I.K. Hart, H.S.R. Coles, J.F. Burne, J.T. Voyvodic, W.D. Richardson, and M.C. Raff. 1992. Cell death and control of cell survival in the oligodendrocyte lineage. Cell 70:31-46.
15. Baum, H.-P., J. Reichrath, A. Theobald, and G. Schock. 1994. Fixation requirements for the immunohistochemical reactivity of PCNA antibody PC10 on cryostat sections. Histochem. J. 26:929-933.
16. Bayer, S.A. 1983. ^{3}H-Thymidine-radiographic studies of neurogenesis in the rat olfactory bulb. Exp. Brain Res. 50:329-340.
17. Belecky-Adams, T., B. Cook, and R. Adler. 1996. Correlation between terminal mitosis and differentiated fate of retinal precursor cells in vivo and in vitro: analysis with the "window-labeling" technique. Dev. Biol. 178:304-315.
18. Beppu, T., Y. Ishida, H. Arai, T. Wada, N. Uesugi, and K. Sasaki. 1994. Identification of S-phase cells with PC10 antibody to proliferating cell nuclear antigen (PCNA) by flow cytometric analysis. J. Histochem. Cytochem. 42:1177-1182.
19. Bernier, P.J. and A. Parent. 1998. Bcl-2 protein as a marker of neuronal immaturity in postnatal primate brain. J. Neurosci. 18:2486-2497.
20. Bonatz, G., J. Luttges, J. Hedderich, D. Inform, W. Jonat, P. Rudolph, and R. Parwaresch. 1999. Prognostic significance of a novel proliferation marker, anti-repp 86, for endometrial carcinoma: a multivariate study. Hum. Pathol. 30:949-956.
21. Bonfanti, L., P. Peretto, A. Merighi, and A. Fasolo. 1997. Newly generated cells from the rostral migratory stream in the accessory olfactory bulb of the adult rat. Neuroscience 81:489-502.
22. Bonfanti, L. and D.T. Theodosis. 1994. Expression of polysialylated neural cell adhesion molecule by proliferating cells in the subependymal layer of the adult rat, in its rostral extension and in the olfactory bulb. Neuroscience 62:291-305.
23. Bonfoco, E., D. Krainic, M. Ankarcrona, P. Nicotera, and S. Lipton. 1995. Apoptosis and necrosis: two distinct events induced, respectively, by mild and intense insults with N-methyl-D-aspartate or nitric oxide/superoxide in cortical cell cultures. Proc. Natl. Acad. Sci. USA 92:7162-7166.
24. Bravo, R. and H. Macdonald-Bravo. 1987. Existence

of two populations of cyclin/proliferating cell nuclear antigen during the cell cycle: association with DNA replication sites. J. Cell Biol. *105*:1549-1554.
25. Bredesen, D.E. 1995. Neural apoptosis. Ann. Neurol. *38*:839-851.
26. Carrio, R., M. Lopez-Hoyos, J. Jimeno, M.A. Benedict, R. Merino, A. Benito, J.L. Fernandez-Luna, G. Nunez, J.A. Garcia-Porrero, and J. Merino. 1996. A1 demonstrates restricted tissue distribution during embryonic development and functions to protect against cell death. Am. J. Pathol. *149*:2133-2142.
27. Celis, J.E., R. Bravo, P.M. Larsen, and S.J. Fey. 1984. Cyclin: a nuclear protein whose level correlates directly with the proliferative state of normal as well as transformed cells. Leukemia Res. *8*:143-157.
28. Chang, L.M.S. and F.J. Bollum. 1986. Molecular biology of terminal transferase. Crit. Rev. Biochem. *21*:27-52.
29. Chun, J. 1998. Apoptotic DNA fragmentation detection using ligation-mediated PCR, p. 23-33. In L. Zhu and J. Chun (Eds.), Apoptosis Detection and Assay Methods. Eaton Publishing, Natick, MA.
30. Clarke, P.G.H. 1990. Developmental cell death: morphological diversity and multiple mechanisms. Anat. Embryol. *181*:195-213.
31. Cunningham, T.J. 1982. Naturally occurring neuron death and its regulation by developing neural pathways. Int. Rev. Cytol. *74*:163-186.
32. Dachary-Prigent, J., J.M. Freyssinet, J.M. Pasquet, J.C. Carron, and A.T. Nurden. 1993. Annexin V as a probe of aminophospolipid exposure and platelet membrane vesiculation: a flow cytometry study showing a role for free sulfhydryl groups. Blood *81*:2554-2565.
33. Daly, J.M., C.B. Jannot, R.R. Beerli, D. Grasus-Porta, F.G. Maurer, and N.E. Hynes. 1997. Neu differentiation factor induces ErbB2 down-regulation and apoptosis of ErbB2-overexpressing breast tumor cells. Cancer Res. *57*:3804-3811.
34. Darzynkiewicz, Z., X. Li, and J. Gong. 1994. Assays of cell viability: discrimination of cells dying by apoptosis. Methods Cell Biol. *41*:15-38.
35. De Bilbao, F., E. Guarin, P. Nef, P. Vallet, P. Giannakopoulos, and M. Dubois-Dauphin. 1999. Postnatal distribution of cpp32/caspase 3 mRNA in the mouse central nervous system: an in situ hybridization study. J. Comp. Neurol. *409*:339-357.
36. DelRio, J.A. and E. Soriano. 1989. Immunocytochemical detection of 5′-bromodeoxyuridine incorporation in the central nervous system of the mouse. Dev. Brain Res. *49*:311-317.
37. Didenko, V.V. and P.J. Hornsby. 1996. Presence of double-strand breaks with single-base 3′ overhangs in cells undergoing apoptosis but not necrosis. J. Cell Biol. *135*:1369-1376.
38. Duchrow, C. Schluter, G. Key, M.H. Kubbutat, C. Wohlenberg, H.D. Flad, and J. Gerdes. 1995. Cell proliferation-associated nuclear antigen defined by antibody Ki-67: a new kind of cell cycle-maintaining proteins. Arch. Immunol. Ther. Exp. *43*:117-121.
39. Fadok, V.A., D. Voelker, P.A. Campbell, J.J. Cohen, D.L. Bratton, and P.M. Henson. 1992. Exposure of phosphatidylserine on the surface of apoptotic lypmphocytes triggers specific recognition and removal by macrophages. J. Immunol. *148*:2207-2216.
40. Fernandez-Alnemri, T., G. Litwack, and E.S. Alnemri. 1994. CPP32, a novel human apoptotic protein with homology to *Caenorhabditis elegans* cell death protein Ced-3 and mammalian interleukin-1-beta-converting enzyme. J. Biol. Chem. *269*:30761-30764.
41. Fernandez, A., J. Kiefer, L. Fosdick, and D.J. McConkey. 1995. Oxygen radical production and thiol depletion are required for $Ca^{(2+)}$-mediated endogenous endonuclease activation in apoptotic thymocytes. J. Immunol. *155*:5133-5139.
42. Fesus, L., Z. Nemes, L. Piredda, A. Madi, M. di Rao, and M. Picentini. 1987. Induction and activation of tissue transglutaminase during programmed cell death. FEBS Lett. *224*:104-108.
43. Frade, J.-M., P. Bovolenta, J.R. Martinez-Morales, A. Arribas, J.A. Barbas, and A. Rodríguez-Tébar. 1997. Control of early cell death by BDNF in the chick retina. Development *124*:3313-3320.
44. France-Lanord, V., B. Brugg, P.P. Michel, Y. Agid, and M. Ruberg. 1997. Mitochondrial free radical signal in ceramide-dependent apoptosis: a putative mechanism for neuronal death in Parkinson's disease. J. Neurochem. *69*:1612-1621.
45. Frankfurt, O.S. 1998. Detection of apoptotic cells with monoclonal antibodies to single-stranded DNA, p. 47-62. In L. Zhu and J. Chun (Eds.), Apoptosis Detection and Assay Methods. Eaton Publishing, Natick, MA.
46. Gage, F.H. 1998. Stem cells of the central nervous system. Curr. Opin. Neurobiol. *8*:671-676.
47. Galli-Resta, L. and M. Ensini. 1996. An intrinsic time limit between genesis and death of individual neurons in the developing retinal ganglion cell layer. J. Neurosci. *16*:2318-2324.
48. Gates, M.A., L.B. Thomas, E.M. Howard, E.D. Laywell, B. Sajin, A. Faissner, B. Gotz, J. Silver, and D.A. Steindler. 1995. Cell and molecular analysis of the developing and adult mouse subventricular zone of the cerebral hemispheres. J. Comp. Neurol. *361*:249-266.
49. Gavrieli, Y., Y. Sherman, and S.A. Ben-Sasson. 1992. Identification of programmed cell death in situ via specific labeling of nuclear DNA fragmentation. J. Cell Biol. *119*:493-501.
50. Gerdes, J., U. Schwab, H. Lemke, and H. Stein. 1983. Production of a mouse monoclonal antibody reactive with a human antigen associated with cell proliferation. Int. J. Cancer *31*:13-20.
51. Gold, R., M. Schmidt, G. Rothe, H. Zischler, H. Breitschopf, H. Wekerle, and H. Lassmann. 1993. Detection of DNA fragmentation in apoptosis: application of in situ nick translation to cell culture systems and tissue sections. J. Histochem. Cytochem. *41*:1023-1030.
52. Hall, P.A., P.J. Coates, R.A. Goodlad, I.R. Hart, and D.P. Lane. 1994. Proliferating cell nuclear antigen expression in non-cycling cells may be induced by growth factors *in vivo*. Br. J. Cancer *70*:244-247.
53. Hines, M.D. and B.L. Allen-Hoffmann. 1996. Keratinocyte growth factor inhibits cross-linked envelope formation and nucleosomal fragmentation in cultured human keratinocytes. J. Biol. Chem. *271*:6245-6251.

54. Holt, S.E. and J.V. Shay. 1999. Role of telomerase in cellular proliferation and cancer. J. Cell. Physiol. 180:10-18.
55. Jin, K., J. Chen, T. Nagayama, M. Chen, J. Sinclair, S.H. Graham, and R.P. Simon. 1999. In situ detection of neuronal DNA strand breaks using the Klenow fragment of DNA polymerase I reveals different mechanisms of neuron death after global cerebral ischemia. J. Neurochem. 72:1204-1214.
56. Kerr, J.F., A.H. Wyllie, and A.R. Currie. 1972. Apoptosis: a basic biological phenome with wide-ranging implications in tissue kinetics. Br. J. Cancer 26:239-257.
57. Kondo, K. and T. Makita. 1996. Electron microscopic immunogold labeling of bromodeoxyuridine (BrdU) in routine electron microscopy. Acta Histochem. Cytochem. 29:115-119.
58. Korhonen, L., S. Hamner, P.A. Olsson, and D. Lindholm. 1997. Bcl-2 regulates the levels of the cysteine proteases ICH and CPP32/Yama in human neuronal precursor cells. Eur. J. Neurosci. 9:2489-2496.
59. Krueger, B.K., J.F. Burne, and M.C. Raff. 1995. Evidence for large-scale astrocyte death in the developing cerebellum. J. Neurosci. 15:3366-3374.
60. Lee, S., S. Christakos, and M.B. Small. 1993. Apoptosis and signal transduction: clues to a molecular mechanism. Curr. Opin. Cell Biol. 5:286-291.
61. Lois, C., J.-M. Garcia-Verdugo, and A. Alvarez-Buylla. 1996. Chain migration of neuronal precursors. Science 271:978-981.
62. Lossi, L., S. Mioletti, and A. Merighi. 2000. Undifferentiated progenitors and granule cell precursors undergo apoptosis within a few hours after their generation, but represent an endogenous source of brain-derived neurotrophic factor for the post-natal cerebellar cortex. Soc. Neurosci. Abstr. 26:598.
63. Lossi, L., D. Zagzag, M.A. Greco, and A. Merighi. 1998. Apoptosis of undifferentiated progenitors and granule cell precursors in the post-natal human cerebellar cortex correlates with expression of BCL-2, ICE and CPP-32 proteins. J. Comp. Neurol. 399:359-372.
64. Luskin, M.B. 1993. Restricted proliferation and migration of postnatally generated neurons derived from the forebrain subventricular zone. Neuron 11:173-189.
65. Luskin, M.B. 1997. Characterization of neural progenitor cells of the neonatal forebrain, p. 67-86. In F.H. Gage and Y. Christen (Eds.), Isolation, Characterization and Utilization of CNS Stem Cells. Springer-Verlag, Berlin.
66. Meisenholder, G.W., S.G. Martin, D.R. Green, J. Nordber, B.M. Babior, and R.A. Gottlieb. 1996. Events in apoptosis. Acidification is downstream of protease activation and bcl-2 protection. J. Biol. Chem 271:16260-12262.
67. Migheli, A., A. Attanasio, and D. Schiffer. 1995. Ultrastructural detection of DNA strand beaks in apoptotic neural cell by in situ end-labelling techniques. J. Pathol. 176:27-35.
68. Miller, M.A., C.M. Mazewski, Y. Sheikh, L.M. White, G.A. Yanik, D.M. Hyams, B.C. Lampkin, and A. Raza. 1991. Simultaneous immunohistochemical detection of IUdR and BrdU infused intravenously to cancer patients. J. Histochem. Cytochem. 39:407-412.
69. Miller, M.W. and R.S. Novakowski. 1988. Use of bromodeoxyuridine-immunohistochemistry to examine the proliferation, migration and time of origin of cells in the central nervous system. Brain Res. 457:44-52.
70. Mitev, V., A. Pauloin, and L.M. Houdebine. 1994. Purification and characterization of two casein kinase type II isozymes from bovine gray matter. J. Neurochem. 63:717-726.
71. Muto, M., M. Utsuyama, T. Horiguchi, E. Kubo, T. Sado, and K. Hirokawa. 1995. The characterization of the monoclonal antibody Th-10a, specific for a nuclear protein appearing in the S phase of the cell cycle in normal thymocytes and its unregulated expression in lymphoma cell lines. Cell Prolif. 28:645-657.
72. Nakajima, T., K. Kagawa, T. Deguchi, H. Hikita, T. Okanoue, K. Kashima, E. Konishi, and T. Ashihara. 1994. Biological role of proliferating cell nuclear antigen (PCNA) expression during rat liver regeneration. Acta Histochem. Cytochem. 27:135-140.
73. Naot, D., R.V. Sionov, and D. Ish-Shalom. 1997. CD44: structure, function, and association with the malignant process. Adv. Cancer Res. 71:241-319.
74. Nicholson, D.W., A. Ali, N.A. Thornberry, J.P. Vaillancourt, C.K. Ding, M. Gallant, Y. Gareau, and P.R. Griffin. 1995. Identification and inhibition of the ICE/CED-3 protease necessary for mammalian apoptosis. Nature 376:37-43.
75. Nicotera, P. and A.D. Rossi. 1994. Nuclear Ca^{2+}: physiological regulation and role in apoptosis. Mol. Cell Biochem. 135:89-98.
76. Nishikawa, S. and F. Sasaki. 1995. DNA localization in nuclear fragments of apoptotic ameloblasts using anti-DNA immunoelectron microscopy: programmed cell death of ameloblasts. Histochem. Cell. Biol. 104:151-159.
77. Oppenheim, R.W. 1985. Naturally occurring cell death during neural development. Trends Neurosci. 8:487-493.
78. Peretto, P., A. Merighi, A. Fasolo, and L. Bonfanti. 1996. Glial tubes in the rostral migratory stream of the adult rat. Brain Res. Bull. 42:9-21.
79. Peretto, P., A. Merighi, A. Fasolo, and L. Bonfanti. 1999. The subependymal layer in rodents: a site of structural plasticity and cell migration in the adult mammalian brain. Brain Res. Bull. 49:221-243.
80. Piacentini, M., L. Fesus, M.G. Farrace, L. Ghibelli, L. Piredda, and G. Melino. 1991. The expression of "tissue" transglutaminase in two human cancer cell lines is related with programmed cell death (apoptosis). Eur. J. Cell Biol. 54:246-254.
81. Polakowska, R.R., M. Piacentini, R. Batlett, L.A. Goldsmith, and A.R. Haake. 1994. Apoptosis in human skin development: morphogenesis, periderm and stem cells. Dev. Dynamics 199:176-188.
82. Raff, M.C., B.A. Barres, J.F. Burne, H.S.R. Coles, Y. Ishizaki, and M.D. Jacobson. 1994. Programmed cell death and the control of cell survival. Philos. Trans. R. Soc. Lond. [Biol] 345:265-268.
83. Rakic, P. 1996. Radial versus tangential migration of neuronal clones in the developing cerebral cortex. Proc. Natl. Acad. Sci. USA 92:11323-11327.
84. Ray, S., T.J. Kelley, S. Campion, A.P. Seve, and S.

Basu. 1991. Developmental expression of embryonic chicken brain DNA polymerase alpha and its binding with monoclonal antibodies against human KB cell DNA polymerase alpha. Cell Growth Differ. *2*:567-573.

85. Reynolds, J.E. and A. Eastman. 1996. Intracellular calcium stores are not required for Bcl-2-mediated protection from apoptosis. J. Biol. Chem. *271*:27739-27743.

86. Rubini, M., C. D'Ambrosio, S. Carturan, G. Yumet, E. Catalano, S. Shan, Z. Huang, M. Criscuolo, M. Pifferi, and R. Baserga. 1999. Characterization of an antibody that can detect an activated IGF-I receptor in human cancers. Exp. Cell Res. *251*:22-32.

87. Rudolph, P., R. Knuchel, E. Endl, H.J. Heidebrecht, F. Hofstader. and R. Parwaresch. 1998. The immunohistochemical marker Ki-S2: cell cycle kinetics and tissue distribution of a novel proliferation-specific antigen. Mod. Pathol. *11*:450-456.

88. Ryder, E.F. and C.L. Cepko. 1994. Migration patterns of clonally related granule cells and their progenitors in the developing chick cerebellum. Neuron *12*:1011-1029.

89. Sadi, M.V. and E.R. Barrack. 1991. Determination of growth fraction in advanced prostate cancer by Ki-67 immunostaining and its relationship to the time of tumor progression after hormonal therapy. Cancer *67*:3065-3071.

90. Sallinen, P., H. Haapasalo, T. Kerttula, I. Rantala, H. Kalimo, Y. Collan, J. Isola, and H. Helin. 1994. Sources of variation in the assesment of cell proliferation using proliferating cell nuclear antigen histochemistry. Anal. Quant. Cytol. and Histol. *16*:261-268.

91. Santisteban, M.S. and G. Brugal. 1994. Image analysis of *in situ* cell cycle related changes of PCNA and Ki-67 proliferating antigen expression. Cell Prolif. *27*:435-453.

92. Sasaki, K., T. Murakami, M. Kawasaki, and M. Takahashi. 1987. The cell cycle associated change of the Ki-67 reactive nuclear antigen expression. J. Cell. Physiol. *133*:579-584.

93. Sato, Y., T. Ito, A. Nozawa, and M. Kanisawa. 1995. Bromodeoxyuridine and iododeoxyuridine double immunostaining for epoxy resin sections. Biotech. Histochem. *70*:169-174.

94. Schlegel, J., I. Peter, S. Orrenius, D.K. Miller, N.A. Thornberry, T.T. Yamin, and D.W. Nicholson. 1996. CPP32/apopain is a key interleukin 1 beta converting-like protease involved in Fas-mediated apoptosis. J. Biol. Chem. *271*:1841-1844.

95. Schluter, C., M. Duchrow, C. Vohlenberg, M.H. Becker, G. Key, H.D. Flad, and J. Gerdes. 1993. The cell proliferation-associated antigen of antibody Ki-67: a very large, ubiquitous nuclear protein with numerous repeated elements, representing a new kind of cell cycle-maintaining proteins. J. Cell Biol. *123*:513-522.

96. Schwartz, L.M., S.W. Smith, M.E. Jones, and B.A. Osborne. 1993. Do all programmed cell death occur via apoptosis? Proc. Natl. Acad. Sci. USA *90*:980-984.

97. Simonati, A., C. Tosati, T. Rosso, E. Piazzola, and N. izzuto. 1999. Cell proliferation and death: morphological evidence during corticogenesis in the developing human brain. Microsc. Res. Tech. *45*:341-352.

98. Sirri, V., P. Roussel, and D. Hernandez-Verdun. 2000. The AgNOR proteins: qualitative and quantitative changes during the cell cycle. Micron *31*:121-126.

99. Soriano, E., J.A. Del Rio, and C. Auladell. 1993. Characterization of the phenotype and birthdates of pyknotic dead cells in the nervous system by a combination of DNA staining and immunocytochemistry for 5′-bromodeoxyuridine and neural antigens. J. Histochem. Cytochem. *41*:819-827.

100. Soriano, E., J.A. Del Rio, F.J. Martinez-Guijarro, I. Ferrer, and C. Lopez. 1991. Immunocytochemical detection of 5′-bromodeoxyuridine in fluoro-gold-labeled neurons: a simple technique to combine retrograde axonal. J. Histochem. Cytochem. *39*:1565-1570.

101. Spreafico, R., P. Arcelli, C. Frassoni, P. Canetti, G. Giaccone, T. Rizzuti, M. Mastrangelo, and M. Bentivoglio. 1999. Development of layer I of the human cerebral cortex after midgestation: architectonic findings, immunocytochemical identification of neurons and glia, and in situ labeling of apoptotic cells. J. Comp. Neurol. *410*:126-142.

102. Srinivasan, A., K.A. Roth, R.O. Sayers, K.S. Schindler, A.M. Wong, L.C. Fritz, and K.J. Tomaselli. 1998. In situ immunodetection of activated caspase-3 in apoptotic neurons in the developing nervous system. Cell Death Differ. *5*:1004-1016.

103. Takahashi, T. and V.S. Caviness, Jr. 1993. PCNA-binding to DNA at the G1/S transition in proliferating cells of the developing cerebral wall. J. Neurocytol. *22*:1096-1102.

104. Tamatani, R., Y. Taniguchi, and Y. Kawarai. 1995. Ultrastructural study of proliferating cells with an improved immunocytochemical detection of DNA-incorporated bromodeoxyuridine. J. Histochem. Cytochem. *43*:21-29.

105. Tampanaru-Sarmesiu, A., L. Stefaneau, K. Thapar, G. Kontogeorgos, T. Sumi, and K. Kovacs. 1998. Transferrin and transferrin receptor in human hypophysis and pituitary adenomas. Am. J. Pathol. *152*:413-422.

106. Tanaka, M. and T. Marunouchi. 1998. Immunohistochemical analysis of developmental stage of external granular layer neurons which undergo apoptosis in postnatal rat cerebellum. Neurosci. Lett. *242*:85-88.

107. Thiry, M. 1992. Highly sensitive immunodetection of DNA on sections with exogenous terminal deoxynucleotidyl transferase and non-isotopic nucleotide analogues. J. Histochem. Cytochem. *40*:411-419.

108. Thiry, M. 1993. Immunodetection of RNA on ultrathin sections incubated with polydenylated nucleotidyl tranferase. J. Histochem. Cytochem. *41*:657-665.

109. Thiry, M. 1995. Nucleic acid compartmentalization within the cell nucleus by in situ transferase-immunogold techniques. Microsc. Res. Tech. *31*:4-21.

110. Thiry, M. and F. Puvion Dutilleul. 1995. Differential distribution of single-stranded DNA, double- stranded DNA, and RNA in adenovirus-induced intranuclear regions of HeLa cells. J. Histochem. Cytochem. *43*:749-759.

111. Torunsciolo, D.R.Z., R.E. Schimdt, and K.A. Roth. 1995. Simultaneous detection of TdT-mediated dUTP-biotyn nick end-labeling (TUNEL) positive

112. Trere, D. 2000. AgNOR staining and quantification. Micron 31:127-131.
113. Tsanaclis, A.M., S.S. Brem, S. Gately, H.M. Shipper, and R. Wang. 1991. Statin immunolocalization in human brain tumors. Detection of noncycling cells using a novel marker of cell quiescence. Cancer 68:786-792.
114. Tsurusawa, M., M. Ito, Z. Zha, S. Kawai, Y. Takasaki, and T. Fujimoto. 1992. Cell-cycle-associated expressions of proliferating cell nuclear antigen and Ki-67 reactive antigen of bone marrow blast cells in childhood acute leukemia. Leukemia 6:669-674.
115. Tsutsui, K., S. Okada, M. Watanabe, T. Shohmori, S. Seki, and Y. Inoue. 1993. Molecular cloning of partial cDNAs for rat DNA topoisomerase II isoforms and their differential expression in brain development. J. Biol. Chem. 268:19076-19083.
116. Ulaner, G.A. and L.C. Giudice. 1997. Developmental regulation of telomerase activity in human fetal tissues during gestation. Mol. Hum. Reprod. 3:769-773.
117. van Dierendonck, J.H., R. Verheijen, H.J. Kuijpers, R. van Driel, J.L. Beck, G.J. Brakenhoff, and F.C. Ramaekers. 1989. Ki-67 detects a nuclear matrix-associated proliferation-related antigen. II. Localization in mitotic cells and association with chromosomes. J. Cell Sci. 92:531-540.
118. van Dierendonck, J.H., J.H. Wijsman, R. Keijzer, C.J. van de Velde, and C.J. Cornelisse. 1991. Cell-cycle-related staining patterns of anti-proliferating cell nuclear antigen monoclonal antibodies. Comparison with BrdUrd labeling and Ki-67 staining. Am. J. Pathol. 138:1165-1172.
119. Villalba, M., J. Boeckart, and L. Journot. 1997. Pituitary adenylate cyclase-activating polypeptide (PACAP-38) protects cerebellar granule neurons from apoptosis by activating the mitogen-activated protein kinase (MAP Kinase) pathway. J. Neurosci. 17:83-90.
120. Wang, X., N.G. Zelenski, Y. Yang, J. Sakai, M.S. Brown, and J.L. Goldstein. 1996. Cleavage of sterol regulatory element binding proteins (SREBPs) by CPP32 during apoptosis. EMBO J. 15:1012-1020.
121. Waseem, N.H. and D.P. Lane. 1990. Monoclonal antibody analysis of the proliferating cell nuclear antigen (PCNA). Structural conversation and the detection of a nucleolar form. J. Cell Sci. 96:121-129.
122. Watanabe, M., K. Tsutsui, K. Tsutsui, and Y. Inoue. 1994. Differential expression of the topoisomerase II alpha and II beta mRNAs in developing rat brain. Neurosci. Res. 19:51-57.
123. Wood, K.A., B. Dipasquale, and R.J. Youle. 1993. In situ labeling of granule cells for apoptosis-associated DNA fragmentation reveals different mechanisms of cell loss in developing cerebellum. Neuron 11:621-632.
124. Wood, K.A. and R.J. Youle. 1995. The role of free radicals and p53 in neuron apoptosis in vivo. J. Neurosci. 15:5851-5857.
125. Wright, W.E., M.A. Piatyszek, W.E. Rainey, W. Byrd, and J.W. Shay. 1996. Telomerase activity in human germline and embryonic tissues and cells. Dev. Genet. 18:173-179.
126. Wyllie, A.H. 1980. Glucocorticoid-induced tymocyte apoptosis is associated with endogenous endonuclease activation. Nature 284:555-556.
127. Wyllie, A.H., R.G. Morris, A.L. Smith, and D. Dunlop. 1984. Chromatin cleavage in apoptosis: association with condensed chromatin morphology and dependence on macromolecular synthesis. J. Pathol. 142:67-77.
128. Xiong, Y., H. Zhang, and D. Beach. 1992. D type cyclins associate with multiple protein kinases and the DNA replication and repair factor PCNA. Cell 71:505-514.
129. Yamaguchi, Y., K.S.E. Nozawa, N. Hayakawa, Y. Nimura, and S. Yoshida. 1998. Change in telomerase activity of rat organs during growth and aging. Exp. Cell Res. 242:120-127.
130. Yanik, G.A., N. Yousuf, M.A. Miller, S.H. Swerdlow, B.C. Lampkin, and A. Raza. 1992. In vivo determination of cell cicle kinetics of non-Hodgkin's lymphomas using iododeoxyuridine and bromodeoxyuridine. J. Histochem. Cytochem. 40:723-728.
131. Yao, R. and G.M. Cooper. 1995. Requirement for phosphatidylinositol-3 kinase in the prevention of apoptosis by nerve growth factor. Science 267:2003-2006.
132. Yuan, J. 1995. Molecular control of life and death. Curr. Opin. Cell Biol. 7:211-214.
133. Zakeri, Z. and R.A. Lockshin. 1994. Physiological cell death during development and its relationship to aging. Ann. NY Acad. Sci. 719:212-229.
134. Zandvilit, D.W., A.M. Hanby, C.A. Austin, K.L. Marsch, I.B. Clak, N.A. Wright, and R. Poulsom. 1996. Analysis of foetal expression sites of human type II DNA topoisomerase alpha and beta mRNAs by in situ hybridization. Biochim. Biophys. Acta 1307:239-247.
135. Zhang, C. 1998. Monoclonal antibody as a probe for characterization and separation of apoptotic cells, p. 63-73. In L. Zhu and J. Chun (Eds.), Apoptosis Detection and Assay Methods. Eaton Publishing, Natick, MA.
136. Zhang, G., V. Gurtu, J.-T. Ma, and S.R. Kain. 1998. Sensitive detection of apoptosis using enhanced color variants of Annexin V conjugates, p. 1-6. In L. Zhu and J. Chun (Eds.), Apoptosis Detection and Assay Methods. Eaton Publishing, Natick, MA.
137. Zhang, G., V. Gurtu, C. Spencer, J.-T. Ma, and S.R. Kain. 1998. Detection of caspase activity associated with apoptosis using fluorimetric and colorimetric methods, p. 7-14. In L. Zhu and J. Chun (Eds.), Apoptosis Detection and Assay Methods. Eaton Publishing, Natick, MA.
138. Zhang, L. and J.E. Goldman. 1996. Generation of cerebellar interneurons from dividing progenitors in the white matter. Neuron 16:47-54.
139. Zurita, F., R. Himenez, R. Diaz de laGuardia, and M. Burgos. 1999. The relative rDNA content of a NOR determines its level of expression and its probability of becoming active. A sequential silver staining and in situ hybridization study. Chromosome Res. 7:563-570.

15 Confocal Imaging of Nerve Cells and Their Connections

Andrew J. Todd
Laboratory of Human Anatomy, Institute of Biomedical and Life Sciences, University of Glasgow, Glasgow, UK, EU

OVERVIEW

This chapter outlines the use of confocal microscopy combined with immunofluorescence staining for examining nerve cells in the central nervous system (CNS) and studying the synaptic connections between them. A method for carrying out immunolabeling with two or three different fluorescent dyes is described, together with modifications that allow this to be combined with tract tracing, intracellular injection, or lectin labeling, and also a new technique for combining confocal and electron microscopy (EM) on sections which have been processed for immunocytochemistry. With these approaches, it is possible to search for and quantify contacts between neurons at the light microscopic level and then examine representative contacts at the ultrastructural level to determine whether synapses are present.

BACKGROUND

Immunocytochemistry has been used extensively in anatomical studies of the CNS. Early investigators used single antibodies, for example, to demonstrate the distribution of neurons which used a particular neurotransmitter, however, more recently, the detection of two or even three different antigens has become common. This approach (multiple immunolabeling) has many applications in neurobiology. For example, it allows the demonstration of two transmitters in a single neuron, which is important for studies of cotransmission, and permits the association between neurotransmitters and their receptors to be examined (20). Multiple immunolabeling has also proved very useful as a way of investigating the connections between nerve cells which form the basis of neuronal circuits within the CNS. If two populations of neurons can be revealed with different antibodies, then immunocytochemistry can be used to determine whether, for example, the axons of one population form synapses with the cell bodies or dendrites of the other.

There are several ways of identifying more than one antigen in a piece of tissue, however the simplest (both conceptually and in practice) is to apply two or more (primary) antibodies raised in different species to a single section, and then reveal them with different labels. These labels (for example fluorescent dyes) are usually attached to secondary antibodies (anti-immunoglobulins) which bind to the primary antibodies when applied to the section. Because secondary antibodies can be made specific for the immunoglobulin of a particular species (e.g., rabbit, mouse, rat), it is possible to reveal each primary antibody with a different label. This approach has become more straightforward over the last few years because of the increased availability of primary antibodies raised in a variety of different species, together with the development of secondary antibodies which show a high degree of species specificity. While it is possible to use this approach with brightfield microscopy, by revealing each antigen with a different colored dye, a more satisfactory method is to use fluorescent dyes, since it is possible to excite these selectively by means of appropriate filter sets, which means that each fluorescent dye can be viewed independently of any others. Confocal microscopy has provided a particularly useful way of examining such sections for several reasons (3). Firstly, it provides extremely good spectral separation of different fluorescent dyes with minimal "bleed-through" fluorescence, which means that each antigen can be identified with little interference from the others. Secondly, the very limited depth of focus permits thin "optical sections" through a thicker (e.g., Vibratome) section to be viewed, which results in a greatly improved image with increased spatial resolution. Thirdly, these optical sections can be reconstructed to provide three-dimensional information about the distribution of immunostaining with each antibody and the relationship between immunolabeled structures.

In studies of neuronal circuitry, it is necessary to use EM in order to confirm that synapses are present at points of contact between pairs of neurons (see also Chapter 10). Although the high resolution of the electron microscope permits unequivocal identification of synapses, the time-consuming nature of the technique together with the small size of the volumes of tissue that can be examined make it difficult to study large numbers of contacts. Thus, although it may be possible to determine whether synapses are present between two neurons, it is not usually feasible to determine their frequency or the spatial distribution on the postsynaptic cell. While confocal microscopy does not allow identification of synapses, it is still extremely valuable in studies of neuronal connections, since it possible to examine entire neurons with the confocal microscope. This means that the density of contacts which these neurons receive from particular types of axon can be determined, together with the distribution of such contacts on different parts of the cell body or dendritic tree of the target neuron. It is clear that confocal microscopy and EM, therefore, have complementary strengths. Confocal microscopy can be used to sample contacts over whole neurons, while EM allows confirmation that synapses are present at these contacts. In principle, there are two ways of designing a study that involves both confocal microscopy and EM of immunostained material: *(i)* one can prepare different tissue for each technique, using immunofluorescent labeling for confocal microscopy and electron-dense markers for EM; or *(ii)* one can attempt to carry out both techniques on the same tissue section. Although the former approach is technically easier, there are some advantages to carrying out combined confocal microscopy and EM on the same tissue (see below), and we have recently developed a method which allows this to be done (14,15,19).

If the pre- and postsynaptic components of a neuronal circuit possess different antigens which allow them to be distinguished, then it is possible to use immunocytochemistry and confocal microscopy to search for contacts between the two neurons without the need for any experimental intervention. Many different types of antibody can be used in studies of this kind, including antibodies directed against neurotransmitters (or their synthetic enzymes), neuropeptides, or receptors. The choice of antigens–antibodies will obviously be determined by the types of neurons that are to be examined, by the subcellular distribution of antigens (for example some antigens may only be present in the axon or in the soma and dendrites of the neuron to be investigated), and also by the availability of antibodies which have been raised in different species and can therefore be used in combination.

Although immunocytochemistry alone can be used to examine some circuits within the CNS, it is often necessary to combine it with other techniques, such as tract tracing methods or intracellular injection. The use of anterograde or retrograde tract tracing allows circuits that involve identified projection neurons to be examined (see Chapters 12 and 13). A simple way of combining tract tracing with immunofluorescence for studies of neuronal circuitry is to use tracers such as unconjugated cholera toxin B subunit (CTb) or *Phaseolus* leucoagglutinin (PHA-L), which can be detected with antibodies (14,15). Dextrans can also be used as tracers in studies of this kind (9). They can either be conjugated directly to a fluorescent dye, or else to biotin, in which case they are subsequently revealed with avidin conjugated to a suitable fluorochrome. By carrying out immunofluorescence on sections which contain neurons that have been labeled by intracellular injection, the synaptic inputs to or outputs from physiologically identified neurons can be examined (5,13). Two compounds that have been used as intracellular markers in studies of this kind are neurobiotin (which is subsequently detected with an avidin–fluorochrome conjugate) (13) and rhodamine–dextran (5). It is also possible to inject the fluorescent dye Lucifer Yellow into neurons (1), however, Lucifer Yellow has very broad excitation and emission spectra, which make it less suitable in combination with immunofluorescence staining.

In certain situations, lectin binding can be used to identify neuronal populations, for example, unmyelinated primary afferents in the spinal dorsal horn are able to bind a variety of lectins, including *Bandeiraea simplicifolia* isolectin B4 (BSI-B4). The lectin, conjugated either to biotin or to a fluorescent dye, can be applied to tissue sections during immunocytochemical processing in order to reveal these afferents (18,21).

PROTOCOLS

Protocol for Double or Triple Labeling Immunofluorescence

Materials and Reagents

- Phosphate buffer (PB), 0.2 M stock: 0.2 M NaH_2PO_4/Na_2HPO_4 mixed to pH 7.4.
- Fixative: 4% freshly-depolymerized paraformaldehyde in 0.1 M PB. Heat 400 mL distilled water to 60°C and add 40 g paraformaldehyde. Add concentrated NaOH dropwise until suspension clears. Add 500 mL 0.2 M PB and make up to 1 L with distilled water. Filter and use within 24 hours.
- Phosphate-buffered saline (PBS): 0.01 M NaH_2PO_4/Na_2HPO_4 to pH 7.4, 0.3 M NaCl. Note that this solution contains a relatively high concentration of NaCl. We have found that for certain antibodies, the higher ionic strength greatly improves immunostaining by reducing background staining, and we now routinely use this recipe.
- PBS with Triton® X-100 (PBST): PBS, 0.3% Triton X-100.

- Appropriate primary and secondary antibodies [we use secondary antibodies raised in donkey (Jackson ImmunoResearch, West Grove, PA, USA), see below].
- Glycerol-based antifade mounting medium (e.g., Vectashield; Vector Laboratories, Peterborough, UK).

Procedure

1. Fix animal by perfusion with 4% formaldehyde under terminal anesthesia. Remove blocks of CNS and store in same fixative for 2 to 24 hours. Some antigens do not tolerate extended fixation, and in cases of poor or absent immunostaining when using a new antibody, it is worth reducing the fixation time to around 2 hours.
2. Rinse blocks in PB.
3. Cut Vibratome sections (40–70 μm) into PB.
4. Place sections into 50% ethanol in distilled water for 30 minutes immediately after cutting (10). This step greatly increases penetration of antibodies into Vibratome sections.
5. Rinse 3 × 5 minutes in PBS.
6. Incubate 18 to 72 hours in cocktail of primary antibodies (each raised in a different species) at appropriate dilutions in PBST.

   Note: Although many published methods recommend adding normal blocking serum (from the species in which the secondary antibodies were raised) to the antibody cocktails in order to reduce background staining, we have not found any evidence that this improves the quality of immunostaining, and so we do not use blocking serum.
7. Rinse 3 × 5 minutes in PBS.
8. Incubate 2 to 24 hours in cocktail of fluorescent species-specific secondary antibodies diluted in PBST. It is vital that the secondary antibodies are raised in different species from those of the primary antibodies. For convenience, we routinely use secondary antibodies raised in donkey. Donkey has become a "universal donor" for secondary antibodies in multiple-immunolabeling studies, since few (if any) primary antibodies are raised in this species. Obviously, the secondary antibodies chosen must be directed against the species in which the primary antibodies were raised, and each must be conjugated to a different fluorescent dye. The choice of fluorescent dyes depends on the wavelength of the laser lines (3). We use fluorescein isothiocyanate (FITC), lissamine rhodamine (LRSC), and cyanine 5.18 (Cy5™), since these are optimally excited by the lines of a Krypton-Argon laser (488 nm, 568 nm, 647 nm, respectively).
9. Rinse 3 × 5 minutes in PBS.
10. Mount sections on glass slides with an appropriate antifade medium. This reduces the rate of photobleaching, which is otherwise a problem, particularly with FITC.
11. Apply a coverslip and seal around the edges with nail varnish.
12. Store slides in a -20°C freezer until needed. Slides prepared in this way can be kept for extended periods (several years).
13. Examine with the confocal microscope.

Notes for Combining Immunofluorescence Staining with Other Techniques

Tract Tracing Experiments (see also Chapter 12)

The following additional reagents may be needed according to the type of tracer employed:

- Rabbit antibodies against CTb and PHA-L (Sigma, St. Louis, MO, USA).

- Goat and rabbit antibodies against PHA-L (Vector Laboratories).
- Goat antibody against CTb (List Biological Laboratories, Campbell, CA, USA).
- Streptavidin–florochrome conjugates (Jackson ImmunoResearch).

If unconjugated tracers (e.g., CTb, PHA-L) have been used, these are detected by including an appropriate antibody against the tracer in the primary antibody cocktail. If a biotinylated tracer (e.g., biotin–dextran) has been used, then a suitable streptavidin–fluorochrome conjugate (diluted 1:1000) is included in the cocktail of fluorescent secondary antibodies.

Intracellular Injection

If a fluorescent dye (e.g., rhodamine–dextran or Lucifer Yellow) has been used, sections are processed exactly as described in the basic protocol. If neurobiotin was injected into the cell, this is detected with avidin conjugated to a fluorescent dye, as described above.

Lectins

A biotinylated lectin [e.g., BSI-B4 (Sigma) 1 µg/mL] can be included in the cocktail of primary antibodies, and this is detected with avidin conjugated to a fluorescent dye.

Protocol for Combined Confocal and Electron Microscopy of Immunostained Sections

General Principles

The initial stages of this procedure are the same as those described in the protocol for double or triple labeling immunofluorescence above, except that *(i)* the fixative is modified to include 0.2% glutaraldehyde, which is necessary to provide satisfactory EM fixation; *(ii)* an additional stage, treatment with $NaBH_4$, is included to reduce the deleterious effects of glutaraldehyde on antigenicity (7); and *(iii)* Triton X-100 is omitted from all diluents, since the use of detergents gives poor ultrastructure.

Materials and Reagents

- Fixative: 4% formaldehyde, 0.2% glutaraldehyde in 0.1 M PB. Heat 400 mL distilled water to 60°C and add 40 g paraformaldehyde. Add concentrated NaOH dropwise until suspension clears. Add 8 mL 25% glutaraldehyde (EM grade) and 500 mL 0.2 M PB. Make up to 1 L with distilled water. Filter and use within 24 hours.
- 3,3′-Diaminobenzidine tetrahydrochloride (DAB) solution: 25 mg DAB, 15 µL 30% H_2O_2, 50 mL PB. Filter and use immediately.

Note: Extreme care must be used in working with DAB, which is a suspected carcinogen. Any contaminated glassware, surfaces, etc., should be treated with bleach to inactivate the DAB.

- Durcupan: 10 g Durcupan ACM resin (Fluka Chemicals, Gillingham, UK), 10 g dodecanyl succinic anhydride (Agar Scientific, Stansted, UK), 0.3 g dibutyl phthalate (Agar Scientific), 0.3 g DMP-30 (2,4,6,tri[dimethylaminomethyl]phenol) (Agar Scientific).

Procedure

1. Fix animal by perfusion with 4% formaldehyde/0.2% glutaraldehyde under terminal anesthesia. Remove blocks of CNS and store in same fixative for 2 to 24 hours.
2. Rinse blocks in PB.
3. Cut Vibratome sections (40–70 µm) into PB.
4. Place sections into 50% ethanol in dis-

tilled water for 30 minutes immediately after cutting.
5. Rinse 3 × 5 minutes in PBS.
6. Place sections in 1% NaBH$_4$ in PBS for 30 minutes. Leave lids off.
   Note: NaBH$_4$ is toxic and potentially explosive.
7. Rinse thoroughly in PBS (e.g., 10× over 90 minutes). The lids can be placed on bottles once bubbles cease to appear.
8. Incubate 18 to 72 hours in cocktail of primary antibodies diluted in PBS.
9. Rinse 3 × 5 minutes in PBS.
10. Incubate 4 to 24 hours in cocktail of fluorescent species-specific secondary antibodies (raised in donkey) diluted in PBS.
11. Rinse 3 × 5 minutes in PBS.
12. Mount sections on glass slides with a glycerol-based antifade mounting medium.
13. Apply a coverslip and seal around the edges with nail varnish.
14. Store slides in a -20°C freezer until needed.
15. Examine sections with the confocal microscope.
16. Remove coverslip and rinse section in PB (3 × 5 min).
17. Incubate 72 hours in species-specific biotinylated secondary antibodies (raised in donkey, diluted 1:200 in PBS; Jackson ImmunoResearch). Choose secondary antibodies directed against those primary antibodies which are to be revealed with EM. Although the primary antibodies already have fluorescent secondaries attached, they are still able to bind biotinylated secondary antibodies.
18. Rinse 3 × 10 minutes PBS.
19. Incubate 72 hours in avidin–peroxidase conjugate (e.g., extravidin–peroxidase from Sigma, diluted 1:1000). We have found that avidin–peroxidase conjugate penetrates further into Vibratome sections than the avidin–biotin–peroxidase (ABC) complex and is, therefore, more suitable for this protocol.
20. Rinse 3 × 10 minutes PBS.
21. Rinse 5 minutes PB.
22. Incubate for approximately 5 minutes in DAB solution.
23. Rinse 3 × 5 minutes PB.
24. Osmicate: 1% OsO$_4$ in PB for 30 minutes.
    Note: OsO$_4$ is highly toxic and volatile, so this must be carried out in a fume cupboard.
25. Rinse 3 × 5 minutes water.
26. Dehydrate in acetone (70%, 90%, 3 × 100%, 10 min each). Block staining in uranyl acetate (30 min in a saturated solution of uranyl acetate in 70% acetone) can be included during the dehydration stage.
27. 1:1 Acetone:Durcupan mixture for 1 hour.
28. Pure Durcupan for 12 to 24 hours.
29. Flat-embed sections from Durcupan between acetate sheets, lay on a flat surface (e.g., microscope slide) with weight on top, and cure at 60°C for 48 hours.
30. Examine section and identify region of interest. Remove 1 acetate sheet and apply to a block of cured resin with a drop of liquid Durcupan. Cure overnight. Trim block, cut ultrathin sections, and mount on Formvar-coated single-slot grids.
31. Stain grids with lead citrate.
32. Examine with electron microscope.

RESULTS AND DISCUSSION

Primary Afferent Input to Neurons in Lamina III and IV of the Spinal Dorsal Horn

The use of the approaches described

above can be illustrated by our recent studies of the primary afferent input to a population of neurons with cell bodies in laminae III and IV of the spinal cord dorsal horn (14,15,18) (Figures 1 and 2). The neurokinin 1 (NK1) receptor, on which substance P acts, is present on certain cells throughout the dorsal horn, including a population of large neurons with cell bodies in lamina III or IV, which have prominent dorsally-directed dendrites that pass up into lamina I, as well as dendrites that arborize in deeper laminae (2,4). On NK1 receptor–immunoreactive neurons in the spinal cord, the receptor is found throughout the plasma membrane of the cell body and dendritic tree, but is not present on their axons. This means that antibodies against the NK1 receptor can be used to reveal the entire dendritic trees as well as the cell bodies of those neurons on which it is present (Figures 1c and 2a).

Since the large NK1 receptor–immunoreactive neurons in laminae III and IV of the dorsal horn have dendritic trees that can extend from lamina I to lamina V, they could potentially receive monosynaptic input from various types of primary afferent, including both unmyelinated (C) fibers, which terminate mainly in laminae I and II, and large myelinated afferents which arborize in laminae III through V (22). Unmyelinated primary afferents can be further subdivided into 2 major classes: *(i)* those which contain neuropeptides and terminate mainly in lamina I and the outer part of lamina II (II_o); and *(ii)* those that do not contain peptides and which end in the inner part of lamina II (II_i). In the rat, all peptidergic C fibers contain calcitonin gene-related peptide (CGRP) and many also contain substance P (8). Although substance P-containing axons in the dorsal horn can originate from various sources (local neurons, descending axons, or primary afferents), CGRP in the dorsal horn is derived exclusively from primary afferents.

This means that substance P-containing primary afferents can be distinguished from other axons which contain the peptide by the presence of CGRP. By carrying out triple labeling immunofluorescence staining with antibodies against NK1 receptor, substance P, and CGRP (14), we were able to demonstrate that the large lamina III/IV NK1 receptor–immunoreactive neurons receive numerous contacts from substance P-containing axons (Figure 2, c–f), and that over 90% of these are of primary afferent origin, since they also contain CGRP (Figure 1a). Interestingly, the substance P-containing afferents not only contact the dendrites of these cells in laminae I and II_o (where the afferents are very numerous), but also pass ventrally along the dendrites in laminae II_i and III (Figure 2, d–f). In order to establish that the substance P-containing axons formed synapses on the dendrites of these cells, we used the combined confocal–electron microscopic technique described above (Figures 3 and 4) (14). Asymmetrical synapses were found at all of the contacts between substance P-immunoreactive primary afferents and dendrites of lamina III/IV NK1 receptor–immunoreactive neurons which were examined (Figure 4).

We had previously found that some of the lamina III/IV NK1 receptor–immunoreactive neurons projected to the thalamus (12). We therefore injected CTb into the thalamus and performed triple immunofluorescence labeling to detect CTb, NK1 receptor, and substance P (Figure 2) and were able to demonstrate that cells of this morphological type, which belonged to the spinothalamic tract, also received dense innervation from substance P-containing axons.

Nonpeptidergic C fibers bind BSI-B4, however the lectin also binds to some peptidergic afferents. Since all peptidergic afferents contain CGRP (6), it is possible to identify nonpeptidergic C fibers with confocal microscopy, since these will bind BSI-B4 but will not be CGRP immunoreactive.

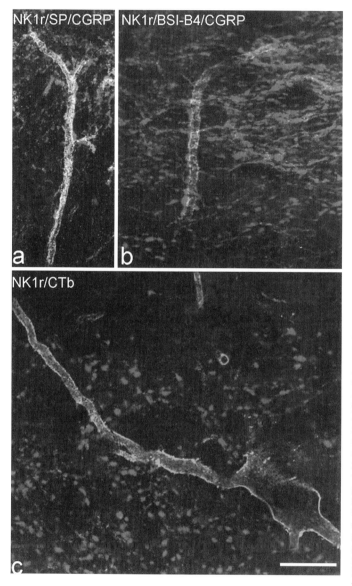

Figure 1. The association between large NK1 receptor–immunoreactive lamina III/IV neurons with long dorsal dendrites and three different types of primary afferent in the dorsal horn of the rat spinal cord, seen in transverse (a) or parasagittal (b and c) sections. Each image contains a dendrite belonging to a NK1 receptor–immunoreactive neuron (green), and in panel c, the cell body of one of these neurons is also seen. (a) A section also reacted with a monoclonal antibody to substance P (blue) and an antiserum against CGRP (red). Most of the immunoreactive axons contain both peptides and therefore appear pink, and where these overlap the NK1 receptor–immunoreactive dendrite, they appear white. The dendrite is associated with many varicosities which contain both substance P and CGRP (i.e., substance P-containing primary afferent terminals). (b) The section has been immunostained with CGRP antiserum (blue) and incubated in biotinylated BSI-B4, which was revealed with avidin–rhodamine (red). Many of the CGRP-immunoreactive axons have also bound the lectin and appear pink or purple. Numerous red axons are present in the lower half of the field: these have bound the lectin but are not CGRP-immunoreactive and, therefore, belong to nonpeptidergic C fibers. Very few of these axons make contact with the NK1-immunoreactive dedrite. (c) Shown is the cell body and proximal dendrites of a lamina III NK1 receptor–immunoreactive neuron from a rat in which CTb had been injected into the sciatic nerve 3 days previously. CTb (red) has been taken up and transported by myelinated afferents belonging to the sciatic nerve, which terminate extensively in lamina III. Although there are many CTb-immunoreactive axons surrounding the cell, very few make contact with it. Panel a was generated from 8 optical sections at 1 μm separation, panel b from 9 optical sections at 0.5 μm separation, while panel c is from a single optical section. Scale bar = 20 μm for each image. Panel b is reprinted with permission from Neuroscience 94, Sakamoto, H., R.C. Spike, and A.J. Todd, Neurons in laminae III and IV of the rat spinal cord with the neurokinin-1 receptor receive few contacts from unmyelinated primary afferents which do not contain substance P, p. 903-908, Copyright (1999), with permission from Elsevier Science. Panel c is reproduced with permission from Reference 15. (See color plate A12.)

Using this method, we found that although numerous nonpeptidergic afferents were present in lamina II$_i$, very few of these made contact with the dorsal dendrites of the lamina III/IV NK1 receptor–immunoreactive neurons which passed through this region (Figure 1b) (18).

Myelinated primary afferent terminals in the dorsal horn can be identified by means of an anterograde (transganglionic) transport technique. If CTb is injected into an intact somatic nerve in the rat, it is taken up almost exclusively by myelinated afferents and transported through the dorsal root ganglion and into their axon terminals within the spinal cord (17). We combined transganglionic transport of CTb injected into the sciatic nerve with immunofluorescence staining and found that although the dendrites of lamina III/IV NK1 receptor–immunoreactive neurons passed through regions that contained numerous terminals belonging to sciatic nerve myelinated afferents, they received only a limited number of contacts from these afferents (Figure 1c). With the combined confocal–electron microscopic technique, we were able to demonstrate that synapses were present at some of these contacts and, therefore, concluded that the cells receive a limited monosynaptic input from myelinated primary afferents (15).

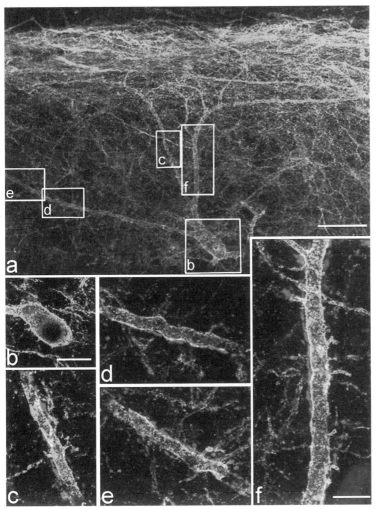

Figure 2. Contacts formed by substance P-containing axons onto a lamina III NK1 receptor–immunoreactive spinothalamic tract neuron. (a) Shown is the cell body and dendrites of the neuron in a parasagittal section scanned to reveal only NK1 receptor-immunostaining (green). Boxes indicate the regions shown in the remaining parts of the figure. (b) The cell body of the neuron contains CTb (blue), which had been injected into the thalamus 3 days previously, thus allowing identification of this as a spinothalamic neuron. (c–f) Different parts of the dendritic tree receive numerous contacts from substance P-immunoreactive axons, which are orientated along the lengths of the dendrites (red). Panel a was obtained from 9 optical sections 1.5 µm apart, panels c and e through f from 4 optical sections, and panel d from 7 optical sections, each 0.5 µm apart. Scale bars: panel a = 50 µm, panel b = 20 µm, panels c through f = 10 µm. Reproduced with permission from Reference 14. (See color plate A13.)

A.J. Todd

These results demonstrate that even though the large lamina III/IV neurons with NK1 receptors have dendrites that pass through several laminae, their primary afferent input is highly selective. They receive a dense innervation from substance P-containing afferents, which frequently run along their dorsal dendrites and form numerous synapses on them (14), however, they are sparsely innervated by myelinated afferents in laminae III and IV (15), and receive very few contacts from nonpeptidergic C fibers (18).

Technical Aspects

Benefits of Triple Fluorescence Labeling

Although it is often possible to investigate neuronal circuits with only two labels (one each for the pre- and postsynaptic components of the circuit), there are many

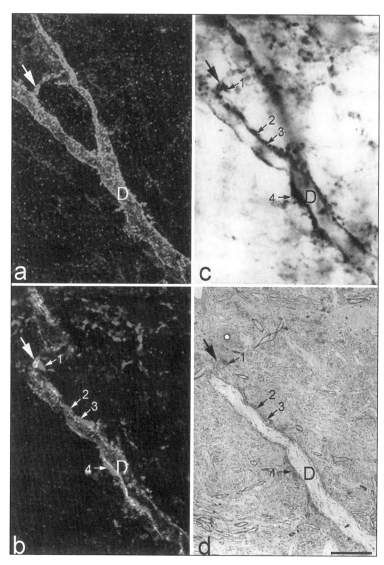

Figure 3. Combined confocal microscopy and EM of a dorsal dendrite belonging to a lamina III NK1 receptor–immunoreactive neuron from a parasagittal section that had been reacted with NK1 receptor, substance P, and CGRP. (a) NK1 receptor immunoreactivity in a series of 7 optical sections at 0.5 µm intervals. The arrow shows a small branch given off from the main dendritic shaft (D). (b) A single confocal section from the series used to generate panel b. In this case, NK1 receptor appears green, substance P is blue, and CGRP is red. Boutons, which contain both peptides, appear pink. Several substance P-immunoreactive varicosities are in contact with the NK1 receptor–immunoreactive dendrite, and 4 of these are indicated with the small numbered arrows. The varicosity numbered 2 is only substance P-immunoreactive, whereas those numbered 1, 3, and 4 have both substance P and CGRP immunoreactivity. (c) The corresponding region seen with light microscopy after the immunoperoxidase reaction and at a focal depth approximately equivalent to the confocal image in panel b. The main dendritic shaft (D) and part of its small branch (arrow) are clearly seen, whereas most of the right-hand branch is out of focus. Many substance P-immunoreactive varicosities are visible, including the ones indicated with arrows in panel b. (d) A low magnification electron micrograph of the corresponding region. The section is at a depth nearly equivalent to the confocal image in panel b. Scale bar = 10 µm. Modified and reproduced with permission from Reference 14. (See color plate A14.)

situations in which it is necessary or desirable to use a third label. Thus, in some cases it is possible to refine the identification of one of the two neurons which are involved in the circuit by the use of two labels. For example, combining either substance P-immunostaining or BSI-B4 binding with detection of CGRP allowed us to identify two different neurochemical types of primary afferent terminal in the superficial dorsal horn (Figure 1, b and c). A third label can then be used to reveal the other component of the circuit (in this case, neurons with the NK1 receptor). In addition, triple labeling can be used to examine inputs to a single neuron from two different axonal populations simultaneously, thus allowing interactions between the two inputs to be examined. This approach is also useful in studies of neurons that have been labeled by intracellular injection, since it increases the amount of data that can be collected from each cell

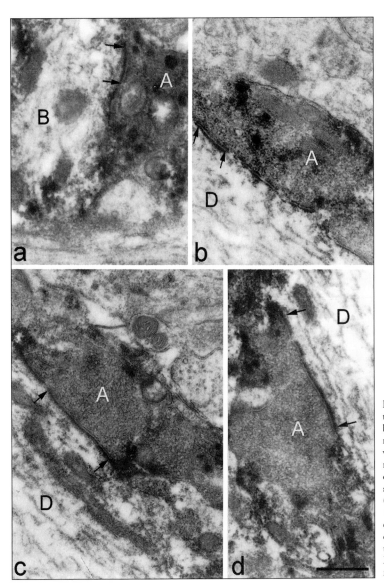

Figure 4. High magnification electron micrographs to show synapses between the substance P-immunoreactive axonal boutons (A), which were indicated with numbered arrows in Figure 3, and the NK1 receptor–immunoreactive dendritic shaft (D) or its small branch (B). (a–d) Shown are boutons numbered 1 through 4, respectively. In each case an asymmetrical synaptic specialization is clearly seen between the arrows. Scale bar = 0.5 µm. Modified and reproduced with permission from Reference 14.

and may therefore reduce the number of neurons that need to be injected.

Another situation in which a third label can be extremely useful is when studying projection neurons that have been identified with retrograde tracers. Even the most sensitive tracers used for retrograde labeling (e.g., CTb) seldom give complete filling of the dendritic trees of labeled neurons, and examining connections formed by axons on the distal dendrites of these cells is therefore usually not possible. If the entire dendritic tree of the retrogradely labeled neuron can be immunostained, for example with an antibody against a membrane protein such as a receptor, then this allows even the distal dendrites to be examined (Figure 2). Pollock et al. (16) have recently refined this approach to allow the simultaneous detection of four different fluorescent labels in the same section, by using the retrograde tracer Fluorogold (which is excited by UV light and was viewed with conventional epifluorescence) and carrying out triple immunofluorescence labeling and confocal microscopy with secondary antibodies conjugated to FITC, LRSC, and Cy5.

Combined Confocal and Electron Microscopy

The advantages of being able to detect three compounds simultaneously led us to develop a method that would allow both confocal and electron microscopic examination of the same immunostained section. With conventional (pre-embedding) immunocytochemistry for EM, if more than one antigen is to be detected, each must be labeled with a different electron-dense marker. Although it is relatively straightforward to label two different antigens, for example with DAB and silver-intensified immunogold (20), triple labeling is much more difficult. In the method which is described in this chapter, only one electron-dense marker (DAB) is used, and the distinction between profiles that were stained with different antibodies is made by correlating the appearance in electron microscope sections with that which had been obtained with the confocal microscope (Figures 3 and 4).

There are two main limitations to this method. Firstly, although penetration of the immunofluorescence staining in the Vibratome sections is generally very good (even without the use of detergents), the penetration of the peroxidase/DAB reaction product is much more limited and usually extends less than 5 µm from the cut surfaces of the section. For this reason, it is necessary to select contacts which lie close to the section surface for analysis. Secondly, ultrastructural preservation in tissue that has been prepared in this way is usually not as good as that seen with conventional electron microscopic methods, presumably due to the large number of stages involved and the prolonged processing time. However, it is good enough to allow identification of synaptic specializations, and this is the main reason for using this method in studies of neuronal circuitry.

Quantification of Contacts

One of the advantages of confocal microscopy for studies of neuronal circuits is that the frequency of contacts which a neuron receives from a particular type of axon can be measured (14,15,18). This means that, for example, the density of contacts formed by a population of axons onto two different groups of target neurons can be compared to determine whether one group is selectively innervated (14). In order to analyze the density of contacts onto the dendritic trees of neurons, it is obviously necessary to measure the lengths of these dendrites from confocal images. We have found that the program Neurolucida for Confocal (MicroBrightField, Colchester, VT, USA) is particularly suitable for this purpose, since it can be used to analyze stacks of confocal images.

Measurements along the z-axis (depth through the section) in series of confocal images are obtained from the focus motor that moves the specimen up and down between scans, and the vertical distance between each image in the series (z separation) is used by the program when it calculates the length of dendrites. However, because of differences in refractive index between the mounting medium of the specimen and the medium which lies between the objective lens and the coverslip (either air or immersion oil, depending on the objective lens), the vertical distance moved by the stage is not exactly the same as the apparent distance moved through the specimen, and a correction factor is required (11). This problem is discussed in detail by Majlof and Forsgren (11), who provide correction factors for various combinations of media. With a glycerol-based mounting medium, the correction factor for a dry lens is 1.47 and for an oil lens it is 0.97. Thus, for example, if a z-series is scanned with a dry lens and the separation between each image is 1 µm (measured from the focus motor), the actual vertical separation between images in the series is 1.47 µm.

In quantitative studies of neuronal circuits, it is obviously essential that penetration of antibodies into the section is complete, otherwise the number of contacts will be underestimated. Since antibodies penetrate from the cut surfaces of a section, the degree of penetration of immunostaining can usually be assessed by scanning through the full thickness of the section. If the density of immunostained profiles is approximately constant throughout the section thickness, then penetration is probably adequate. If there is a reduction in the number of stained profiles towards the center of the section thickness, then this clearly suggests that penetration is not uniform. We have found that, for most of the antibodies which we use, pretreatment with ethanol (10) and the use of detergents (e.g., Triton X-100) in the incubation buffers results in complete penetration of immunostaining. However, with certain antibodies it is not possible to achieve satisfactory penetration into the sections, and these antibodies are therefore not suitable for use in quantitative studies.

ACKNOWLEDGMENTS

I thank Dr. S. Vigna for the gift of NK1 receptor antiserum, Drs. M. Naim, R.C. Spike, S.A.S. Shehab, and H. Sakamoto for their participation in the experiments described in this chapter, Mrs. C. Watt and Mrs. M.M. McGill for expert technical assistance, and the Wellcome Trust for financial support.

REFERENCES

1. **Belichenko, P.V. and A. Dahlström.** 1995. Confocal laser scanning microscopy and 3-D reconstructions of neuronal structures in human brain cortex. Neuroimage *2*:201-207.
2. **Bleazard, L., R.G. Hill, and R. Morris.** 1994. The correlation between the distribution of NK1 receptor and the actions of tachykinin agonists in the dorsal horn of the rat indicates that substance P does not have a functional role on substantia gelatinosa (lamina II) neurons. J. Neurosci. *14*:7655-7664.
3. **Brelje, T.C., M.W. Wessendorf, and R.L. Sorenson.** 1993. Multicolor laser scanning confocal immunofluorescence microscopy: practical applications and limitations, p. 97-181. *In* B. Matsumoto (Ed.), Cell Biological Applications of Confocal Microscopy. Acadamic Press, San Diego.
4. **Brown, J.L., H. Liu, J.E. Maggio, S.R. Vigna, P.W. Mantyh, and A.I. Basbaum.** 1995. Morphological characterization of substance P receptor-immunoreactive neurons in rat spinal cord and trigeminal nucleus caudalis. J. Comp. Neurol. *356*:327-344.
5. **Jankowska, E., D.J. Maxwell, S. Dolk, and A. Dahlström.** 1997. A confocal and electron microscopic study of contacts between 5-HT fibres and feline dorsal horn interneurons in pathways from muscle afferents. J. Comp. Neurol. *387*:430-438.
6. **Ju, G., T. Hökfelt, E. Brodin, J. Fahrenkrug, J.A. Fischer, P. Frey, R.P. Elde, and J.C. Brown.** 1987. Primary sensory neurons of the rat showing calcitonin gene-related peptide immunoreactivity and their relation to

substance P-, somatostatin-, galanin-, vasoactive intestinal polypeptide- and cholecystokinin-immunoreactive ganglion cells. Cell Tissue Res. *247*:417-431.

7. Kosaka, T., I. Nagatsu, J.-Y. Wu, and K. Hama. 1986. Use of high concentrations of glutaraldehyde for immunocytochemistry of transmitter-synthesizing enzymes in the central nervous system. Neuroscience *18*:975-990.

8. Lawson, S.N. 1992. Morphological and biochemical cell types of sensory neurons, p. 27-59. *In* S.A. Scott (Ed.), Sensory Neurones: Diversity, Development and Plasticity. Oxford University Press, New York.

9. Li, J.-L., T. Kaneko, S. Nomura, Y.-Q. Li, and N. Mizuno. 1997. Association of serotonin-like immunoreactive axons with nociceptive projection neurons in the caudal spinal trigeminal nucleus of the rat. J. Comp. Neurol. *384*:127-141.

10. Llewellyn-Smith, I.J. and J.B. Minson. 1992. Complete penetration of antibodies into Vibratome sections after glutaraldehyde fixation and ethanol treatment: light and electron microscopy for neuropeptides. J. Histochem. Cytochem. *40*:1741-1749.

11. Mailof, L. and P.-O. Forsgren. 1993. Confocal microscopy: important considerations for accurate imaging, p. 79-95. *In* B. Matsumoto (Ed.), Cell Biological Applications of Confocal Microscopy. Acadamic Press, San Diego.

12. Marshall, G.E., S.A.S. Shehab, R.C. Spike, and A.J. Todd. 1996. Neurokinin-1 receptors on lumbar spinothalamic neurons in the rat. Neuroscience *72*:255-263.

13. Mason, P., S.A. Back, and H.L. Fields. 1992. A confocal laser microscopic study of enkephalin-immunoreactive appositions onto physiologically identified neurons in the rostral ventromedial medulla. J. Neurosci. *12*:4023-4036.

14. Naim, M., R.C. Spike, C. Watt, S.A.S. Shehab, and A.J. Todd. 1997. Cells in laminae III and IV of the rat spinal cord which possess the neurokinin-1 receptor and have dorsally-directed dendrites receive a major synaptic input from tachykinin-containing primary afferents. J. Neurosci. *17*:5536-5548.

15. Naim, M., S.A.S. Shehab, and A.J. Todd. 1998. Cells in laminae III and IV of the rat spinal cord which possess the neurokinin-1 receptor receive monosynaptic input from myelinated primary afferents. Eur. J. Neurosci. *10*:3012-3019.

16. Pollock, R., R. Kerr, and D.J. Maxwell. 1997. An immunocytochemical investigation of the relationship between substance P and the neurokinin-1 receptor in the lateral horn of the rat thoracic spinal cord. Brain Res. *777*:22-30.

17. Robertson, B. and G. Grant. 1985. A comparison between wheatgerm agglutinin and choleragenoid-horseradish peroxidase as anterogradely transported markers in central branches of primary sensory neurones in the rat with some observations in the cat. Neuroscience *14*:895-905.

18. Sakamoto, H., R.C. Spike, and A.J. Todd. 1999. Neurons in laminae III and IV of the rat spinal cord with the neurokinin-1 receptor receive few contacts from unmyelinated primary afferents which do not contain substance P. Neuroscience *94*:903-908.

19. Todd, A.J. 1997. A method for combining confocal and electron microscopic examination of sections processed for double- or triple-labeling immunocytochemistry. J. Neurosci. Methods *73*:149-157.

20. Todd, A.J., C. Watt, R.C. Spike, and W. Sieghart. 1996. Colocalization of GABA, glycine and their receptors at synapses in the rat spinal cord. J. Neurosci. *16*:974-982.

21. Vulchanova, L., M.S. Riedl, S.J. Shuster, L.S. Stone, K.M. Hargreaves, G. Buell, A. Surprenant, R.A. North, and R. Elde. 1998. P2X3 is expressed by DRG neurons that terminate in inner lamina II. Eur. J. Neurosci. *10*:3470-3478.

22. Willis, W.D. and R.E. Coggeshall. 1991. Sensory Mechanisms of the Spinal Cord. Plenum, New York.

16. Confocal Imaging of Calcium Signaling in Cells from Acute Brain Slices

Wim Scheenen[1] and Giorgio Carmignoto[2]
[1]Department of Cellular Animal Physiology and Institute of Cellular Signalling, University of Nijmegen, Nijmegen, The Netherlands, EU; [2] Department of Experimental Biomedical Sciences and CNR Center for the Study of Biomembranes, University of Padova, Padova, Italy, EU

OVERVIEW

Measurements of intracellular calcium concentrations ($[Ca^{2+}]_i$) using intracellularly trapped fluorescent dyes have become a popular way to determine changes in $[Ca^{2+}]_i$ in individual cells. Techniques for performing such experiments are well documented (5,9,14,15). More recent advancements in confocal laser scanning microscopy (CLSM) have made it possible to perform these measurements even within tissue slices. Preparation of the CLSM tissue and performing an experiment on a tissue slice is slightly different than that for single cells. In this chapter, we provide an overview of the basics of confocal microscopy and subsequently describe how to set up and perform an experiment on brain slices. We also show that neurons and astrocytes can be differentially loaded with the Ca^{2+} indicator indo-1 depending on the solution used during slice cutting and dye loading procedures.

BACKGROUND

CLSM is now established as a valuable tool for obtaining high resolution images and 3D reconstructions of a variety of biological specimens. With the use of specific Ca^{2+}-sensitive dyes, this technique can also be used to obtain reproducible results on Ca^{2+} dynamics. Here, a general description of CLSM will be given.

In CLSM, a laser light beam is expanded to make optimal use of the optics in the objective. This beam is turned into a scanning beam and focused to a small spot onto a fluorescent specimen by an objective lens. The mixture of reflected light and emitted fluorescent light is captured by the same objective and is directed through a dichroic mirror (beam splitter). The reflected light is deviated by this mirror while the emitted fluorescent light passes through in the direction of the detector, a photomultiplier tube. In this way, only fluorescence emission coming from the tissue is recorded. A confocal aperture (pinhole) is placed in front of the photodetector, such that the fluorescent light from points on the specimen that are not within the plane of focus (the so-called out-of-focus light) will be largely obstructed by the pinhole. By optimizing the size of the confocal pinhole, out-of-focus light (both above and below the focal plane) is greatly reduced. This becomes especially important when dealing with thick specimens as brain slices. Generally, the plane of focus (Z-plane) is selected by a computer-controlled stepping motor that allows the experimenter to focus through the specimen, thereby selecting the proper Z-plane for the experiments.

The optical section, which is a 2D image of a small partial volume of the specimen centered around the focal plane is generated by performing an XY scan with the focused laser spot over the specimen at that focal plane. As the laser scans across the specimen, the fluorescent light detected by the photomultiplier is converted into a digital signal. This will lead to a pixel-based image that typically is displayed on a computer monitor attached to the CLSM. The relative intensity of the fluorescent light emitted from the specimen will therefore correspond to the grayscale value of the resulting pixel in the image. An experiment in which time effects need to be observed can be performed by repeatedly scanning the same XY plane with a fixed time interval, leading to an XYt scan. This XYt scanning will typically be performed when imaging Ca^{2+} dynamics in brain slices.

Considerations on the Experimental Setup

One of the physical properties of fluorescent dyes is bleaching; upon a certain number of excitations, a fluorescent molecule cannot be exited anymore and becomes nonfluorescent. In generating a confocal image, spatial resolution, image intensity (or brightness), scanning speed, and dye bleaching are interconnected to each other. In other words, if one wants a very bright and sharp confocal image, one should use a small pinhole, a slow scanning speed combined with a high laser excitation intensity, and should allow for bleaching to occur. However, performing an experiment that follows Ca^{2+} dynamics in living cells requires that processes like bleaching should be kept to an absolute minimum. Therefore, the laser power used should be set as low as possible. This can be done by directly reducing the laser output power or by introducing neutral density filters in the optical path. As a consequence of the reduced excitation light, the intensity of the obtained fluorescence signal becomes very low in comparison to system noise, i.e., one gets a low signal-to-noise ratio (S/N). Increasing the size of the pinhole will increase the S/N. However, by increasing the size of the pinhole, light from a thicker plane is allowed to pass to the detector, leading to a reduced optical sectioning. For brain slices, one can be interested in: *(i)* different cells within a slice; or *(ii)* intracellular differences, for example between the processes of a given cell. For the first example, a thick optical section will be sufficient, and a good S/N can be obtained by increasing the confocal pinhole. For the second example, i.e., intracellular Ca^{2+} dynamics in neuronal processes, a thin confocal section is required, hence the confocal pinhole should be small, and intrinsically, these images will have a lower S/N and look "noisy".

Another option in obtaining better signals is to average sequential images. The disadvantage of this option is the reduction of image acquisition speed, and as a consequence, fast Ca^{2+} changes will be missed.

The detector gain should be set as low as possible to minimize the contribution of system noise. However, care has to be taken that anticipated changes in fluorescence intensity fall within the dynamic range of the detector. Therefore, for increases in fluorescence, the gain should not be set so high that intensity values reach the maximum 8-bit value of 255, and decreases in fluorescence should be possible without the pixel values reaching 0. Most systems have a special color lookup table that will indicate when too many pixels fall out of the dynamic range of the detector. In experiments where the ratiometric dye indo-1 is used, special care has to be given to this topic to allow reliable ratio values to be obtained (see below).

Relation Between Speed of the Experiment and Scanning Speed of the Equipment

The theory described in the previous section applies to a point scanning microscope. Point scanning CLSM can be either a fast or slow process depending on the equipment used. For most types of microscopes commercially available today, it generally takes a few seconds to scan one image at full screen resolution. It will be clear that for most time-related Ca^{2+} dynamics, a time resolution of 2 to 5 seconds is an absolute minimum, and often, experiments with faster acquisition speed are required. A way to increase the sampling speed is to reduce the scanning box.

In practice, this means that the spatial resolution is decreased since the same information is presented in fewer pixels. A disadvantage of reducing the screen size is that the spatial resolution can get too low (i.e., only a few pixels per cell). This disadvantage can be overcome by performing an optical zoom by which the area that is scanned by the laser is reduced. However, one has to take into consideration that bleaching becomes a more serious problem when optical zooming is performed. From these statements it will become clear that each experimental situation and tissue will need its own settings. As a general rule for beginning experimenters to set up a proper protocol, keep in mind that dye bleaching and phototoxicity-related problems are the biggest enemies of confocalists working on living tissue.

Another development in confocal microscopy has led to a faster scanning type. This faster scanning is achieved by scanning the specimen with the laser, which is transformed into a line, or by using point scanning with fast scanning optics. Generally, time-related problems do not occur when the experiments can be performed on such fast scanning equipment.

A full description of all confocal microscopes is beyond the aim of this chapter and can be found in *The Handbook of Biological Confocal Microscopy* (Reference 11; see also Reference 14).

PROTOCOLS

Protocol for Preparing Acute Central Nervous System Slices for Real Time Ca^{2+} Imaging with CLSM

Materials

All chemicals are from Sigma, (St. Louis, MO, USA) unless otherwise stated.

- Solutions for slice cutting and dye incubation (see Table 1).

Table 1. Solution N. 1

Chemical	MW	mM	g/L	Stock (M)	mL/L
NaCl	58.4	120.0	7.0	2.0	60.0
KCl	74.6	3.2	0.23	2.0	1.6
$CaCl_2$	147.0	1.0	0.15	1.0	1.0
$MgCl_2$	203.3	2.0	0.4	1.0	2.0
KH_2PO_4	136.1	1.0	0.14	1.0	1.0
$NaHCO_3$	84.0	26.0	2.2	0.5	52.0
Glucose	180.2	2.8	0.5		
The following antioxidants can be added:					
Ascorbic Acid	176.1	0.57	0.1		
NaPyruvate	110.0	2.0	0.2	0.5	4.0

- Solutions for slice cutting and dye incubation (see Table 2).

Table 2. Solution N. 2

Chemical	MW	mM	g/L	Stock (M)	mL/L
CholineCl	139.6	110.0	15.4	2.0	55.0
KCl	74.6	2.5	0.19	2.0	1.25
$CaCl_2$	147.0	0.5	0.07	1.0	0.5
$MgCl_2$	203.3	7.0	1.42	1.0	7.0
NaH_2PO_4	137.1	1.2	0.14	1.0	1.0
$NaHCO_3$	84.0	25.0	2.1	0.5	50.0
Ascorbic acid	176.1	1.3	0.23		
NaPyruvate	110.0	2.4	0.26	0.5	4.8
Glucose	180.2	20.0	3.6		

In both solutions N. 1 and N. 2, add $NaHCO_3$ at pH 7.4 (with 5% CO_2-95% O_2).

- Indo-1/AM (Molecular Probes, Eugene, OR, USA).
- Dimethyl sulfoxide (DMSO).
- Pluronic acid F-127.
- Sulfinpyrazone.
- Vibrating microtome (Electron Microscopy Science, Fort Washington, PA, USA).
- Disposable 16-well culture plates (Nalge Nunc International, Roskilde, Denmark).

Procedure

Slice Preparation

The following is a typical procedure for cutting slices from developing rats (from postnatal day 5 to 20).

1. Rats are deeply anesthetized with a

lethal dose of either ketamine or diethyl ether. Cryoanesthesia can be used for very young animals. After decapitation, the brain is rapidly removed and placed in ice-cold physiological saline (solution N. 1 or 2).

2. The tissue is immobilized with an acrylic glue on the dish of the Vibratome at orientation and angle for obtaining appropriate slices from the region of interest. The tissue has to be submerged in the same ice-cold solution under continuous bubbling with 5% CO_2-95% O_2. Slices of the desired thickness are obtained. A thickness between 200 and 300 µm is optimal for most of the experimental approaches. No support is, in general, necessary for cutting slices from the brain. In contrast, for obtaining transverse slices from the spinal cord, it is mandatory to stabilize the tissue, for example by attaching the segments of the spinal cord (4–6 mm long) to a shaped block of silicon (small quantity of glue has to be used).

3. Slices are allowed to recover for at least 45 minutes at 37°C in the same physiological saline under continuous bubbling with 5% CO_2-95% O_2. In the case that slices have to be incubated with dyes, recovery periods shorter than 45 minutes have to be used.

Dye-Loading

The dye loading described in this section is for using indo-1. For a discussion on the use of other probes, see the Results and Discussion section.

4. Incubating slices in the physiological saline added with the cell permeant acetoxymethyl derivative of indo-1 (indo-1/AM; 20 µM) and 0.02% pluronic acid F-127 in a water bath at 37°C for 40 to 50 minutes under continuous influx of the gas mixture (5% CO_2-95% O_2).

Notes:

a. Continuous mild stirring is crucial for optimizing the loading of the dye. We found it very convenient to perform loading in a 16-well plate. By inflating the gas mixture through an 18 to 20 gauge needle onto the surface of the incubating medium, one can ensure both a proper oxygenation of the medium and a slow continuous movement of the slices.

b. Since AM esters of the dyes do not readily dissolve in water, the stock solution prepared in DMSO will form small crystals when diluted in medium. In order to prevent this, and to optimize the actual loading concentration, a small amount of the detergent pluronic acid F-127 (12) can be added to the loading medium.

c. Although the use of dye AM should ideally yield a strict cytoplasmic localization of the dye, cellular activity such as that of organic anion transporters might cause uptake of the dye in intracellular organelles or extrusion of the dye. In order to inhibit organic anion transporter activity, sulfinpyrazone or probenicid can be added to the loading medium and/or experimental solution (3). However, take into consideration that blockers of intracellular compartmentalization can affect cellular functions. For example, probenicid has been shown to enlarge the duration of spontaneous Ca^{2+} oscillations in pituitary melanotrope cells (16).

d. For AM loading, a loading concentration of 5 to 20 µM is sufficient to get a reasonable dye concentration in the tissue. Loading times vary, but for slices, an incubation in the loading solution, i.e., medium containing the dye, should be between 30 and 60 minutes at 37°C. Note that the viability of neurons can be significantly increased by adding to the incubation solution antioxidants agents (see P-rotocols) as well as antagonists of the N-methyl-D-aspartate receptor (NMDAR), such as 10 µM D-2-amino-5-phosphonopentanoic acid. After dye loading, the slices should be kept at 20°C in constantly oxygenated medium for at least 30 minutes to allow complete de-esterification of the dye.

In the experiments that will be described below, loading of cells is compared in slices incubated with indo-1 either in the presence or in the absence of antioxidant agents.

Confocal Analysis

5. Slices are mounted in a chamber which is placed on the stage of an inverted microscope (Diaphot 300; Nikon, Melville, NY, USA) equipped with a 40× water immersion objective (NA = 1.1; Nikon) connected with a real time confocal microscope (RCM8000; Nikon). The 351-nm band of the argon ion laser is used for excitation, and the emitted light, separated into its 2 components (405 and 485 nm) by a dichroic mirror at 455 nm, is collected by 2 separate photomultiplier tubes. The ratio of the intensity of the light emitted at the 2 wavelengths (R405/485) is displayed as a pseudocolor scale using a linear lookup table. Time series can be acquired with a frame interval of 1, 2, or 3 seconds, and 8 or 16 images are averaged for each frame. During recordings, slices are continuously perfused (3 mL/min) with physiological saline of the following composition (in mM): NaCl, 120; KCl, 3.1; NaH_2PO_4, 1.25; $NaHCO_3$, 25; dextrose, 5; $MgCl_2$, 1; $CaCl_2$, 2; at pH 7.4 with 5% CO_2-95% O_2.

Note: The R405/485 in basal conditions was observed to vary little in different cells. Occasionally, a slight decrease could be observed in R405/485 basal levels. Indeed, prolonged UV irradiation of indo-1 can cause overall photobleaching and photodamaging, a conversion to a fluorescent, but Ca^{2+} insensitive, species (17).

Protocol for Performing Extracellular Calibrations

After an experiment is done, one can get an idea of the actual free Ca^{2+} concentrations that were obtained by performing a calibration. Calibration methods have been described in great detail elsewhere (5,9,15). Since, for brain slices, an intracellular calibration is extremely difficult to accomplish, we will only describe how to perform extracellular calibrations based on experiments using indo-1. However, we would like to emphasize that extreme care has to be taken in interpreting the measured values as absolute Ca^{2+} values. The most important pitfall that occurs when performing extracellular calibrations is that the solutions used to perform the calibration do not reflect the composition of the cytoplasm. For example, the calibration solution lacks proteins present in the cytoplasm that may alter the Ca^{2+} binding characteristics of the dye. It is, therefore, always wise to consider whether calibrated Ca^{2+} values are absolutely necessary. In fact, for most experiments, the ratio or intensity values alone will be sufficient. For a detailed description of pitfalls in the calibration procedure we would like to refer to Reference 15.

1. Add 100 µM of the dye in its free acid form to 200 µL of Ca^{2+}-free medium added with 10 mM EGTA. The composition and pH of this medium have to be as close as possible to those of the cytosol.
2. Make a small well on a coverslip in which you place this dye solution.

3. Measure the minimal ratio obtained in this way (Rmin) and the fluorescence intensity at 485 nm (sf).
4. Without changing the focus of the microscope, add to the well 5 μL 1 M Ca^{2+} solution to obtain the high Ca^{2+} medium.
5. Measure the maximal ratio value obtained (Rmax) and the fluorescence intensity in this condition at 405 nm (sb).

Calculating Absolute Ca^{2+} Values

After obtaining the Rmin, Rmax, sf, and sb, the intracellular concentration can be calculated using the following formula (4):

Ca^{2+} = [(R-Rmin)/(Rmax-R)] × (sf/sb) × Kd

in which R is the ratio value for which the concentration is desired, and Kd is the dissociation constant of the dye (250 nM for indo-1).

RESULTS AND DISCUSSION

Which Ca^{2+}-Sensitive Dye to Use?

In order to detect Ca^{2+} dynamics, a brain slice needs to be loaded with a fluorescent dye. This dye loading can be done through microinjection, through a patch pipet, or through loading with membrane-permeable forms of the dye, the acetoxymethyl-esters (dye AM). For dye AM loading, each dye and each tissue will give specific loading results like a high or a low loading, shallow or deep penetration in the tissue, etc. It will be up to the experimenter to select the most useful dyes. However, the choice of dye also depends on the type of laser available.

Basically, the available dyes can be divided into two groups, one showing only a shift in fluorescence intensity upon Ca^{2+} binding, the other also showing a spectral shift, either in excitation or in emission. Since for the latter group a change in Ca^{2+} can be expressed as a change in ratio between the fluorescence intensities at two selected wavelengths upon Ca^{2+} binding, this group is referred to as ratiometric dyes. The ratiometric dye that can be used in CLSM experiments is indo-1, and a working and thoroughly tested protocol has been described in the section Dye Loading. This dye can be excited in the UV at 351 nm, and its emission maximum in the Ca^{2+}-bound form lies at 405 nm and in its Ca^{2+}-free form at 485 nm. Hence, a ratio of the 405 nm emission over the 485 nm emission will give an optimal change in intensity upon Ca^{2+} changes.

Unfortunately, most confocal laser scanning microscopes are equipped with only visible wavelength lasers, preventing the use of indo-1. All dyes that can be used for CLSM equipped with a visible-wavelength laser are single wavelength dyes like fluo-3, Calcium Green, Oregon Green, and Fura Red. They are called single wavelength because, as mentioned before, they will only change their fluorescence intensity at one wavelength upon Ca^{2+} binding. All of the single wavelength dyes mentioned, except Fura Red, have a higher fluorescence intensity in their Ca^{2+} bound form. Fura Red, in contrast, decreases its fluorescence intensity upon Ca^{2+} binding.

As mentioned before, which dye is the best to use depends largely on the tissue used and needs to be tested for each individual case. For single cells, the use of two single wavelength dyes, Fura Red in combination with either fluo-3 or Oregon Green, have been described, leading to "pseudo-ratio" images that overcome some of the problems of single wavelength dyes (7,8). In theory, this technique should also be applicable to brain slices, although in practice, loading characteristics of the two dyes may be too different to obtain reliable results.

Although both groups of dyes are useable, ratiometric dyes are clearly preferable since the measured signal, being expressed as a ratio value, will be insensitive to factors such as dye concentration, bleaching, uneven distribution through the tissue, or movement of the slice in the chamber. Using pseudo-ratio imaging will overcome problems with movement of the slice, but uneven distribution through the slice and different bleaching rates will become more of a problem. For example, in a pilot study using thalamic tissue, the bleaching rate for Fura Red was found to be significantly higher than that of Oregon Green, preventing proper use of the obtained pseudo-ratio values (Scheenen, unpublished result). In practice, the use of two single wavelength probes is useful to check for possible movements of the slice. Final analysis of the results should, however, be performed on one of the two dyes that has the lowest amount of bleaching, shows the least amount of uptake into intracellular compartments, and has a reliable loading profile.

Antioxidant Agents Help Cell Viability and Dye Loading

In general, optimal staining can be observed in neurons from slices obtained

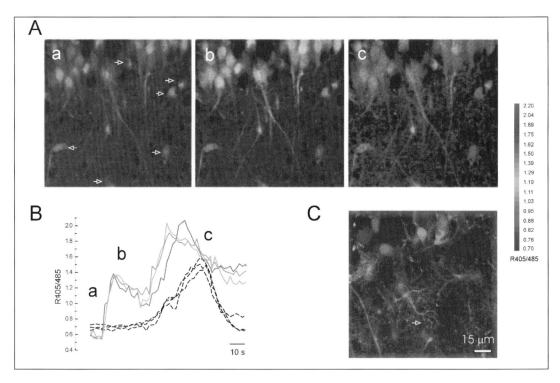

Figure 1. Effects of antioxidant agents in loading of cells from acute brain slices with Ca^{2+} fluorescent dyes. (A) Indo-1 loaded cells from the CA1 hippocampal region of a young rat at postnatal day 10 following slicing and incubation procedures in the presence of antioxidant agents. Most of the neuronal dendrites resulted well loaded. The time series of pseudocolor images illustrates the $[Ca^{2+}]_i$ changes occurring in these neurons following perfusion of the slice with 60 mM KCl. The sequence shows the $[Ca^{2+}]_i$ increase in pyramidal neurons occurring several seconds before that in astrocytes (open arrows). The R405/485 is displayed as a pseudocolor scale. Sampling rate 2 seconds. The stimulation with high K^+ extracellular solution was obtained by iso-osmotic replacement of Na^+ with K^+. (B) Kinetics of the $[Ca^{2+}]_i$ changes in the neurons (continuous lines) and astrocytes (dotted lines), indicated by arrows in panel A, following KCl stimulation, as expressed by the ratio between indo-1 emission wavelength at 405 and 485 nm. Letters a through c correspond to images a through c in panel A. (C) Indo-1 loaded cells from the CA1 hippocampal region of a young rat at postnatal day 10 following slicing and incubation procedures in the absence of antioxidant agents. Under these conditions, dendrites are less loaded with indo-1 while astrocyte processes (arrow) are, in general, more loaded with the dye. Furthermore, several dying neurons can be observed scattered in the CA1 pyramidal layer. (See color plate A15.)

from rats during the first 2 postnatal weeks. The number of labeled neurons decreases, however, in slices from older animals, and the quality of the staining also changes dramatically depending on the age of the animal.

A considerable improvement can be achieved by using antioxidant agents during slice cutting and dye loading procedures. Antioxidant agents such as ascorbic acid can protect neurons from the degeneration caused by the slicing procedures (13). In our experience, antioxidant agents can increase both the number of loaded neurons and, more importantly, the quality of the loading. This holds for cells in slices obtained from different brain regions, such as neocortex, hippocampus, and spinal cord. Figure 1 illustrates the advantage of using antioxidant agents such as ascorbic acid. Images illustrate two typical examples of cells from the CA1 hippocampal region loaded with indo-1 after cutting and incubating procedures are performed in the presence (Figure 1A) or absence (Figure 1B) of antioxidant agents. When antioxidant agents are used, dendrites of pyramidal neurons result well loaded. In some cases, by increasing the laser power and focusing on different planes, dendrites can be followed for most of their extension. The sequence of pseudocolor images of Figure 1A illustrates the transient increase in the $[Ca^{2+}]_i$ occurring in these neurons upon slice perfusion with 60 mM KCl. All neurons in the field are able to respond properly to activation of voltage-activated Ca^{2+} channels by K^+-induced depolarization, thus providing a clue of their healthy conditions (Figure 1A).

When antioxidant agents are not included, good loading of the dye was, instead, obtained in very few neurons. Note also that in Figure 1B, the very thin processes of the astrocyte are well loaded, while in Figure 1A, only the astrocyte cell bodies are discernible. The possibility of monitoring the spatial and the temporal features of $[Ca^{2+}]_i$ changes at the level of the processes following different stimuli can provide crucial information for a better understanding of Ca^{2+} signaling in astrocytes. Indeed, highly localized $[Ca^{2+}]_i$ elevations in discrete regions along the distal processes of hippocampal astrocyte were observed following stimulation of Schaffer collaterals (10). These $[Ca^{2+}]_i$ elevations can either remain confined to the process or propagate intracellularly to the soma and other processes, suggesting that each astrocyte process represents an independent compartment of $[Ca^{2+}]_i$ signaling.

Under conditions where antioxidant agents are absent, the number of dying neurons is also greatly increased. In Figure 1B, several neurons are well loaded with the dye but display high $[Ca^{2+}]_i$ and are most likely destined to die in a short time. Curiously, small cells with a mean soma diameter of 5 to 10 μm are, in general, less loaded in the presence than in the absence of antioxidant agents. These cells are, most likely, astrocytes. On the basis of pure morphological criteria, however, astrocytes can be hardly distinguished from small neurons with a stellate or bipolar shape. We thus developed a new experimental approach to functionally distinguish each astrocyte and neuron present in the recording field. The experiment illustrated in Figure 1 refers to the CA1 hippocampal region, though identical results are obtained in the visual cortex. We took advantage of the observation that in mixed neuron–astrocyte cultures, while neurons, as identified by immunocytochemical criteria, responded promptly with a $[Ca^{2+}]_i$ increase to depolarization induced by 60 mM K^+, none of the immunocytochemically identifiable astrocytes were sensitive to this treatment. We thus applied the same protocol to brain slices in order to analyze whether neurons and presumed astrocytes responded differently, as in culture, to stimulation with 60 mM K+. All cells display a significant [Ca^{2+}]$_i$ increase upon depolarization with high extracellular K+. The onset of the response from pyramidal neurons and smaller cells (arrows) is, however, clearly different. In pyramidal neurons, the perfusion with high K+ induces a prompt [Ca^{2+}]$_i$ increase, which is followed by a further [Ca^{2+}]$_i$ elevation. In contrast, small-sized cells with astrocyte-like morphology initially fail to respond and display a clear [Ca^{2+}]$_i$ increase several seconds after that of pyramidal neurons. This response occurred at approximately the same time of the second [Ca^{2+}]$_i$ peak in the neurons. The delayed [Ca^{2+}]$_i$ increase in these cells is, to a large extent, attributable to the release of glutamate and/or other neurotransmitters by depolarized synaptic terminals and to the subsequent activation of metabotropic receptors linked to inositol–trisphosphate (IP$_3$) production (10). The remarkable delay that occurs in the response of astrocytes upon a depolarizing stimulus with 40 to 60 mM KCl with respect to the prompt [Ca^{2+}]$_i$ increase in neurons can be used as a functional tool to identify astrocytes from acute brain slices (1,10). This type of response represents also the first of a series of observations that led us to conclude that astrocytes in acute brain slices, at least from the visual cortex and the CA1 hippocampal region of young rats at postnatal day 5 to 18, do not significantly express functional voltage-operated Ca^{2+} channels (1).

Future Directions for Ca^{2+} Imaging in Brain Slices

Although a wealth of information can be obtained using indo-1 in combination with CLSM, two major disadvantages need to be taken into consideration. The first is that using UV excitation light may cause damage to the tissue, and the second is that, since UV light has a short wavelength, major light scattering will occur. The result of this last point is that imaging deep into the tissue will become practically impossible. Both disadvantages are overcome by the development of multiphoton excitation microscopy (MPE) (for overview see Reference 2). In this microscopy mode, fast pulses of laser light are phased into the same plane of focus in the tissue, and 2 or 3 individual photons will cause dye excitation within the same excitation event. As a consequence, laser light with a considerably longer wavelength and lower energy is used in this technique. Generally, MPE microscopes are equipped with near infrared lasers. Therefore, no UV light will damage the specimen, and near infrared light will undergo less scattering, making imaging deeper in the tissue possible. An additional advantage is that only the plane of focus is illuminated, and photobleaching, photodamaging, and photon toxicity of the out-of-focus field (which is the majority of the slice) is avoided.

A limitation of the use of CLSM in the study of [Ca^{2+}]$_i$ homeostasis in cells from acute brain slices is the insufficient penetration of the fluorescent dye in cells from slices of fully mature animals. Recently, a new procedure (6) resulted in a great improvement of viability and dye loading of cells from acute brain slices. After deep anesthesia, rats are perfused through the heart with cold physiological saline (see solution N. 2). After removal, the brain is processed as usual. Slices obtained by using this protocol could be maintained in excellent conditions for at least 8 hours, and cells are more easy to patch. In transverse spinal cord slices, we recently noticed that the number of substantia gelatinosa neurons displaying good loading with indo-1 is substantially increased. Even more importantly, under these conditions patch clamp recordings can be successfully performed in cells from slices obtained from animals of 2 months of age. Preliminary observations suggest that load-

16. Confocal Imaging of Calcium Signaling in Cells from Acute Brain Slices

ing with indo-1 in these cells is also greatly improved, thus allowing the use of the confocal microscope approach in future experiments also for the study of calcium signaling in slices from fully developed brain.

ACKNOWLEDGMENTS

This work was supported by the Armenise-Harvard University Foundation and Telethon-Italy Grant No. 1095 to G.C. We thank Cristina Fasolato and Tullio Pozzan for helpful discussion and critical reading of the manuscript.

REFERENCES

1. Carmignoto, G., L. Pasti, and T. Pozzan. 1998. On the role of voltage-dependent calcium channels in calcium signaling of astrocytes in situ. J. Neurosci. 18:4637-4645.
2. Denk, W. and K. Svoboda. 1997. Photon upmanship: why multiphoton imaging is more that a gimmick. Neuron 18:351-357.
3. Di Virgilio, F., T.H. Steinberg, and S.C. Silverstein. 1990. Inhibition of fura-2 sequestration and secretion with organic anion transport blockers. Cell Calcium 13:313-319.
4. Grynkiewicz, G., M. Poenie, and R.Y. Tsien. 1985. A new generation of Ca^{2+} indicators with greatly improved fluorescence properties. J. Biol. Chem. 260:3440-3450.
5. Hofer, A.M. and W.J.J.M. Scheenen. 1999. Imaging calcium in the cytoplasm and in organelles with fluorescent dyes: general principles, p. 53-91. In R. Rizzuto and C. Fasolato (Eds.), Imaging Living Cells. Springer-Verlag, Berlin.
6. Hoffman, D.A. and D. Johnston. 1998. Down regulation of transient K^+ channels in dendrites of hippocampal CA1 pyramidal neurons by activation of PKA and PKC. J. Neurosci. 18:3521-3528.
7. Koopman, W.J.H., M.A. Hink, A.J.W.G. Visser, E.W. Roubos, and B.G. Jenks. 1999. Evidence that Ca^{2+}-waves in *Xenopus* melanotropes depend on calcium-induced calcium release: a fluorescence correlation microscopy and linescanning study. Cell Calcium 26:59-67.
8. Lipp, P. and E. Niggli. 1993. Ratiometric Ca^{2+} measurements with visible wavelength indicators insolated cardiac myocytes. Cell Calcium 14:359-372.
9. Nuccitelli, R. (Ed.). 1994. A practical guide to the study of calcium in living cells. In Methods in Cell Biology, Vol. 40. Academic Press, New York.
10. Pasti, L., A. Volterra, T. Pozzan, and G. Carmignoto. 1997. Intracellular calcium oscillations in astrocytes: a highly plastic, bidirectional form of communication between neurones and astrocytes in situ. J. Neurosci. 17:7817-7830.
11. Pawley, J.B. (Ed.). 1995. Handbook of Biological Confocal Microscopy. Plenum Press, New York.
12. Poenie, M., J. Alderton, J. Steinhart, and R.Y. Tsien. 1986. Calcium rises abruptly and briefly throughout the cell at the onset of anaphase. Science 233:886-889.
13. Rice, M.E., M.A. Pérez-Pinzón, and E.J.K. Lee. 1994. Ascorbic acid, but not glutathione, is taken up by brain slices and preserves cell morphology. J. Neurophysiol. 71:1591-1560.
14. Rizzuto, R. and C. Fasolato (Eds.) 1999. Imaging Living Cells. Springer-Verlag, Berlin.
15. Scheenen, W.J.J.M., A.M. Hofer, and T. Pozzan. 1998. Intracellular measurements of calcium using fluorescent probes, p. 363-374. In J.E. Celis (Ed.), Cell Biology: A Laboratory Handbook. Academic Press, New York.
16. Scheenen, W.J.J.M., B.G. Jenks, E.W. Roubos, and P.H.G.M. Willems. 1994. Spontaneous calcium oscillations in *Xenopus laevis* melanotrope cells are mediated by ω-conotoxin sensitive calcium channels. Cell Calcium 15:36-44.
17. Scheenen, W.J.J.M., L.R. Makings, L.R. Gross, T. Pozzan, and R.Y. Tsien. 1996. Photodegradation of Indo-1 and its effects on apparent Ca^{2+} concentration. J. Chem. Biol. 3:765-774.

Index

A
ABC, 71–72, 162–163, 178, 191, 195, 197, 210, 214, 224, 226, 228, 264
Acetylation, 136, 152–154, 157
Acetylcholinesterase, 119, 176
AChe, 119, 176
Acridine Orange, 89
Acrolein, 20, 25, 165
Acrylamide, 245
Acrylic embedding media, 166
Adaptor proteins, 1, 12
   Grb2, 1, 10–11
   Shc, 1–2, 12
Adeno-associated virus, 83
Adenoviral vectors, 51
Adenovirus, 36, 68
Adrenergic receptor(s), 54
Agarose gel electrophoresis, 41–42, 87, 96, 100, 106, 242
AgNORs, 238
Alkaline phosphatase. See AP
Alkaline phosphatase streptavidin. See AP-streptavidin
Amacrine cell(s), 86–87, 91, 107
Amino acid(s), 167, 176, 178, 179, 182, 205
   anterograde tracing with, 176, 178, 180, 203, 205, 207–208, 211, 224–225, 228, 231–233
   isotope-conjugated neurotransmitters, 204
   isotope labeled amino acids, 205
   tritiated amino acids, 205
7A6 monoclonal antibody, 240
AMPA, 65–66, 87, 107, 201
Amplicon, 146, 149, 156
Amyloid precursor protein. See APP
Annealing temperature, 98, 156

Annexin V, 237, 239, 255, 258
   annexin V-Cy3, 57, 92–94, 100, 105
   annexin V-FITC, 19, 54, 57, 125, 127–128, 210, 239, 262, 270
   binding assays, 237, 239
Anterograde degeneration, 203
Anterograde labeling. See Anterograde tracing
Anterograde tracers. See Anterograde tracing; Tracer(s)
Anterograde tracing, 176, 178, 180, 203–205, 207–208, 211, 212, 216–218, 224–225, 228, 231–233
   BDA, 205–207, 224–225, 227–230, 232
   biocytin, 182, 191–193, 198, 201, 205, 207, 210, 212, 224–225, 232–233
   carbocyanine derivatives, 205, 210, 216
   CTb, 205–208, 223–231, 233, 261 263, 265–267, 270
   DiA, 205, 207, 211
   DiI, 205, 207, 211, 215
   DiO, 205, 218
   Fast DiI, 205
   Fluoro-Emerald, 205–207
   Fluoro-Ruby, 205–207, 219
   HRP, 128, 162, 176, 178, 183–186, 188, 198, 205, 208, 210–214, 216, 219, 223, 231, 234
   Lipophilic carbocyanine dyes, 205
   neurobiotin, 182, 191–192, 195, 198, 201, 205, 207, 224–225, 232, 261, 263
   PHA-L, 205–208, 211–212, 218, 224–225, 232–234, 261–263
Antigenicity, 130, 136, 162, 165–168, 173, 179, 187, 254, 263
   antigen-antibody reaction, 161
   antigenic site, 197

Index

antigen(s), 3, 8, 12, 17, 19, 87, 90, 120–121, 128, 136, 141, 143, 161, 162, 165, 167, 170–180, 187, 201, 216, 234, 236, 238, 240, 246, 248, 249, 251–252, 254–262, 270
  masking of antigenic sites, 137
  spatial distribution of antigen, 173, 260
  subcellular localization of antigens, 4, 65, 138, 142, 166, 170, 173
Antisense primers, 98
AP, 49, 51, 85, 107–108, 120, 122, 124–125, 130, 142, 145, 149–151, 153, 163, 176, 177, 178, 226, 229, 231, 244, 248, 260, 272
  streptavidin, 226, 229
ApoHRP
  WGA, 205
Apopain. See Caspase 3
Apoptosis, 100, 103, 106, 161, 235, 237, 239–243, 245, 247–259
  apoptotic bodies, 240, 251
  apoptotic cells, 235, 237, 239–241, 244, 248–255, 257–258
  apoptotic nuclei, 240–241
  apoptotic program, 237, 240
  biochemical assays, 235, 241
  in situ detection, 141, 158, 239, 242, 250, 256
APP, 33–36
  anti-human, 33
  distribution in axons, 34–35
  distribution in dendrites, 34–35
Argyrophilic nucleolar organizer region proteins. See AgNORs
Ascorbic acid, 276, 281, 283
Astrocyte(s), 2, 31, 36, 39, 76, 78, 170, 173, 256, 273, 280–283
Asymmetrical synapses, 265
Autofluorescence, 208, 213, 215, 232
Autolysis, 241
Autoradiographic procedures, 137, 139, 205, 218, 254
Autoradiography, 16, 49, 99, 122, 132, 178, 205, 208, 236, 249
  silver grains, 26, 124, 139, 236, 249
  tritiated amino acids, 205
  tritiated tymidine, 236, 248–249
Avidin-biotin-peroxidase complex. See ABC
Avidin-fluorochrome, 261
Axon(s)
  axonal boutons, 201, 229, 269
  axonal tracers, 204, 215
  axonal transport, 203, 205, 209, 212, 218
  filling in ultrastructural tracing studies, 223
  silver impregnation of degenerating axons, 203, 218–219

B

Background, 2, 12, 16, 26, 29, 37, 39, 48, 53, 57, 67, 85, 120, 135–136, 145, 149, 154, 156, 161, 170, 173, 175, 181, 193, 198, 204, 213, 221, 228, 231–232, 235, 243, 247, 250, 259, 261–262, 273
  fluorescent tracers, 205, 208, 212–215, 218, 223–224
  subtraction (in confocal microscopy), 57, 193, 195, 198, 200, 216
  in ultrastructural tracing studies, 223
BamHI, 55
Banderaea simplicifolia. See BSI-B4; Lectins
BCIP, 122, 130, 132, 150, 156, 244
Bcl-2, 240, 248, 254, 256
BDA, 176, 205–207, 224–225, 227–230, 232
BDHC, 222
Benzamidine, 12
Benzidine dihyrochloride. See BDHC
Benzodiazepine, 54, 58
BET, 47
Bicuculline, 53, 62
Bi-directional tracing with CTb, 226, 228
Bi-directional transport of tracers, 216, 225
Binding studies, 140
Biocytin, 182, 191–193, 198, 201, 205, 207, 210, 212, 224–225, 232–233
Biolistics, 67–69, 71, 73, 76–78, 82–84
  gene gun, 67–69, 73–76, 78, 83–84
  tungsten particles, 68
Biotin, 121, 125, 127, 130, 132, 134, 147, 162, 191, 195, 206–208, 210, 214, 217, 224, 241, 261, 263–264
  biotin-based tracers, 205, 208, 210, 214, 224–225
Biotinylated dextran amine. See BDA
Biotinylated nucleotides, 242
Bisbenzimide, 237
Bleaching, 274–275, 280
Brain slices, 15–16, 18–19, 21, 23, 25, 69, 83–84, 88, 273–275, 277–283
BrdU, 174, 236, 238, 242–243, 246–250, 253, 256
5-bromodeoxyuridine, 236, 247
BSI-B4, 261, 263, 265–266, 269
B-tubulin CMV, 33, 35, 40, 78

C

CA1 hippocampal region, 280–282
$[Ca^{2+}]i$ increase, 65, 84, 107, 239, 256, 273–275, 277–283
Calbindin D28k, 69
Calcitonin gene-related peptide, 174, 179, 265, 271
Calcium, 12, 29–30, 34–36, 54, 56, 67–69, 76, 78, 83, 88, 91, 95, 111, 239, 257, 273, 275,

Index

277, 279, 281, 283
binding proteins, 69, 83, 258
[Ca2+]i increase, 65, 84, 107, 239, 256, 273–275, 277–283
channels, 54, 60, 65, 83, 86–88, 90, 107–108, 111, 182, 281–283
Green, 35, 39, 53, 57, 65–66, 69, 84, 90, 107–108, 124, 127, 137, 206–207, 217, 237, 256, 266–268, 279–280
phosphate DNA co-precipitation. See Calcium phosphate precipitation
phosphate precipitation, 30, 56
Calcium indicators, 239
  calcium green, 279
  calibration, 278
  Fluo-3, 239, 279
  Fura-2, 239, 283
Calibration
  calculating absolute [Ca2+] values, 279
  extracellular calibration, 278, 280, 282
  intracellular calibration, 278
  solution, 276–280, 282
Carbocyanine derivatives, 205, 210, 216
CARD, 124, 128, 140–142
Casein-kinase 2a, 238
Caspase 3, 240, 242, 255–258
Caspase(s), 240, 242–243, 255, 258
  caspase 3, 240, 242, 255, 257
  caspase-9, 2
CAT, 3, 6, 29, 39–40, 45–46, 51, 78–82, 177–180, 182–183, 196, 201, 232–233, 238, 272
Catalase(s), 212
Catalyzed reporter deposition. See CARD
Catecholamines, 91
Cationic lipid transfection, 29
Cationic polymers, 38
CD 44 cell surface adhesion molecule, 240
CDNA, 17–19, 22–24, 34, 36, 56, 58–61, 85–88, 90, 92, 95–96, 98–100, 102, 104, 107–108, 147, 149, 152–153, 155–158
CDNA template(s), 147
Cell, 2, 5, 7, 9–10, 15–19, 21–25, 27–29, 32–34, 36, 38, 43, 45, 48, 51, 53–56, 58–62, 65, 67–69, 73–74, 76–93, 95–98, 100, 102–104, 106–108, 109–111, 115, 121, 124, 132, 135, 138–139, 141–143, 145–146, 152, 154, 157–158, 161, 171, 173–183, 186–190, 192–201, 203–205, 208, 215–218, 224, 230, 232–243, 245–260, 263, 265–267, 269, 271–272, 274–275, 277, 280–281, 283
  compartment-specific (cellular) antigens, 19
  cycle, 99–100, 155, 235–236, 238, 247, 254–257
  death, 2, 224, 235, 242, 254–258

death apoptotic. See Apoptosis
death autophagic, 239
death non-lysosomal vesiculate, 239
division, 236–238, 247, 249
grafting, 38
lines, 38, 51, 54, 74, 76–77, 89, 107, 256, 262, 280
lysates, 5, 9–10, 33, 36
membrane asymmetry, 239
proliferation, 1–2, 12, 55, 161, 235–239, 241, 243, 245–249, 251, 253–257
proliferation markers, 236
ultrastructure, 28, 161–162, 165–166, 171, 178–180, 186, 231, 233–234, 240, 263
Cerebellum, 69–70, 109–114, 146, 180, 245–248, 250, 252–254, 256–258
  cerebellar cortex, 174, 246, 249, 254, 256
  cerebellar interneurons, 246, 258
  cerebellar neurons, 2, 53–54, 60, 63
  EGL, 246–250, 252
  granule cell precursors, 174, 246, 249–250, 256
  granule cells, 56, 65, 82, 88, 109–110, 112, 174, 234, 237, 246, 247, 249–250, 256, 257–258
  IGL, 246–248, 250, 253
  ML, 4–8, 11, 17, 20, 30–34, 40–41, 44–46, 55–57, 70–72, 75, 93–97, 100, 121, 150–154, 156, 185–188, 215, 227–229, 243–248, 261, 263, 276, 278
  Purkinje cells, 69, 83, 89, 107, 146, 254
C fibers, 265–266, 268
C-fos, 1, 215, 218
CGRP, 124, 174, 200, 231, 265–266, 268–269
Chemiluminescence. See ECL
Chemiluminescent detection system. See ECL
Chloramphenicol acetyl transferase. See CAT
Chloride, 33, 53–54, 60, 91, 150, 152, 156, 165, 183, 187, 191, 193, 205, 226–227, 244–245, 263
  current, 53, 56, 59, 158, 161, 186, 193, 200, 211, 227
  ions, 91, 95
Choleragenoid. See CTb
Cholera toxin. See CTb
Cholera toxin B. See CTb
Chromatin, 237, 240–241, 251–253, 258
  condensation, 37, 41, 47, 240, 251–252
  eterochromatin, 249–250
Chromatography, 47, 94–95, 100
C-jun, 8
Clathrin -coated pits, 114
CLSM. See Confocal microscope
CMV-CAT, 45–46
CMV promoter, 33, 35, 40, 78

Index

CNTF, 8–9
Colocalization, 54, 57, 61, 63, 79–80, 82, 142, 179, 234, 272
Combination of transgenic technology with single-cell RT-PCR, 101
Combined confocal and electron microscopy, 263, 270
Combined primer annealing, 156
Compartment-specific cellular antigens, 19
  lamp proteins, 19
  rab proteins, 19
  syntaxin, 6, 23–24
Concavalin A, 95–96, 97
Cone bipolar(s), 86, 179
Confocal laser scanning microscope. See Confocal microscope
Confocal microscope, 16–19, 22, 28, 120, 122, 127, 138, 141, 193, 194, 195, 198, 200, 216, 218, 233, 237, 259–260, 262, 264, 265, 268, 270–275, 278, 279, 282, 283
  confocal microscopy of retrogradely labeled neurons, 223, 225
  correlated studies of confocal microscopy and electron microscopy, 222
  image acquisition speed, 275
  photobleaching, 54, 262, 278, 282
  photodamaging, 278, 282
  pinhole, 274–275
  point scanning, 275
  ratiometric dyes, 279–280
  ratio value, 279–280
  refractive index, 271
  sampling speed, 275
  signal-to-noise ratio, 194, 214, 274
  spatial resolution, 26, 173–174, 260, 274–275
  stacks of confocal images, 271
  three-dimensional reconstruction of neurons, 223
  UV, 41, 98, 167, 270, 278–279, 282
Confocal microscopy. See Confocal microscope
Correction factors, 271
Cortex, 49, 69–70, 83, 86, 109–114, 146, 174, 183, 191, 200–201, 205, 217, 234, 246, 249, 254, 256–257, 271, 281–282
  cortical neurons, 36, 55–56, 60–61, 63–64, 201
  pyramidal neurons, 183, 191, 280–283
COS-7 cells, 17–19, 22–24, 26, 28
Cotransfection, 34–35, 67, 82
CP, 29–30, 34–35
CPP32. See Caspase 3
CREB, 8, 84
Cresyl violet, 214, 218
Cross-contamination, 106
Cross-reactivity in ultrastructural tracing studies, 223, 225, 230

Cryoprotectant medium, 71
Cryoprotection, 195
Cryostat sections, 46, 100, 121–122, 130, 132, 135–136, 138, 152, 165, 254
Cryosubstitution, 162
CTb, 205–208, 223–234, 261–263, 265–267, 270
  bidirectional tracing, 221, 226, 228
  conjugate with saporin, 224
  gold, 208, 211, 217–218, 223, 231, 233, 243–244, 250–251, 253, 255
  neuronal uptake of CTb tracers, 223
  saporin-CTb conjugate, 224
  studying projection neurons, 270
  unconjugated CTb, 223–225, 229
Curing, 166, 173, 189
Cy3.5, 57, 92–94, 100, 105
Cyanine, 262
Cy3-conjugated antibody, 90
Cytokines, 1–2
  interferon, 1, 115
  interleukin, 1, 257

D

DA, 28, 49, 51, 85, 91–92, 95, 100, 102, 105–106, 107–108, 110–111, 142, 178–179
  biosynthesis, 85, 107, 196
  transporter. See DAT
DAB, 71–73, 125, 128, 130, 134, 163, 165, 172–173, 175, 183, 187–188, 195–196, 210, 214–217, 222, 227–230, 233, 263–264, 270. See also HRP substrate(s)
DAB reaction product, 187, 215, 270
DAPI, 89, 237
Dark-field observation (of tracers), 218
DAT, 49, 51, 115, 179, 191, 193, 201, 277
DATP, 241
DBH, 227, 229–230
Dead cells. See Apoptosis
Degeneration methods, 176, 180–181
  anterograde degeneration, 203
  retrograde degeneration, 203
  Wallerian degeneration, 203
Degeneration techniques. See Degeneration methods
Dehydration in ultrastructural tracing studies, 223
Dendrites, 23, 27, 30, 34–35, 57, 61–62, 69, 83, 95, 139, 142, 196–197, 201, 204, 206, 216–217, 223, 230, 233–234, 259, 261, 265–268, 270–272, 280–281, 283
  dendritic processes, 76
  dendritic synthesis, 141
  density of contacts onto the dendritic tree, 260, 270
  filling in dendritic tree, 62, 200, 223, 225, 260,

Index

265, 267, 270
  filling in ultrastructural tracing studies, 223
  length, 205, 236, 271
Densitometry, 242
Density filters, 274
Desensitization, 25, 58
Detector gain, 275
Detergents in ultrastructural tract tracing studies, 222–223
DEVD tetrapeptide, 243
Dextran(s), 121, 176, 205–207, 211, 217, 224, 232–234, 261, 263
Diabete, 68
  diabetic states, 68
4',6-diamidino-2-phenylindole, 237
Diamidino yellow, 206–207
3,3' diaminobenzidine, 71, 128, 263
Dichroic mirror, 274, 278
DIG, 121–122, 125, 147, 151–152, 154–156, 241
  digoxigenin labeled oligoprobes, 121
  Digital camera, 57, 187, 189, 214. See also Image analysis
  Digoxigenin. See DIG
  Dilution of primary antibodies, 19–20, 168, 170
  Dimethyl-chloro-sylane, 96–97
DiO, 205, 218
Direct delivery of genes, 38
Direct in situ PCR, 146–147, 149, 158
Direct in situ RTPCR, 146–147, 149–150
DNA, 2, 4, 9, 29–32, 36–39, 41–42, 47–48, 51, 55, 65, 67–68, 73–74, 76, 82–84, 87, 98–100, 106, 121, 142, 145–147, 149, 151–154, 156–159, 236–242, 247, 249–251, 253, 255–258
  agarose gel electrophoresis, 41–42, 87, 98–100, 106, 121, 142, 242
  annealing temperature, 98, 156
  condensation, 37, 41, 47, 240, 251–252
  fragmentation, 196, 239–242, 251, 253, 255, 258
  modifying enzyme(s), 241, 253
  nick(s), 241, 255, 257
  nick translation, 241, 255
  nuclease(s), 241
  nucleosome(s), 241
  polymerase, 55, 85, 88, 99, 103, 140, 145, 151, 153–154, 156, 158, 238, 241–242, 251, 256–257
  polymerase a, 238
  thermal denaturation, 240
DNAse, 96–97, 121, 150–154, 157
Dopamine. See DA
Dopamine-beta-hydoxylase. See DBH
Dopaminergic amacrine cell(s), 91
Dorsal horn, 111, 113–114, 174, 178–180, 182, 190, 200–201, 229–233, 261, 264–267, 269, 271
Dorsal root ganglion, 29, 111, 113–114
Double face method. See Multiple immunolabeling
Double fluorescent detection, 142
Double fluorescent ISH, 122–123, 128, 138
Double immunogold staining, 170. See Multiple immunolabeling
Double ISH at the electron microscope level, 125, 129, 138
Durcupan, 263–264
DUTP, 146, 241
Dye(s), 87, 88, 89, 90, 92–94, 103, 120, 137, 182, 189, 193, 204–205, 212–214, 216, 217–218, 233, 237, 239, 245–246, 253, 259–263, 273–283. See also Fluorescent dyes
  intra-axonal injection, 204
  intracellular injection, 184, 192, 259, 261, 263, 269
  lipophilic dyes, 204
Dying cells, 237, 239, 241, 252. See also Apoptosis
Dying neurons, 280–281. See also Apoptosis
Dynamic of neurotransmitter receptors, 54

E

E6, 90, 92–93, 100, 103, 105
ECL, 246
EcoRi, 55
E6-Cy3, 85, 90, 92–93, 95, 97, 100, 103, 105
EGF, 1, 11
EGFP-N1, 35–36, 55–56
EGL, 246–248, 250, 252, 257
E6-Hy hybridoma cells, 92, 100
Electron microscope, 18, 20, 89–90, 91, 119–120, 125, 127, 128–129, 135–140, 142, 161, 164, 166, 169, 177–180, 182, 189, 191, 195, 197, 200, 205, 213, 216, 219, 221–222, 229, 230–234, 241, 243–244, 249, 251, 256, 259, 260, 263, 264, 270, 272, 276
  correlated light and electron microscopy analysis, 222
  correlated light and ultrastructural observations in ultrastructural tracing studies, 223
  structural preservation, 167, 222
Electron microscopy. See Electron microscope
Electrophoresis, 5, 7–8, 12, 33, 41–42, 47, 87, 94, 96, 100, 106, 156–157, 242, 245–246, 251
  agarose gel, 41–42, 87, 96, 100, 106, 242
Electroporation, 32, 35–36, 67–68, 76
ELISA, 39, 242
Embedding, 28, 120, 138, 162–163, 166–167, 173, 175–176, 178, 180, 188, 197, 229, 231, 249

Index

Embedding (continued)
    acrylic embedding media, 166
    durcupan, 263–264
    epon, 21–22, 127, 130, 166–167, 179, 185, 188–191, 195, 198, 227, 229
    epoxy resins, 166, 171
    low temperature, 167, 180, 196
    paraffin, 32, 138, 157
    post-embedding immunocytochemistry, 90, 166, 168–169, 177, 179, 197
    post-embedding immunogold, 120, 163, 171, 173–176, 179, 189, 197, 201, 249
    pre-embedding immunocytochemistry, 162, 175, 197
    pre-embedding immunogold, 15, 19, 171, 197
    resin infiltration, 166
Embryos, 10–11, 141, 211
    embryonic development, 235, 255
Emission, 44, 90, 237, 239, 261, 274, 279–280
End labeling, 241
Endocytosis, 16, 23, 25–29, 90, 232
    inhibitors, 10, 12, 136, 245, 254
Endonuclease(s), 255, 258
    enzyme histochemical techniques, 176
Endosome(s), 16, 23, 28
Epidermal growth factor. See EGF
Episomally-located sequences, 51
Epitope(s), 2, 19–20, 34, 173, 238
Etching, 166–167, 171, 177, 189
Ethanol, 5, 7, 41, 44, 70, 75, 121, 153–155, 167, 169, 185, 191, 210, 213, 222, 225, 227–229, 232, 262, 271–272
    to increase tissue penetration, 163, 171–172
EthD-2, 237
Ethidium homodimer-2. See EthD-2
Evans blue, 205–207
Excitation, 57, 90, 98, 208, 214, 237, 261, 274, 278–279, 282
    emission filters, 44
Excitatory Synapses, 190
Expression library, 36
Expression vector(s), 32, 35, 47, 51, 53–56
    pCDM8, 55
    pEGFP-N1, 35–36, 55–56
External granular layer. See EGL
Extracellular recording, 182
Extravidin-peroxidase, 264
Ex vivo gene transfer, 68

F

Fab' fragments, 165, 174, 244
Fast Blue, 205–207, 227, 229–231
Fast DiI, 205

Fast red, 122, 125, 128
FDG, 45
Ferritin, 28, 222
FGF, 1
Fiber(s) en (of) passage, 203, 205
Fibroblast Growth Factor. See FGF
Filter(s), 10–11, 44, 55, 57, 70–73, 150, 169–170, 210, 214–215, 243, 260–261, 263, 274
FITC, 19, 44–45, 54, 57, 90, 121, 125, 127–128, 147, 210, 237, 239, 262, 270
Fixation, 21–22, 25, 39, 44–46, 91, 120, 127, 130, 132, 135–136, 140, 150, 152–154, 157, 161, 162–163, 165–169, 171, 172, 173, 175–176, 178, 186–187, 193, 200, 205–208, 210, 212, 213, 222–225, 227–228, 232, 237–238, 249, 254, 261–263, 272
    acrolein, 20, 25, 165
    formaldehyde, 46, 57, 71–72, 91, 135, 141, 150–152, 166, 169, 175, 193, 213, 261–263
    glutaraldehyde, 20–21, 25, 91, 135, 165–170, 172, 175–178, 183, 186–187, 191, 193, 206–207, 210, 213, 222–225, 227–228, 232, 244–245, 249, 263, 272
    glutaraldehyde as a fixative for ultrastructural tracing studies, 223
    immersion, 135, 152, 187
    osmium ferrocyanide, 167, 169
    osmium in ultrastructural tracing studies, 222–223
    osmium tetroxide, 20–21, 25, 161, 166–169, 183, 191, 222, 243, 249
    paraformaldehyde, 17–18, 25, 33, 44, 122, 125, 127, 130, 132, 150, 152–153, 157, 165, 167–169, 183, 186, 191, 193, 206–207, 210, 213, 222, 227–228, 263
    paraformaldehyde as a fixative for ultrastructural tracing studies, 223
    picric acid, 135, 165, 168, 191, 193
    post-fixation, 22, 25, 45, 137, 162, 166–167, 173, 187, 222–223, 228
    slides, 44, 132, 135–136, 149–150, 152–156, 165, 210, 262, 264
    tannic acid, 166, 180
    uranyl acetate, 167, 169–170, 189, 245, 264
Fixative. See Fixation
Flow cytometry, 159, 239, 255
Fluo-3, 239, 279
Fluorescein. See FITC
Fluorescein isothiocyanate, 19, 54, 127–128, 210, 262
Fluorescence, 45, 57, 60–63, 90, 127, 137, 143, 189, 206–207, 210, 213–215, 217–218, 237, 239, 241, 260, 268, 274–275, 279, 283
    intensity, 57, 60–61, 274–275, 278–279

Index

microscope, 45, 210, 214
Fluorescence microscope. See Fluorescence
Fluorescen-conjugated dextran. See Fluoro Emerald
Fluorescent bead. See Fluorescent microspheres
Fluorescent clusters, 53
Fluorescent dye(s), 88, 182, 189, 237, 239, 253, 260–262, 279, 282
   acridine Orange, 89
   annexin V-Cy3, 57, 92–94, 100, 105
   annexin V-FITC, 19, 54, 57, 125, 127–128, 210, 239, 262, 270
   bisbenzimide, 237
   bodipy, 19
   Calcium Green, 35, 39, 53, 57, 65–66, 69, 84, 90, 107–108, 124, 127, 137, 206–207, 217, 237, 256, 266–268, 279–280
   Cy3.5, 57, 92–94, 100, 105
   cyanine, 262
   DAPI, 89, 237
   Dextran(s), 121, 176, 205–207, 211, 217, 224, 232–234, 261, 263
   Diamidino yellow, 206–207
   EthD-2, 237
   Ethidium bromide, 42, 47, 99–100
   Evans blue, 205–207
   Fast Blue, 205–207, 227, 229–231
   Fast DiI, 205
   Fast red, 122, 125, 128
   FITC, 19, 44–45, 54, 57, 90, 121, 125, 127–128, 147, 210, 237, 239, 262, 270
   Fluo-3, 239, 279
   Fluorogold, 206–207, 223–224, 231–234, 270
   Fura-2, 239, 283
   Fura Red, 279–280
   Hoechst 33342, 237
   Hoechst dyes, 237
   Indo-1, 273, 275–283
   Indocarbocyanine, 55
   Lipophilic carbocyanine dyes, 205
   Lissamine rhodamine (TRITC), 210, 262
   Lucifer yellow, 88, 191–193, 198, 200, 263
   mBBr, 239
   mBCl, 239
   Merocyanine 450, 239
   Mito Tracker Red CMX Ros, 239
   Oregon green, 279–280
   SNARF-1, 239
   SYTO stains, 237
   SYTOX Green, 237
   Texas Red, 22, 217, 241
   TRITC, 210
   YO-PRO(tm)-1 iodide, 205, 237
Fluorescent ligands, 15–16, 28, 140
Fluo-NT, 17, 22–25
Fluo-SRIF, 17, 22–24
Fluorescent microspheres, 205–208
Fluorescent stains. See Fluorescent dye(s)
Fluorescent tracer(s), 205, 208, 212–215, 218, 223–224. See also Fluorescent dye(s)
   carbocyanine derivatives, 205, 210, 216
   CTb, 205–208, 223–234, 261–263, 265–267, 270
   DiA, 205, 207, 211
   diamidino yellow, 206–207
   Evans blue, 205–207
   Fast Blue, 205–207, 227, 229–231
   Fast DiI, 205
   Fast red, 122, 125, 128
   Fluoro-Emerald, 205–207
   Fluoro-Ruby, 205–207, 219
   lipophilic carbocyanine dyes, 205
   nuclear yellow, 206–207
   PI, 205–206, 237
Fluorochrome(s), 92, 134, 198, 206–208, 215, 223–224, 241, 261, 263
   fluorochrome-conjugated secondary antibodies, 210, 214
   fluorochrome-conjugated streptavidin, 206–207, 263
   fluorogenic substrate(s), 242
   wavelength, 208, 214, 262, 279–280, 282–283
Fluoro-Emerald, 205–207
Fluorogold, 206–207, 223–224, 231–234, 270
Fluoro-Ruby, 205–207, 219
Focus, 103, 120, 192, 204, 260, 268, 271, 274, 279, 282
Forkhead factors, 2
Formaldehyde, 46, 57, 71–72, 91, 135, 141, 150–152, 166, 169, 175, 193, 213, 261–263
Freeze-thawing, 138, 165, 171, 196
Frontal cortex, 70, 200
Fura-2, 239, 283
Fura Red, 279–280

G

GABAA, 53–55, 57–61, 63, 65–67, 92, 102, 201
   clusters, 32, 53, 57, 61–65, 162–163, 166, 174, 190
   colocalization, 61, 63, 79–80, 82
   gabaergic presynaptic terminals, 61
   gabaergic synapse(s), 53, 65, 178, 201
   GFP, 35–36, 39, 48–49, 53–55, 57–63, 65, 69, 78, 80–82, 90
   receptor, 1–2, 8, 34, 36, 51, 63, 65–66, 83, 86–88, 88, 92, 95, 102, 107, 109–110, 115, 119, 124, 138, 158–159, 165, 181, 201,

Index

GABAA (continued)
   215, 223, 233, 234, 238, 240, 257, 259, 261, 278, 282
   receptor clusters, 53, 61, 65
   receptor subunits, 53–54, 57–58, 61, 92, 102, 107, 141
   subunits, 53–66
GAD, 61, 63, 65, 87, 115, 178, 183
Gal, 113, 115, 142
   construct, 30–31, 38, 41–42, 45
   polyclonal antibody, 33, 46, 48–49, 77, 245–246
   Galactosidase. See Gal
Galanin, 115, 124, 127, 132, 138, 142–143, 272
Gelatin, 20, 34, 45, 125, 127, 130, 138, 168, 191, 193, 213
Gel-based DNA fragmentation assay, 242
Gene, 1, 29, 32, 36–40, 46–51, 55, 67–69, 71, 73–89, 91–93, 95, 97, 99, 101–103, 105, 107–108, 109, 124, 138–142, 149, 155, 158, 215, 276
   expression, 1, 28, 30, 32–33, 35–39, 45–47, 49–51, 53–56, 65–69, 76, 78, 80–89, 91–93, 95, 97, 99, 101–103, 105, 107–108, 109, 138–139, 141–142, 158
   Gene Gun, 67–69, 73–76, 78, 83–84
   product, 45, 46, 49, 55, 72–73, 86, 88–91, 98–102, 106–107, 109, 139–141, 149, 155–157, 214–216, 219, 270
   transfer, 12, 31, 33, 37–40, 50–51, 67–69, 71, 73, 75, 77, 79–81, 83–84, 97
General lysis buffer, 245
Genomic probe(s), 147
GFAP, 39, 78, 113–115, 170–172
GFAP monoclonal antibody, 170
GFP, 39, 53, 65–66, 69, 90, 107–108
Glia, 30, 78, 110, 257
   Glial cell, 39, 78, 109–112, 114, 146, 212, 230, 250, 253
   glial filaments, 171–173
Glial cell(s). See Glia
Glial fibrillary acidic protein. See GFAP
Glucose oxidase, 71–72, 228, 231
GluR1, 34
GluR2, 88, 115
Glutamate receptor 1. See GluR1
Glutamate receptor 2. See GluR2
Glutamate receptors. See Receptors
Glutamic acid decarboxylase. See GAD
Glutaraldehyde, 20–21, 25, 91, 135, 165–170, 172, 175–177, 183, 186–187, 191, 193, 206–207, 210, 213, 222–225, 227–228, 232, 244–245, 249, 263, 272. See also Fixation
Glycine, 5, 54, 65, 93–94, 142, 178, 183, 234, 245–246, 272
   glycine receptor, 54, 65, 142. See also Receptor
GM1 ganglioside receptor, 223
Gold, 20, 23, 26–27, 67–68, 73–75, 78–79, 82, 120, 127–128, 130–132, 137, 139, 161–163, 165–167, 169–180, 197, 201, 208, 211, 217–218, 223, 231, 233, 243–244, 250–251, 253, 255. See also Immunogold
   clusters, 32, 53, 57, 61–65, 64, 162–163, 166, 174, 190
   colloidal gold particles, 162–163, 176, 253
   IgG gold conjugates, 171, 243–244
   Nanogold, 163, 166, 168, 171, 174, 178
   particles, 22–24, 26–27, 67–68, 73, 75, 78–79, 82, 128, 130–131, 137, 150, 162–163, 171, 173–176, 178, 208, 216, 223, 244, 250, 253
   protein A-gold, 169–170, 173, 177, 180
   protein G-gold, 177
   suspension, 7, 75, 95, 97, 210–211, 261, 263
   toning, 223
   ultra small gold, 120
Golgi impregnation, 86, 176, 178, 180, 203–204
Golgi method. See Golgi impregnation
Golgi staining. See Golgi impregnation
Gomori technique, 91
Gonadotropin-releasing hormone neurons, 90, 108
GPCR, 16
Grb2, 1, 10–11
Green fluorescent protein. See GFP
Growth factor(s), 1–2, 8, 10–11, 38, 149, 240, 255, 258
   CNTF, 8–9
   EGF, 1, 11
   FGF, 1
   NGF, 2, 142
   PDGF, 1

H

Halogenated indolyl derivative, 90
Hapten-labeled nucleotide(s), 155, 157
Hapten-labeled primer(s), 146, 149, 157
Heating, 135, 142, 150, 180, 240–241, 250
Heterooligomeric proteins, 54
HindIII, 55
Hippocampus, 2, 39, 69, 86, 108–114, 170–172, 200–201, 281
   hippocampal neurons, 28–30, 34–36, 65, 82, 142
   hippocampal slice, 79
   hippocampal slice-explant culture, 84
   primary cultures, 17, 53–55, 65, 87–88
   pyramidal neurons, 183, 191, 280–283
Histochemical procedures. See Histochemistry
Histochemical reactions. See Histochemistry

Index

Histochemistry, 39, 48, 71, 119, 135, 141–143, 153, 158, 177–178, 180, 182, 206–208, 210, 213–214, 219, 222, 223, 229, 233–234, 257
HNP/Fast red, 122, 125, 128
HNPP, 122, 125, 127–128, 137
Hoechst 33342. See Bisbenzimide
Hoechst dyes, 237
HRP, 28, 72, 120, 127–128, 132, 134, 139–140, 162, 176, 178, 180, 183–186, 188, 195, 198, 205–214, 216, 219, 222–224, 223, 228, 230–231, 234, 248, 264, 268, 270, 272
   based tracers, 205, 214, 223
   crystals, 5, 211, 229–230, 277
   gelfoam, 211
   histochemistry, 71, 119, 141–143, 177–178, 180, 182, 206–208, 210, 213–214, 219, 223, 229, 233–234
   modified TMB stabilization, 214
   Nickel-intensified DAB reaction, 214
   penetration of peroxidase/DAB reaction product, 187, 215, 270
   proteolytic degradation by lysosomal enzymes in ultrastructural tracing studies, 223
   ultrastructural tracing studies, 223
HRP substrate(s)
   BDHC, 222
   DAB, 71–73, 125, 128, 130, 134, 163, 165, 172–173, 175, 183, 187–188, 195–196, 210, 214–217, 222, 227–230, 233, 263–264, 270
   modified TMB stabilization, 214
   Nickel-intensified DAB reaction, 214
   Pyronin, 229–230
   TMB, 210, 214, 219, 222, 227–231
   Vector VIP, 222, 234
Human placental alkaline phosphatase. See PLAP
2-hydroxy-3-naphtoic acid-2'-phenylanilide phosphate, 122, 128
Hypothalamus, 50–51, 70, 76–77, 83, 111, 113–114, 124, 142–143, 232
   hypothalamic cultures, 78, 81
   hypothalamic magnocellular neurons, 127, 132
   hypothalamic neurons, 69, 78, 80
   hypothalamic slice-explant culture, 83–84
   hypothalamo-neurohypophysial system, 76
Hypothermia, 43

I

ICC. See Immunocytochemistry
IdU, 236, 243, 246–250
IgG, 3, 46, 55, 71–72, 93–94, 103, 169, 171, 183, 188, 191, 193, 195, 243–244
IgG gold conjugate(s), 171, 243–244
IgG-gold technique(s), 173
IGL, 246–248, 250, 253
Image acquisition speed, 275
Image analysis, 46, 178, 180, 196–197, 200, 214, 238, 257
Immunoautoradiographic method. See Immunoautoradiography
Immunoautoradiography, 39
Immunoblotting, 3
Immunocytochemical labeling. See Immunocytochemistry
Immunocytochemical staining. See Immunocytochemistry
Immunocytochemical techniques. See Immunocytochemistry
Immunocytochemistry, 3, 4, 8, 18, 22–23, 33, 39, 48–49, 57, 63, 71, 78–79, 85, 87, 90, 91, 92, 102–103, 119–120, 121, 123, 124, 125, 127, 129, 130–143, 145, 147, 157, 161–180, 187–189, 188, 195, 196, 197, 200–201, 204–205, 208, 210, 215-218, 231–234, 236, 237, 240, 247, 248, 249, 252–254, 257, 259–262, 271, 272
   immunodetection of cancer cells, 90
   post-embedding immunocytochemistry, 90, 166, 168–169, 177, 179, 197
   pre-embedding immunocytochemistry, 162, 175, 197
Immunoenzymatic techniques, 162, 173
   immunofluorescence, 3, 10, 30, 33–35, 206–208, 210, 224, 226, 259, 261–263, 265, 270–271
   immunoperoxidase methods in ultrastructural studies, 221–224
   immunoperoxidase reaction(s), 226–227
   photoconversion, 215, 218–219, 224, 231, 233
Immunogold, 15, 19, 25–27, 120, 163, 166, 169–171, 173–180, 189, 197, 201, 249, 251–252, 256, 270
   IgG gold conjugate(s), 171, 243–244
   IgG-gold technique(s), 173
   post-embedding immunogold protocols, 120
   pre-embedding, 15, 19, 171, 197
   Protein A-gold, 169–171, 173, 177, 180
   Protein G-gold, 177
   silver intensification, 20–21, 162, 171, 175, 178, 208, 210, 217, 223
   silver intensified immunogold. See Silver intensification
   single immunogold staining, 169–170
   sodium metaperiodate, 166–167, 169–171, 180, 189, 243
   spatial resolution, 26, 173–174, 260, 274–275
   transferase-immunogold technique(s), 257
   ultra small gold, 120

Index

Immunogold (continued)
  ultrathin frozen sections, 162, 175, 178
Immunohistochemistry. See Immunocytochemistry
Immunolabeling. See Immunocytochemistry
Immunoprecipitation, 3–8, 10–12
  assay, 4, 6–8, 10, 39, 43, 45–48, 51, 71, 77–78, 146, 237, 239, 242, 245–246, 255, 258
  of protein kinase, 1, 4, 7–8, 238, 258
Immunostaining. See Immunocytochemistry
Impermeant dyes. See Fluorescent dye(s)
Indirect in situ PCR, 146–147
Indirect in situ RT-PCR, 146–149, 152, 154
Indo-1, 273, 275–283
Indocarbocyanine, 55
Inhibitory post-synaptic potential(s). See IPSPs
Inhibitory synapses, 61
Inositol-triphosphate. See IP3
In situ detection, 141, 158, 239, 242, 250, 256
In situ end-labeling techniques. See ISEL techniques
In situ hybridization, 71, 86, 88, 119, 121, 123, 125, 127, 129, 131, 133, 135, 137, 139, 141–143, 145–147, 152, 156–159, 161, 204, 240, 255, 258
  detection sensitivity, 146
  posthybridization washing, 121–122, 125, 127, 132, 136
  radioactive and enzymatic double ISH, 121, 123, 130, 133, 138–139
  riboprobe(s), 121–122, 136–138, 141
  triple labeling combining ICC and a radioactive and enzymatic double ISH, 133
In situ nick translation, 241, 255
In situ PCR, 142, 145–147, 149–150, 152, 155–158
  acetylation, 136, 152–154, 157
  amplicon, 146, 149, 156
  cDNA template(s), 147
  combined primer annealing, 156
  direct, 146–147, 149, 158
  genomic probe(s), 147
  hapten-labeled nucleotide(s), 155, 157
  hapten-labeled oligonucleotide primer(s), 154
  hapten-labeled primer(s), 146, 149, 157
  indirect, 146–147
  internal oligonucleotide probe(s), 149
  labeled oligonucleotide, 48, 147, 154
  non-specific nuclear staining, 154, 156–157
  random oligo (dT) primer(s), 155
  thermocyclers, 152
In situ polymerase chain reaction. See In situ PCR
In situ RT-PCR, 154, 156–157
Insulin, 32, 56, 61, 63–66
Insulin-like growth factor receptor, 240
Interferon(s), 1, 115

Interleukin(s), 1, 257
Internal granular layer. See IGL
Interplexiform cell(s), 91
In toto X-gal revelation, 45
Intracellular calcium concentration, 30, 257, 273, 275, 277, 279, 283
Intracellular dye filling, 204
Intracellular pH, 239
Intracellular recording, 181–183, 200
Intraventricular delivery, 38
Intron(s), 78, 80, 154, 156
In vivo gene transfer, 40, 51
5-iododeoxyuridine, 236
Ion channels, 86–87
Iontophoresis, 56, 182, 186, 201, 206–209, 211, 225
  iontophoretic injection, 182, 186
  microionophoretic filling of single neurons, 176
IP3, 282
IPSP(s), 186, 193
ISEL techniques, 241, 251
Isotope-conjugated neurotransmitters, 204
Isotopic methods, 147, 236, 249
Isotopic protocols. See Isotopic methods

J

JAK, 2, 4, 12–13
Janus Kinase(s), 2
Jurkat cells, 240
Juxtacellular labeling, 182

K

Ki-67, 236, 238, 246, 248, 255, 257–258
Kinases, 1–2, 4, 8, 10, 12, 258–259
  JAK, 2, 4, 12–13
  Janus, 2
  MEK, 1
  PKA, 283
  PKB, 2, 8
  PKC, 8, 115, 283
  Protein kinase assay, 6–7
  Raf1, 1
  RSK, 145, 149, 154, 156–158
  Serine/Threonine, 2, 10, 238
  Src, 1, 8
Kinases inhibitors, 12
  okadaic acid, 10
  sodium fluoride, 6, 10
Ki67 nuclear antigen, 236, 238, 246, 248
KiS2, 238, 257
KiS5, 238
Kozak sequence, 90
KpnI, 55

Index

Krox-24 gene, 51

L
Labeled oligonucleotide(s), 147, 154
LacZ, 40, 69, 78, 89, 180
Large dense-cored vesicles. See LGVs
Laser, 194, 232–233, 237, 262, 271–275, 278–279, 281–282
  light, 28, 39, 43, 45, 47, 70, 89–92, 98, 130, 139, 142–143, 153, 156–157, 161, 163, 165, 167–169, 176–183, 187–190, 195–197, 199–201, 203, 205–211, 213–217, 219, 221–222, 224–226, 228, 230–235, 241, 244, 246–247, 249, 259, 268, 270, 272–274, 278, 282
  power, 77, 79, 171–172, 214, 216, 229, 274, 281
  scanning confocal microscopy. See Confocal microscope
  scanning cytometry, 237
Laser scanning confocal microscopy. See Confocal microscope
Laser Scanning Microscope, 194
Lead citrate, 91, 107, 127, 130, 167, 170, 189, 229, 233, 245, 264
Lectins, 176, 224, 261, 263
  binding to oligosaccharides in axonal membranes, 223–224
  BSI-B4, 261, 263, 265–266, 269
  Concanavalin A, 95–96, 97
  degenerating neuronal profiles, 223
  inactivation of protein synthesis, 139, 142, 205, 224
  internalization, 15–16, 18, 23–25, 28, 65, 142, 180
  neuronal death induced by, 224, 254–255
  PHA-L, 205–208, 211–212, 218, 224–225, 232–234, 261–263
  ricin, 224, 234
  suicide tracers in ultrastructural tracing studies, 224
  volkensin, 224
  WGA, 205, 208, 223
  WGA-gold, 223
  WGA-HRP, 178, 208
  WGA-HRP-gold, 223
Lesion(s) of neuronal pathways, 224
Leupeptin, 4–5, 7, 12, 245
LGV(s), 173–174
Ligand-gated channel(s), 102
Ligand(s), 9, 15–19, 22–24, 28, 54, 87, 119
Ligation-mediated PCR, 242, 255
Ligation-mediated polymerase chain reaction. See Ligation-mediated PCR

Light microscope, 39, 89, 91, 139, 163, 168–169, 169, 176, 182, 187, 188, 189, 196, 200, 205–208, 210, 213, 216, 221, 228, 244, 246, 249, 268
  correlated light and electron microscopy analysis, 222, 224–225
Light microscopy. See Light microscope
Lipid-mediated cell transfection, 76
Lissamine rhodamine (TRITC), 210, 262
Low temperature, 167, 180, 196
Luciferase, 38–40, 43, 45–46, 48–51, 69, 78
Luciferin, 43
Lucifer yellow, 88, 191–193, 198, 200, 263
  photoconversion, 215, 218–219, 224, 231, 233
Luteinizing hormone receptor, 142
Lymphocyte precursor(s), 241
Lysosome, 212, 223

M
MAP2, 34, 36, 113
Markers, 18, 36, 39, 76, 86–89, 106, 112–113, 120, 127, 138–139, 176, 201, 208, 215–216, 222, 224, 230, 233, 236–238, 240, 245–246, 249, 253, 258, 260–261, 272
  of apoptosis, 100, 103, 106, 237, 239, 242, 257
    single stranded DNA monoclonal antibodies, 237, 240
  of cell proliferation, 235–236, 238, 243, 246–250, 254, 257
    BrdU, 174, 236, 238, 242–243, 246–250, 253, 256
    Casein-kinase 2a, 238
    IdU, 236, 243, 246–250
    Ki67 nuclear antigen, 236, 238, 246, 248
    KiS2, 238, 257
    KiS5, 238
    Mib-1, 238
    Th-10a, 238, 256
    Topoisomerase IIa, 238
    transferrin receptor, 16, 238, 257
  proliferating cells. See Markers of cell proliferation
MBBr, 239
MBCl, 239
Melanotrope cells, 277, 283
Meninges, 70, 152, 245
Merocyanine 450, 239
Messenger molecule(s), 139
Metabotropic glutamate receptor. See MGluR2
MGluR2, 34, 115
Mib-1, 238
Microinjection, 37–38, 43, 67, 279
Microtubule-Associated Protein2. See MAP2
Microvilli, 90

295

Index

Microwave, 135, 142
Migration, 23, 42, 157, 225, 246–247, 254, 256–257
 migratory events, 247
 migratory phenomena, 248
Miniature synaptic currents, 194
Mitochondria, 68, 83, 224, 229, 240
 mitochondrial membrane antigen, 240
 mitochondrial membrane depolarization, 239
 oxidative activity, 239
Mitosis, 51, 254
Mitotic index, 235
Mito Tracker Red CMX Ros, 239
ML, 246–248, 254
Modified calcium phosphate transfection, 34–35
Molecular layer. See ML
Monoamine neurons, 119
Monobromobimane. See MBBr
Monochlorobimane. See MBCL
Monoclonal antibodies, 6, 10, 12, 45, 83, 90, 107, 188, 196–197, 200, 236–237, 240, 255, 257–258
 7A6, 240
 anti-HA, 33
 anti-histone, 242
 anti-phosphoserine/threonine, 2, 3, 10, 238
 anti-phosphotyrosine, 3, 6, 10–11
 BrdU, 174, 236, 238, 242–243, 246–250, 253, 256
 E6, 90, 92–93, 100, 103, 105
 GFAP, 39, 78, 113–115, 170–172
 IdU, 236, 243, 246–250
 Ki-67, 236, 238, 246, 248, 255, 257–258
 KiS2, 238, 257
 KiS5, 238
 luciferase, 38–40, 43, 45–46, 48–51, 69, 78
 Mib-1, 238
 NeuN, 39
 single stranded monoclonal antibodies, 237, 240
 Th-10a, 238, 256
MRNA, 86, 92, 95, 99–100, 102–103, 106, 109, 124, 127, 130, 132, 135–143, 145–147, 149, 152–159, 238, 255
Multi-photon excitation microscopy, 282
Multiple immunolabeling, 170, 259
 double immunogold staining, 170
 simultaneous labeling, 174–175
 triple labeling combining ICC and a radioactive and enzymatic double ISH, 133
Multiple labeling, 120, 130, 137–138, 141, 175, 208, 214–215, 223–224, 230
Multiple staining. See Multiple immunolabeling
Multiplex semi-nested RT-PCR, 102–103

Myelinated afferents, 265–268

N

NADPH-d, 176, 180
N-(2-aminoethyl) biotinamide hydrochloride. See Neurobiotin
Nanogold, 163, 166, 168, 171, 174, 178
NBT, 91, 122, 130, 132, 244
NBT/BCIP, 122, 130, 132, 150, 156, 244
Necrosis, 115, 240–241, 254–255
Neocortex, 86, 109-110, 112-113, 183, 191, 281. See Cortex
NeuN, 39
Neurobiotin, 182, 191–192, 195, 198, 201, 205, 207, 224–225, 232, 261, 263
Neurofilament, 39
Neurogenesis, 2, 13, 247, 254
Neurokinin 1. See NK1
Neuronal circuits, 221–222, 224, 259, 268, 270–271
Neuronal connection(s), 161, 175–176, 221–224, 232, 260
Neurons, 2, 28–36, 39, 53–57, 60–61, 63–65, 67–69, 76–78, 80, 82–92, 95–98, 100–105, 107–108, 119, 124, 127, 132, 139, 141–143, 146, 161–162, 175–183, 185–187, 189, 191, 193, 195, 197–201, 203–204, 212–213, 216–218, 221–225, 229–235, 247, 250, 253, 255–261, 264–273, 278, 280–283
Neuropeptide receptor antibodies, 15
Neuropeptide(s), 15–16, 18–19, 28, 115, 124, 138, 141–143, 233
Neuropil, 25, 171, 216
Neurosteroid(s), 53, 58–59, 66
Neurotensin, 15, 28, 142. See NT
Neurotensin receptor 1. See NT1 receptor(s)
Neurotrophins, 1, 83
 NGF, 2, 142
Neurotropic virus(es), 204
Neutral Red, 214
NGF, 2, 142
NheI, 55
Nickel grids, 169, 243
Nick translation, 241, 255
Nicotinamide adenine dinucleotide phosphate-diaphorase. See NADPH-d
Nissl staining, 132, 203, 218
Nitric oxide, 115, 176, 254
Nitro blue tetrazolium/bromochloro-indolylphosphate. See NBT/BCIP
Nitro blue tetrazolium chloride. See NBT
NK1, 28, 265–269, 271
NMDA, 36, 54, 65–66, 201, 278
NMDAR2A subunit, 107

Index

NMDAR1 subunit, 109
NO, 3, 6, 9, 12, 16, 24, 38, 45, 47–49, 54, 68, 70, 72–73, 78–79, 82, 89–90, 99, 102, 104, 106, 122, 127, 135–137, 150, 154, 156–157, 166, 176–177, 187, 198, 210, 232, 237, 249, 254, 265, 267, 269, 272, 277, 282–283
Non-isotopic methods, 146, 236, 249
Non-isotopic protocols. See Non-isotopic methods
Non-specific nuclear staining, 154, 156-157. See in situ RT-PCR
Non-specific staining, 57, 92, 103, 163
Non-viral vectors, 38
Northern blot(ting), 86, 88, 145
NSE, 78, 80, 113–115
NT, 15, 17–18, 22, 24, 28, 99, 142, 172
NT1, 22–24
NT1 receptor(s), 22–24
Nuclear staining, 154, 156–157, 218, 253
Nuclear yellow, 205–207
Nucleic acid(s), 73, 142, 145–146, 152–154, 157–158, 237, 240, 251, 257
Nucleic acid thermal denaturation, 240
Nucleosome(s), 241
Nucleus, 1–2, 9–10, 22, 24, 29, 57–58, 70, 76, 79, 82, 109, 111–114, 124, 127, 132, 141, 177–180, 216–217, 226, 231–234, 240, 250, 257, 271–272
  chromatin, 237, 240–241, 251–253, 258
  chromatin condensation, 37, 41, 47, 240, 251–252
  eterochromatin, 249–250
  nuclear staining, 154, 156–157, 218, 253

O

OF-1 mouse, 42
Okadaic acid, 10
Olfactory lobes, 70
Oligo(dT) primer(s), 98, 137, 151, 153, 155
Oligomer(s), 241
Oligonucleotide(s), 38, 47, 51, 99, 143, 145, 147, 149, 152–154
Oligoprobe(s), 121, 136, 138
Oligosaccharide in axonal membrane, 224
Opioid(s), 15, 17, 28, 177–178, 201
Optical section(s), 266, 274
Optic nerves, 70
Oregon green, 279–280
Organotypic culture, 71, 73–81, 83–84
Organotypic slice, 69, 73–75, 83
Oscillations, 277, 283
Osmicated material. See Osmium
Osmication. See Osmium

Osmium, 20–21, 25, 125, 130, 161–162, 166–169, 171, 173, 175, 183, 188, 191, 195, 197, 222, 224–225, 243, 249
  ferrocyanide, 44, 167, 169
  post-fixation, 22, 25, 45, 137, 162, 166–167, 173, 187, 222–223, 228
  tetroxide, 20–21, 25, 161, 166–169, 183, 191, 222, 243, 249
  in ultrastructural tracing studies, 162, 198, 205, 221–224, 253
Oxidative activity, 239
Oxygenation, 277
Oxytocin (OT), 76, 83, 113, 115, 124, 127, 142–143

P

Pancreatic islets, 68, 83, 178
PAP, 162–163, 178, 196, 197, 227–228
Papain, 95–97, 103–104
Paraffin, 32, 138, 157
Paraformaldehyde, 17–18, 25, 33, 44, 122, 125, 127, 130, 132, 150, 152–153, 157, 165, 167–169, 183, 186, 191, 193, 206–207, 210, 213, 222, 227–228, 263
  as a fixative for ultrastructural tracing studies, 221–224
  vapors, 175
Paraventricular nucleus (PVN), 76–77, 80–81
PARP, 242–243, 245–246, 251
Patch clamp, 56, 87–88, 90, 181–182, 189, 192, 194, 199–200, 282
PC12 cells, 2
PCD, 235, 239–240, 242, 254–257. See also Apoptosis
Pcis-CMV-CAT, 41
PCMV-luciferase, 40
PCMV(nls)-Lac-Z, 40, 69
PCNA, 236, 238, 254–258
PCR, 55, 85, 87, 96, 98–100, 102, 104, 106–108, 140, 142, 145–147, 149–153, 155–159, 242, 255
  antisense primers, 98
  multiplex semi-nested RT-PCR, 102–103
  oligo(dT) primer(s), 98, 137, 151, 153, 155
  promoter constructs, 78
  single round PCR, 98
  two rounds PCR, 99
PDGF, 1
PEG, 138, 209, 211
PEGFP-N1, 35–36, 55–56
PEI, 37–43, 39, 41, 43, 45–47, 49, 51

Index

Penetration, 68, 82, 135, 140, 162–163, 165, 171–172, 196–197, 211, 222, 225, 232, 234, 262, 270–272, 279, 282
   of antibodies, 162, 196–197, 232, 262, 271–272
   of peroxidase/DAB reaction product, 183, 195, 210, 214, 222, 228, 264
   of reagents for immunocytochemical studies, 161–164, 171, 195–196, 232, 281
   of reagents for ultrastructural tracing studies, 162, 198, 222
Pepsin, 150–151, 153, 157
Pepstatin, 4–5, 7, 12
Peptidase inhibitor(s), 17
   aprotinin, 4–5, 7, 245
   leupeptin, 4–5, 7, 12, 245
Peptide(s), 3–4, 12, 15–17, 76, 84, 87, 115, 119, 124, 132, 139, 141–142, 173–174, 179, 265–266, 268, 271
Perfusion, 44–45, 56, 62, 98, 135, 186–187, 210, 212–213, 222, 224, 227–228, 262–263, 280–282
Peroxidase. See HRP
Peroxidase-anti-peroxidase. See PAP
Peroxide toxicity, 241
Phagocytes, 239–240, 247, 252
PHA-L, 205–208, 211–212, 218, 224–225, 232–234, 261–264
Phaseolus leucoagglutinin. See PHA-L
Phaseolus vulgaris-leucoagglutinin. See PHA-L
Phenilmethylsulfonyl fluoride, 4, 6, 10, 245
Phosphatase(s), 5, 7, 49, 85, 91, 107–108, 120, 122, 124–125, 130, 142, 145, 149–151, 153, 176, 178, 226, 229, 231, 244, 248
Phosphatidylserine, 239, 255
Photobleaching, 54, 262, 278, 282
Photoconversion, 215, 218–219, 224, 231, 233
Photodamaging, 278, 282
Photoreceptor, 86
PI, 205, 237
Picric acid, 135, 165, 168, 191, 193
Pinhole, 274–275
PKA, 283
PKB, 2, 8
PKC, 8, 115, 283
Plancental alkaline phosphatase. See APP
PLAP, 85, 90–92, 95, 100, 103–105, 107
Plasmid(s), 30, 34, 37–38, 40–42, 46–57, 61, 65, 67, 73–74, 82, 83
   DNA, 2, 4, 9, 29–32, 36–39, 41–42, 47–48, 51, 55, 65, 67–68, 73–74, 76, 82–84, 87, 98–100, 106, 121, 142, 145–147, 149, 151–154, 156–159, 236–242, 247, 249–251, 253, 255–258

pcis-CMV-CAT, 41
pCMV-luciferase, 40
pEGFP-N1, 35–36, 55–56
plasmid-based genes, 37
Platlet-derived growth factor. See PDGF
Pluronic acid, 276–277
Point scanning, 275
Poly-ADP-ribose polymerase. See PARP
Polyclonal antibodies, 12, 44, 210
   anti-CTb, 227–228
   anti-DBH, 229
   anti-human APP, 33
   anti-MAP2, 33
   anti-PHA-L, 210
   to PAP, 162–163, 178, 196, 197, 227–228
Polyethylene glycol. See PEG
Polyethylenimine. See PEI
Poly-L-lysine, 55
Polymerase a, 238
Polymerase chain reaction. See PCR
Post-embedding, 90, 120, 162–163, 166–177, 179, 189, 196–197, 201, 249
Post-mitotic cell(s), 247
Postmitotic primary neurons, 29
Postsynaptic receptor genes, 102
Pre-embedding, 15, 19, 137, 162–163, 165–166, 170–173, 175–177, 179–180, 188, 196–197, 201, 233–234, 270
Primary fixation. See Fixation
Primary hippocampal neurons, 29
Primer(s), 87–88, 98–100, 102, 104, 106, 146–147, 149, 151-157
   antisense, 98
   custom designed forward, 155
   flanking introns, 154
   hapten-labeled, 146, 149, 157
   random oligo (dT), 155
Progenitor cells, 1–2, 8–12, 246, 249, 256
Programmed cell death. See PCD
Proliferating cell nuclear antigen. See PCNA
Proliferating index, 235–236
Promoter, 2, 33, 35, 37–40, 47, 50–51, 66, 68, 76, 78, 80, 83, 85, 88–90, 92, 104, 108
   constructs, 32, 35, 37–38, 57–58, 61, 67–68, 73, 76, 78, 80–82
   regulation, 2–4, 12, 28–29, 37–39, 47, 50–51, 58, 60–61, 63, 65, 69, 88, 141–142, 149, 158, 255–256, 258, 283
Propidium iodide. See PI
Protease digestion, 150, 153–154, 157
Protease inhibitor(s), 10, 12, 245
   aprotinin, 4–5, 7, 245
   benzamidine, 12

Index

leupeptin, 4–5, 7, 12, 245
pepstatin, 4–5, 7, 12
phenilmethylsulfonyl fluoride, 4, 6, 10, 245
Proteinase K, 150, 154, 244, 249
Protonation, 38
PS, 77, 239–240
PVN, 76–77, 80–81
Pyramidal cells. See Pyramidal neurons
Pyramidal neurons, 89, 109, 183, 191, 200, 233, 280–283
Pyronin, 229–230

Q

Quantification, 39, 138, 174, 178, 180, 238, 258, 270
Quantitative analysis, 26, 84, 178, 201
    quantification of immunostaining, 174
    quantification of the radioactive signal, 138
    quantitative expression of transgene, 83, 180
    quantitative studies, 49, 271

R

Rab, 19, 23, 46
Radioactive procedures, 32, 39, 99, 120–124, 127, 130, 133, 136–139, 141–142, 236, 249
    isotopic methods, 236
    non-isotopic methods, 146, 236, 249
    radioactive and enzymatic double ISH, 121, 123, 130, 133, 138–139
    radioimmunological procedures, 162
    radioisotopes, 242
    radiolabeled nucleotides, 2, 88, 143, 145, 176, 196–197
    tritiated tymidine, 236, 248–249
Raf1, 1
Random examers, 98
Ras, 1–2
Ratiometric dye(s), 275, 279–280
Ratio value, 279–280
Receptors, 1–2, 8, 15–20, 22–28, 34, 36, 51, 53–55, 57–61, 63, 65–66, 83, 86–88, 92, 95, 102, 107, 109–110, 115, 119, 124, 138, 139–143, 158–159, 165, 177–179, 181, 201, 215, 223, 233, 234, 238, 240, 257, 259, 261, 265–272, 278, 282
    adrenergic receptor(s), 54
    channel complex, 58
    cluster colocalization, 57
    cluster(s), 32, 53, 57, 61–65, 162–163, 166, 174, 190
    epitope-tagged, 18–19
    extrasynaptic clusters, 61
    for ganglioside GM1, 223

GluR1, 34
GluR2, 88, 115
glycine receptor, 54, 65, 142
GPCR, 16
insulin-like growth factor receptor, 240
ligand-induced internalization, 16
mGluR2, 34, 115
neuropeptide(s), 15–16, 18–19, 28, 115, 124, 138, 141–143, 233
NK1, 28, 265–269, 271
NMDA, 36, 54, 65–66, 201, 278
NMDAR1 subunit, 109
NT1, 22–24
receptor-ionophore complex, 88
receptor-ligand complex, 16
resensitization, 16
sequestration, 16, 283
sst2A, 22, 24–25, 27–28
subunits, 53–54, 57–58, 61, 92, 102, 107, 141
TR, 50, 122
transferrin, 16, 28, 238, 257
TrkA, 2
Reciprocal circuits, 225
Recombinant virus, 29–30
Reduced glutathione, 239
Refractive index, 271
Reporter-based tracers, 204
Reporter gene(s), 38, 40, 48–49, 89, 176
Repp86, 238
Resin infiltration, 166
Retina, 69, 84–87, 90–92, 95, 97–98, 100, 102–104, 107–110, 112, 178–179, 255
    amacrine cells, 86–87, 91, 107
    bipolar cells, 107, 110
    cone bipolar(s), 86, 179
    dopaminergic cells, 90, 107
    ganglion cells, 86, 95, 100, 102, 110, 232, 272
    interplexiform cell(s), 91
    Photoreceptor, 86
    rod bipolar, 89, 91, 95, 110
Retrograde degeneration, 203
Retrograde labeling. See Retrograde tracing
Retrograde tracing, 176, 204–206, 208, 212, 216–218, 223, 224–225, 231–233, 270
    with colloidal gold particles, 162–163, 176, 253
    CTb-HRP, 205–208, 223
    diamidino yellow, 206–207
    Evans blue, 205–207
    Fast Blue, 205–207, 227, 229–231
    fluorescent counterstains, 215
    fluorescent tracers, 205, 208, 212–215, 218, 223–224
    HRP-based tracers, 205, 214, 223

Index

Retrograde tracing (continued)
  with HRP conjugated to plant lectins, 176
  Nuclear yellow, 205–207
  PI, 205, 237
  radiolabeled amino acids, 176
  retrograde degeneration, 203
  retrograde transport of tracers, 89, 232, 234
  WGA-apoHRP, 205
Retrovirus, 107
Reverse transcriptase. See RT
Reverse transcriptase polymerase chain reaction. See RT-PCR
Rhodamine. See TRITC
Ribosomal S6 kinase. See RSK
Ribosome(s), 240
Ricin, 224, 234
RNA, 35, 41, 83, 86, 88, 95–96, 98, 103–104, 108, 116, 121, 135, 141–143, 147, 150–151, 158, 257
  mRNA, 86, 92, 95, 99–100, 102–103, 106, 109, 124, 127, 130, 132, 135–143, 145–147, 149, 152–159, 238, 255
  mRNA by single-cell RT-PCR, 92
  reverse transcriptase, 88, 96, 99, 104, 147, 151–152, 155, 157, 238
  RNAse inhibitor(s), 96, 100, 136
Rod(s), 89, 91, 95, 110
RSK, 145, 149, 154, 156–158, 158
RSV, 78
RT, 88, 96, 98, 99, 104, 147, 151–152, 155, 157, 238
RT-PCR, 85, 87–88, 92, 95–96, 98–104, 106–107, 109, 140, 142, 145–158

S
SacI, 99–101, 106
Sampling speed, 275
Saporin, 224
Sciatic nerve, 266–267
Sepharose, 3, 12, 93–94
Sequestration, 16, 283
Shc, 1–2, 12
Signal-to-noise ratio, 194, 214, 274
Signal transducer and activator of transcription. See Transcription
Silver, 20–21, 26, 124–125, 127, 130, 137, 139, 162, 171, 174–175, 177–180, 201, 203, 208, 210, 216–219, 223, 227, 236, 238, 249, 255, 258
  enhanced gold labeling, 137, 161, 166, 175–178, 251
  grains, 26, 124, 139, 236, 249
  impregnation of degenerating axons, 203, 218

intensified immunogold. See Immunogold
Simultaneous labeling. See Immunogold
Single immunogold staining. See Immunogold
Single-stranded DNA. See ssDNA
Skin, 34, 67, 155, 185, 204, 256
Slice preparation(s), 107, 176, 182, 191, 199, 237, 276
SNARF-1, 239
Sodium borohydride, 183, 187, 191, 193, 222, 225, 228, 232
Sodium fluoride, 6, 10
Sodium metaperiodate, 166–167, 169–171, 180, 189, 243
Somatostatin. See SRIF
Southern blot(ting), 99, 145
Spatial distribution, 173, 260
Spatial resolution, 26, 173–174, 260, 274–275
SpeI, 55
Spermidine, 74
Spinal cord, 29, 111, 113–114, 166, 170, 174, 177–180, 182–183, 185, 200–201, 217, 222, 230–234, 265–267, 271–272, 277, 281–282
  neurons, 28–36, 119, 124, 175–183, 185–187, 189, 191, 193, 195, 197–201, 203–204, 212–213, 216–218, 221–225, 229–235, 247, 264–273, 278, 280–283
  spinal dorsal horn, 180, 229, 231, 261, 264
  spinothalamic tract, 267
  substantia gelatinosa, 177–179, 271, 282
  in ultrastructural tract tracing studies, 162, 198, 205, 221–224, 253
Spontaneous inhibitory postsynaptic currents, 61
Src, 1, 8
SRIF, 15, 17, 22, 23, 26–27, 28, 115, 272
SsDNA, 237, 240–241, 255, 257
STAT, 2, 4, 9, 12–13
Statin, 238, 258
STM buffer, 244
Stratacooler, 104
Streptavidin, 125, 127, 130, 134, 198, 226, 229, 263
Structural preservation, 167, 222
Substance P, 28, 115, 174, 179, 187, 201, 233, 265–266, 268, 271–272
Substantia nigra, 47, 49, 110–112, 114, 217
Suicide tracers in ultrastructural tracing studies, 224
Survival, 1–2, 32, 70, 206–207, 212, 236, 246–247, 250, 253–254, 256
  post injection survival, 207, 212
Synapse(s), 28, 53, 60–61, 65, 161, 166, 171, 178–179, 190, 197, 198–201, 204, 224, 226, 230, 234, 259–260, 265, 267–269, 272
  in degenerating neurons, 178, 203, 218, 224
  gabaergic, 53, 65, 178, 201

Index

monosynaptic connections, 204
polysynaptic connections, 204
retrograde transsynaptic tracers, 204
synaptic circuitry, 181, 221, 232
synaptic connection(s), 86, 89–92, 177, 179–180, 259
synaptic input to neurons in ultrastructural tracing studies, 223
synaptic specialization(s), 221, 269, 270
synaptic vesicles, 172–174, 201, 229
Synaptophysin, 53
Synthetic vectors, 38
SYTO stains, 237
SYTOX Green, 237

T

Tannic acid, 166, 180
Taq DNA polymerase, 99, 151, 153–154, 156
T4 DNA ligase, 241, 250
TdT, 241–242, 244, 251
Telomerase, 238, 256, 258
Template(s), 88, 99–100, 106, 121, 147, 156
Terminal deoxynucotidyl transferase. See TdT
Tetramethyl rhodamine-dextran amine. See Fluoro-Ruby
Texas Red, 22, 217, 241
TH, 56, 85, 89, 91–92, 99–100, 102, 104, 106, 107, 112–113, 115, 124, 178
  gene, 55, 67–69, 71, 73–89, 91–93, 95, 97, 99, 101–103, 105, 107–108, 109, 124
Th-10a, 238, 256
Thalamus, 217, 265, 267
Three-dimensional reconstruction, 223
Thyroid, 50–51
Thyroid hormone receptor, 51. See TR
Thyrotropin Releasing Hormone, 51. See TRH
Time lapse video microscopy. See Video microscopy
TMB, 210, 214, 219, 222, 227–231
Topoisomerase IIa, 238
Toxin(s), 205–208, 223–224, 231–234, 261
  CTb, 205–208, 223–234, 261–263, 265–267, 270
  saporin, 224
TR, 50, 51, 122
Tracer(s), 176–177, 204–219, 221–234, 236, 261, 262–263, 270
  anterograde tracing, 176, 178, 180, 203–205, 207–208, 211, 224–225, 228, 231–233
  axonal tracers, 204, 215
  axonal transport, 203, 205, 209, 212, 218
  BDA, 176, 205–207, 224–225, 227–230, 232
  bi-directional labeling. See Bidirectional transport
  bi-directional transport, 216, 225
  biocytin, 182, 191–193, 198, 201, 205, 207, 210, 212, 224–225, 232–233
  biotin-based tracers, 205, 208, 210, 214, 224–225
  combined retrograde and anterograde labeling, 212
  CTb-HRP, 206–208, 223
  cytoplasmic labeling, 216, 218
  dark-field observation, 218
  Diamidino yellow, 206–207
  Evans blue, 205–207
  Fast Blue, 205–207, 227, 229–231
  fluorescent tracer(s), 205, 208, 212–215, 218, 223–224
  gold-conjugated tracers, 208, 210–211, 213, 216
  HRP-based tracers, 214
  isotope-conjugated neurotransmitters, 204
  juxtacellular labeling, 182
  labeled neurotransmitters, 178, 223
  lipophilic dyes, 204
  Neurobiotin, 182, 191–192, 195, 198, 201, 205, 207, 224–225, 232, 261, 263
  neurotropic virus(es), 204
  nuclear staining, 154, 156–157, 218, 253
  Nuclear yellow, 205–207
  PI, 205, 237
  post-injection survival, 207, 212
  reporter-based tracers, 204
  retrograde tracing, 176, 204–206, 208, 212, 216–218, 223, 224–225, 231–233, 270
  retrograde transsynaptic tracers, 204
  transganglionic tracing, 204
  transmembrane dye diffusion, 216
  in ultrasonic tracing studies, 223
  ultrastructural visualization, 171–172, 176, 179, 243, 249–250, 252
Tracing studies, 221–223, 225, 232
Tract-tracing methods, 176, 218
Tranferase(s), 257
  TdT, 241–242, 244, 251
  Transcription factor(s), 1–2, 4, 8, 47, 138–139, 158
Transfection, 15–17, 19, 29–36, 39, 45, 47, 49–51, 53–57, 59–63, 65, 67–69, 71, 73, 75–85
  Biolistics, 67–69, 71, 73, 76–78, 82–84
  of cDNA in cultured mammalian cells, 29
  COS-7 cells, 17–19, 22–24, 26, 28
  efficiency, 29–30, 33, 35, 38–39, 45, 47–48, 54, 65, 68–69, 73, 82–83, 90, 92, 102, 135, 137, 142, 154, 156, 166
  Gene Gun, 67–69, 73–76, 78, 83–84
  heterologous (transfection) systems, 15–16, 124, 139
  hyperexpression, 82

Index

Transfection (continued)
  integrated sequences, 51
  integrated transcriptional responses, 37
  lipid-mediated cell transfection, 76
  modified calcium phosphate transfection, 34–35
  transfected cell lines, 51
  viral transfection, 60
  in vivo gene transfer, 40, 51
Transfection methods. See Transfection
Transferase-immunogold technique(s), 257
Transferrin receptor, 16, 28, 238, 257
Transganglionic tracing, 204
Transganglionic transport, 267
Transgene(s), 38–39, 45–46, 48, 51, 82–83, 89, 108, 141, 180
  transgene expression, 38, 45, 83, 108
  transgenesis, 37–38
  transgenic mouse (mice), 66, 76, 78, 82, 90, 92, 102, 103, 107–108, 180
  transgenic retina, 100
Transmembrane dye diffusion (in tracing studies), 216
TRH, 50–51
TRH promoter, 50–51
Triple labeling combining ICC and a radioactive and enzymatic double ISH, 130, 133, 139
TRITC, 210, 239, 261–263, 266
Tritiated tymidine, 236, 248–249
Triton, 5–6, 20, 33, 71–72, 138, 165, 168–171, 193, 195, 222, 243, 245, 261, 263, 271
TrkA, 2
Trophic factor(s), 235
Tubulin CMV, 33, 35, 40, 78
TUNEL, 241, 248–253, 257
Tungsten particles, 68
Two rounds PCR, 99
Tymidine, 236, 247–249
Tyramide, 128, 132, 134, 140, 142
  amplification, 132, 134, 140–142
Tyrosine hydroxylase. See TH
Tyrosine kinase receptor(s), 8. See also Receptor(s)

U

Ultra small gold, 120
Ultrastructural studies, 28, 161–162, 165–166, 171, 178–180, 186, 198, 205, 221–224, 231, 233–234, 240, 253, 263
  analysis, 161–163, 193–197, 221–222, 253–255
  neuronal connection(s), 161, 175–176, 221–224, 232, 260
  preservation, 162–163, 196–200, 222–225
  preservation in tracing studies, 223
Ultrastructural visualization, 171–172, 176, 179, 243, 249–250, 252
Ultrastructure. See Ultrastructure studies
Ultrathin frozen sections, 162, 175, 178
Unmodified T7 polymerase, 241
Unmyelinated fibers, 261, 265–266, 272
Uranyl acetate, 167, 169–170, 189, 245, 264
UV, 41, 98, 167, 270, 278–279, 282

V

Vaccinia virus, 38
Vascular perfusion, 222, 224
Vasopressin. See VP
Vector VIP, 113, 115, 222, 234
Vibrating microtome. See Vibratome
Vibratome, 18, 21, 39, 44–45, 49, 95, 125, 136, 138, 165, 168–169, 177, 189, 191, 193, 196–197, 210, 213, 227–228, 232, 260, 262–264, 270, 272, 276
  antifreeze mixtures for vibratome sections, 228
  sections, 17–18, 20, 22, 24–25, 44–46, 49, 95, 124–125, 135–136, 138, 161–163, 165–171, 173, 175–180, 187–190, 193–199, 201, 213–216, 228–232, 257–264, 266–268, 270–272
Video microscopy, 35
Viral messengers, 120
Viral transfection methods, 60
Viral vectors, 68, 76, 81–82
  adeno-associated virus, 83
  for gene delivery, 51
  recombinant virus, 29–30
  retrovirus, 107
  vaccinia virus, 38
Visual cortex, 69, 83, 281–282
Visual perception, 85
VLM, 226, 229–230
VOC(s), 54, 87–88, 90, 158, 282
Volkensin, 224
Voltage-gated calcium channel(s). See VOC
Voltage-operated calcium channel(s). See VOC
VP, 76, 78, 83, 115, 124, 127, 132, 138, 142–143

W

Wallerian degeneration, 203
Wavelength, 208, 214, 262, 279–280, 282–283
Western blot(ting), 3, 8, 10, 15, 30, 32, 33, 35–36, 243, 246, 251
WGA, 205, 208, 223, 272
  apoHRP, 205
  CTb, 205–208, 223–234, 261–263, 265–267, 270
  CTb-HRP, 206–208, 223
  gold, 208, 223

Index

HRP, 178, 208, 223
HRP-gold, 223
Wheat-germ agglutinin. See WGA
White matter, 246–248, 252–253, 258

X

Xenopus oocyte(s), 37, 47, 54, 83, 142, 283
X-gal, 44–45, 48, 69

Y

Yama. See Caspase 3
YO-PRO(tm)-1 iodide, 205, 237

Z

Zeta-sizing, 47
Zinc, 58
Zolpidem, 53, 56, 58–59

DATE DUE

FAC MAY 16 2005		
FAC MAY 14 2007		
MAY 4 2007		
DEC 19 2008		
JAN 0 8 REC'D		
GAYLORD		PRINTED IN U.S.A.

SCI QP 356.2 .C45 2002

Cellular and molecular
methods in neuroscience